The DYNAMIC ANALYSIS of Automatic Process Control

IRP *Instructional Resource Package*

STUDENT TEXT

Developed by
THE EDUCATION DEPARTMENT
Instrument Society of America

Ralph L. Moore

The DYNAMIC ANALYSIS of Automatic Process Control

INSTRUMENT SOCIETY OF AMERICA

THE DYNAMIC ANALYSIS OF
AUTOMATIC PROCESS CONTROL

Printed in the United States of America.

The Instrument Society of America
67 Alexander Drive, P.O. Box 12277
Research Triangle Park, NC 27709

ISBN 0-87664-817-0

The Instrument Society of America wishes to acknowledge the cooperation
of those manufacturers, suppliers, and publishers
who granted permission to reproduce material herein.
The Society regrets any omission of credit that may have occurred
and will make such corrections in future editions.

First printing April 1985

CONTENTS

Figures ix
Tables xiii
Preface xv

Section 1
INTRODUCTION TO THE STUDENT

1-1 Automatic Control 1
1-2 Course Content 2
1-3 Prerequisites 2
1-4 Learning Objectives 2
1-5 Definition of Terms 3

Section 2
JUSTIFICATION FOR AUTOMATIC CONTROL

2-1 Introduction 7
2-2 Home Heating System — On-off Control 7
2-3 Home Heating System — On-off plus Derivative Control 9
2-4 Operating an Automobile — Manual Control 9
2-5 The Water Closet — Proportional Control 10
2-6 Tank Level — Manual Control 10
2-7 Tank Level — Proportional Control 10
2-8 Tank Level — Sampled Data 11
2-9 A Shower Bath — Dead Time 12
2-10 System Variables — Cause and Effect 13
2-11 Block Diagrams 14
2-12 Block Diagram — Self-Contained Pressure Regulator 15
2-13 Control Dynamics 16
2-14 Mathematical Modeling 17
2-15 Effects of Measurement Accuracy and Speed 17
2-16 Control System Design 18
2-17 Justification of Control 20
Problems 21

Section 3
ELEMENTS OF PROCESS DYNAMICS

3-1 Commentary 25
3-2 Introduction 26
3-3 The Basic Concepts 27
3-4 Laplace Notation 27
3-5 Transfer Functions 28
Problems 29
3-6 Capacitance and the Unsteady-State Equation 31
3-6-1 Electrical System 31
3-6-2 Thermal Systems 32
Problems 33
3-6-3 Incompressible Fluid Systems 34
3-6-4 Compressible Fluid Systems 35
Problems 36
3-6-5 Compressible Fluid Systems 37
Problems 39
3-6-6 Composition Systems 40
3-6-7 Mass Transfer Systems 41
3-6-8 Summary 41
Problems 42
3-7 Resistance and the Driving-Force Equation 43
3-7-1 Electrical System 43
3-7-2 Thermal Systems 43
Problems 46
3-7-3 Fluid Flow Systems 47
Problems 49
3-7-4 Mass Transfer Systems 51
3-7-5 Summary 51
Problems 52

Section 4
SELF-REGULATING PROCESS SYSTEMS

4-1 Introduction 53
4-2 Nonself-Regulating Systems 53
4-3 Self-Regulating Systems 54
Problems 56
4-4 First-Order Systems 57
4-4-1 Electrical Systems 57
4-4-2 Thermal Systems 58
Problems 60
4-4-3 Incompressible Fluid Systems 62
4-4-4 Compressible Fluid Systems 63
Problems 65
4-4-5 Composition and Mass Transfer — Composition Systems 66

4-4-6 Mass Transfer Systems 67
4-5 Dead Time 69
Problems 70

Section 5
CONTROL VALVES

5-1 Introduction 73
5-2 Inherent Valve Flow Characteristic 73
5-3 Installed Valve Flow Characteristic 74
5-4 Valve Gain 74
5-5 Open-Loop Gain 75
Problems 76

Section 6
SIGNAL FLOW DIAGRAMS

6-1 Introduction 79
6-2 Signal Flow Diagram Elements 79
6-3 Controller Error Measurement 80
6-4 Open Loop versus Closed Loop 81
6-5 Signal Flow Diagrams 82
6-6 Open-Loop Gain 84
Problems 85
6-7 Open- and Closed-Loop Concepts 88
6-7-1 The Open Loop: Allowable Loop Gain 88
6-7-2 The Closed Loop 89
6-8 Rule for Combining Block Diagram 89
Problems 92

Section 7
SINUSOIDAL ANALYSIS

7-1 Introduction 97
7-2 Sinusoidal Testing 97
7-3 Sinusoids 100
7-4 Frequency Response 100
Problems 104
7-5 Controller Actions 105
7-5-1 Derivative Action (Lead-Lag Network) 105
7-5-2 Reset Action (Lead-Lag Network) 106
7-5-3 Dead Time 108
Problems 108
7-6 Control Loop Dynamics 111
7-6-1 General Concepts 111
7-6-2 Dead Time 112
Problems 115
7-7 Dynamic Elements in the Loop 117
7-7-1 Dead Time and First-Order Lag 117
Problems 119
7-7-2 Combinations of First-Order Lags 121
7-7-3 Derivative Action 122
7-7-4 Summary 123
Problems 125
7-8 Controller Tuning 126
7-8-1 Preliminary Estimation of Controller Settings 126
7-8-2 Final Adjustment of Controller Settings 127
7-8-3 General Comments 128
Problems 130

Section 8
DYNAMIC RESPONSE DETERMINATION BY EXPERIMENTAL TESTING

8-1 Introduction 133
8-2 Frequency Response Testing 134
8-3 Transient Response Testing — Step Analysis 139
8-4 Pulse Response Testing 144
Problems 145

Section 9
SELECTION OF CONTROLLER MODES

9-1 Introduction 147
9-2 System Analysis 147
9-3 Positive Feedback 148
9-4 Reactor Control System 148
9-5 Proportional-Only Control 150
9-6 Proportional-Reset Control 150
9-7 Proportional-Reset-Derivative Control 153
9-8 Systems with Dead Time 153
9-9 Floating (Integral-Only) Control 154
9-10 Cascade Control 154
Problems 155

Section 10
FLOW CONTROL

10-1 Introduction 159
10-2 Process Capacitance 159
10-3 Instrument Time Constants 160
10-4 Effect of Pneumatic Transmission Lines 160
10-5 Effect of Nonlinearities 161
10-6 Controller Functions 165
10-7 Conclusions 165
10-8 Problems 166

Section 11
MASS BALANCE CONTROL

11-1 Introduction 167
11-2 Application of Mass Balance Controls 167
11-3 Compressible Fluid Mass Balance 167
11-4 Liquid Mass Balance 169
11-5 Level Control — High Pressure Separator 169
11-6 Level Control — Distillation Column Bottoms 171

11-7 Pressure Control — Distillation Column Overhead 172
11-8 Pressure Control — Steam Heat Exchangers 175
11-9 Pressure Control — Fired Vessels 176
11-10 Pressure Control — Centrifugal Compressor 177
Problems 179

Section 12
TEMPERATURE CONTROL

12-1 Introduction 181
12-2 On-Off Control 181
12-3 Proportional Control 182
12-4 Proportional-Reset (PI) Control 183
12-5 Proportional-Reset-Derivative (PID) Control 184
12-6 Floating (I) Control 185
12-7 Problem 186

Section 13
ADVANCED CONTROL TECHNIQUES: CASCADE AND FEEDFORWARD

13-1 Introduction 187
13-2 Multiple Time Constants 187
13-3 Flow Cascade 188
13-4 Benefits of Cascade Control 189
13-5 Cascade Controller Tuning 191
13-6 Cascade of pH Control on pH Control 191
13-7 Expected Performance Improvement with Cascade Control 192
13-8 The Concept of Feedforward 192
13-9 Applications of Feedforward 193
13-10 Basic Concepts 194
13-11 Three-Element Feedwater Control 195
13-12 Distillation Column 196
13-13 Impulse Feedforward 197
13-14 Problems 199

Section 14
STATE SPACE TECHNIQUES

14-1 Introduction 201
14-2 Control Strategy 201
14-3 Optimal Controller 203
14-4 Adaptive Control 203
14-5 Pairing of Variables for Control 206

Appendix A
ELECTRIC CIRCUIT ANALYSIS BY CLASSICAL METHOD 209

Appendix B
DYNAMICS OF A TEMPERATURE SENSOR 211

Appendix C
RESPONSE OF MULTI-CAPACITY SYSTEMS 215

Appendix D
CHARACTERIZATION OF A LIQUID LEVEL PROCESS CONTROL SYSTEM 219

Appendix E
DYNAMIC ANALYSIS OF A FLOW CONTROL SYSTEM 231

Appendix F
NOMENCLATURE AND DEFINITIONS 235

Appendix G
REFERENCES 237

Appendix H
ANSWERS TO PROBLEMS 243

INDEX 253

FIGURES

1-1. Elements of Controlled System 1
2-1. Home Heating System 7
-2. Automatic Home Heating System 8
2-3. Steering an Automobile 9
2-4. Water Closet Level 10
2-5. Manual Tank Level Control 11
2-6. Proportional Tank Level Control 11
2-7. Tank Level Control, Multiple Tanks 12
2-8. Mixing of Hot and Cold Water 13
2-9. Feedback Control System 13
2-10. Control System Schematic 14
2-11. Simple Proportional Control 15
2-12a. Self-Contained Pressure Regulator 15
2-12b. Block Diagram of Regulator 16
2-13. Basic Response Curves 16
2-14. Variation of *r* as Joint Function of Dead Time *a* and Sampling Time *z* 18
2-15. Effect of Change in Process Throughput on Controller Tuning 19
2-16. Sensitivity of System IAE Performance to Valve Dynamics 19
2-17. Hookup of Valve with Positioner plus Booster 20
2-18. Controlled Variable 20
3-1. Overlap 25
3-2. Liquid Level System 27
3-3. Generalized Heat Exchanger 28
3-4. Block Diagram 28
3-5. Electrical Circuit 31
3-6. Heat Transfer Process 32
3-7. Block Diagram — Heat Transfer 32
3-8. Heat Exchanger Tube Wall 32
3-9. Liquid Level Process 34
3-10. Gas Pressure Process 35
3-11. Vapor Pressure Process 37
3-12. Polymer System 38
3-13. Composition Process 40
3-14. Batch Composition Process 40
3-15. Multicomponent Batch Composition Process 40
3-16. Distillation Column Tray 41
3-17. Adsorption Process 41
3-18. Temperature Sensor 44
3-19. Insulated Pipe 45
3-20. Square Law Nonlinearity 48
3-21. Vapor Diffusion Process 51
4-1. Nonself-Regulating Systems 53
4-2. Graph of Integrating Process 53
4-3. Self-Regulating Systems 54
4-4. First-Order Response 55
4-5. Self-Regulation Diagram 55
4-6. Amplifier with Feedback 55
4-7. Electrical System 57
4-8. Response of Self-Regulating System 58
4-9. Conductive Thermal System 58
4-10. Block Diagram, Conductive Thermal System 58
4-11. Thermal Sensor Response 59
4-12. Thermal Radiation System 59
4-13. Liquid Level Process 62
4-14. Gas Pressure Process 63
4-15. Gas Pressure Block Diagram 64
4-16. Composition System 66
4-17. Recycle Composition Process 67
4-18. Counterdiffusion Process 67
4-19. Distillation Column 68
4-20. Distillation Column Block Diagram 69
4-21. Dead Time Process 69
5-1. Basic Inherent Valve Characteristics 73
5-2. Installed Flow Characteristics 74
5-3. Nonlinear Equal Percentage Valve Characteristic and Square Law Pressure Drop Relationships 75
6-1. Generalized Block Diagram 79
6-2. Signal Flow Diagram Symbols 80
6-3. Generalized Block 80
6-4. Linkage 80
6-5. Displacement Block 80
6-6. Force Block 80
6-7. Controller Linkage 81
6-8. Controller Error Diagram 81
6-9. Open-Loop System 81
6-10. Closed-Loop System 82
6-11. Typical Block 82
6-12. Liquid Level Control System 82
6-13. Summing Junction 83

6-14. Process Block Diagram 83
6-15. Process and Controller Diagram 83
6-16. Control System Signal Flow Diagram 84
6-17. Liquid Level System Signal Flow Diagram 84
6-18. Open-Loop Signal Flow Diagram 88
6-19. Control System Performance Based on Minimum IAE Deviations 88
6-20. Generalized Closed-Loop Diagram 89
6-21. Combined Signal Flow Diagram 89
6-22. Inner Loop Signal Flow Diagram 90
6-23. Consolidated Signal Flow Diagram 90
6-24. Consolidated Block Diagram 90
6-25. Typical Signal Flow Diagram 90
6-26. Unity Feedback Diagram 91
6-27. Block Manipulation 91
7-1. Generation of Sinusoid 97
7-2. Frequency and Period 98
7-3. Phase Angle 98
7-4. Output Variation with Sinusoidal Variation of Input 99
7-5. Frequency Response Test Arrangement 99
7-6. Frequency Response 100
7-7. Fourier Series Representation 100
7-8. Magnitude Ratio and Phase Angle 102
7-9. First-Order Lag 103
7-10. Lead-Lag Network 106
7-11. Lag-Lead Network 107
7-12. Dead Time 111
7-13. Reactor Control 112
7-14. Reactor Signal Flow Diagram 112
7-15. Combined Dead Times 113
7-16. Dead Time and First-Order Lags 117
7-17. Reactor Signal Flow Diagram 118
7-18. First-Order Lag Signal Flow Diagram 121
7-19. Two First-Order Lags 122
7-20. Phase and Gain Margin Signal Flow Diagram 126
7-21. Bode Diagram for Phase and Gain Margin 127
7-22. Process Reaction Curve 129
8-1. Process Transfer Function 133
8-2. General Transfer Function 133
8-3. Testing Configuration 133
8-4. Single-Capacity System 134
8-5. Frequency Response Analysis 135
8-6. Amplitude Ratio 136
8-7. Straight Lines Approximate Data 138
8-8. Static Gain 139
8-9. Step Response, Single-Capacity System 140
8-10. Non-Interacting Time Lags — Percent Incomplete Curves 141
8-11. Response of Second-Order System 142
8-12. Determination of Two Equivalent Time Constants 142
8-13. Simplified Process Dynamics Signal Flow Diagram 142
8-14. Response of Processes with Several Equal Time Constants in Series 143
8-15. Multi-Time Constant Approximation Method 143
8-16. Higher-Order System Approximation 144
8-17. Unit Pulse 144
8-18. Tank Agitation Testing 145
8-19. Impulse Response 145
8-20. Time Constant from Impulse Test 145
9-1. Generalized Block Diagram 147
9-2. Water Jacketed Chemical Reactor with Temperature Control 148
9-3. Combining the Block Diagram for a Reactor 149
9-4. Commercially Available Controller Modes 151
9-5. Response of a Second Order System to a Step Input 152
9-6. Response of a System to a Step Disturbance with Controller Modes as Shown 153
9-7. Relationship of Time Domain Parameters to Damping Coefficient 154
9-8. Chemical Reactor with Cascade Control System 155
10-1. Physical Diagram for Flow Control 159
10-2. Proportional Control with Two Time Constant System 160
10-3. Frequency Response of Pneumatic Transmission Lines 162
10-4. Time Constants of Pneumatic Tubing 163
10-5. Flow Loop Recovery 163
10-6. Effect of Transmission Line Length on Open-Loop Gain 164
10-7. PI Controller Reset Time and Closed-Loop Resonant Frequency (CPM) versus Transmission Distance and Tubing Diameter — 2-Pipe Configuration 164
10-8. Block Diagram of Nonlinearities 164
10-9. Flow System Nonlinearities 165
11-1. Separator Level Control 171
11-2. Level Control with Reboilers 173
11-3. Overhead Control System for Vacuum Distillation Column 174
11-4. Steam Heater with Temperature Control 175
11-5. Vaporizer Pressure Control 176
11-6. Centrifugal Compressor Controls 177
11-7. Compressor Map 178
11-8. Compressor Surge 179
12-1. Temperature Process, Large Capacitance 181
12-2. Switching System 182
12-3. Switching System 182
12-4. Time Proportioning Avoids Large Overshoot of Set Point Temperature 182

12-5. Process Characteristics for a Proportional Controller 183
12-6. Proportional-Only Response for a Temperature System 183
12-7. Temperature Response to Reset Adjustment 184
12-8. Steam Heater 184
12-9. Distillation Column Condenser 185
12-10. Frequency Response of a Floating Controller 185
13-1. Control of a Process with Two Time Constants 187
13-2. Chain-Like Plant Structure 188
13-3. Position System Cascade Control 188
13-4. Temperature Controlled Furnace, Single Loop 189
13-5. Temperature Controlled Furnace, Cascade Loop 190
13-6. Block Diagram of Cascade Control with Secondary Control of Fuel to a Furnace 190
13-7. Temperature Cascade Signal Flow Diagram 191
13-8. Avoiding Primary Controller Windup in Cascade Systems with External Reset Conditions 191
13-9. Cascade pH Control 192
13-10. Relative Improvement of Cascade Control 193
13-11. Feedback Control 193
13-12. Feedforward Control 193
13-13. Block Diagram Used to Determine Feedforward Function 194
13-14. Three-Element Boiler Feedwater Control 195
13-15. Feedforward Control Applied to a Distillation Column 196
13-16. Impulse Function 197
13-17. Impulse Function Response 197
13-18. Pressure Control with Impulse Feedforward 198
13-19. Pressure Controller Response 198
14-1. Representative Distillation Column 201
14-2. Control Strategy Matrix 202
14-3. Signal Flow Diagram Steam Boiler 202
14-4. Optimal Controller 203
14-5. Distillation Column Adaptive Control Advantages 204
14-6. Experimental Titration Curves 205
14-7. Adaptive Control System 205
14-8a. Independent Control Loops 206
14-8b. Matrix Notation for Control Loops 206
14-9a. Conventional Control of an Interacting Process 206
14-9b. Matrix for for Control System of Figure 14-9a 206
14-10. Completely Decoupled Control Scheme 207

TABLES

2-1. Process Properties 19
7-1. Application of Derivative Action in Control of Various Processes 124
7-2. Classification of Controller Actions for Process Characteristics 129
10-1. Variation in Controller Gain with Flow System Nonlinearities 165
13-1. Advantages of Feedforward Control 193

PREFACE

Instruments and regulators were designed primarily by innate ingenuity prior to the 1930's. At that time, it was realized that control systems could be described by mathematical equations. Great advances in this concept occurred during World War II, in the design of servomechanisms for ordnance control. Advances included the use of heat, mass, and force-balance equations to describe physical systems; the use of differential operators, e.g., Laplacians, to more easily handle the resulting differential equations; and the use of the theory of complex variables, with all the power of its inherent mathematical concepts, to assess stability. The 1950's saw the publication of a number of basic books that applied analytical techniques to the regulation of fluid processing systems. This book serves as a compilation of those techniques, many of which, through the years, have led to simplified guidelines for good control applications.

Since control systems lend themselves so well to mathematical analysis, an immense amount of mathematical development has been published in the ensuing years. It has included new signal flow diagrams, new transforms, new graphical techniques, and new computer algorithms. That effort has culminated in the modern control theory described in the final section of this book.

The material herein is not new. It has been drawn from personal experience, from the authors in the list of *References,* and from associates within the DuPont Company. Significant contributions were made by the late Joseph F. Coughlin, ISA Fellow and Principal Consultant in the DuPont Engineering Department, who conceived portions of Sections 3, 4, and 7 as a training manual; and the late Professor C. R. Otto, of Cornell University and the DuPont Engineering Department, who wrote much of Appendix C. I appreciate the cooperation of the DuPont Company in authorizing publication of this information.

Ralph L. Moore
New Castle, Delaware
December 15, 1984

Dedication

To a good friend and devoted co-worker, Mrs. Beverle Eastburn.

SECTION 1

INTRODUCTION TO THE STUDENT

1-1 AUTOMATIC CONTROL

Automatic process control is widely used because it results in economy of operation of industrial processes that more than pays for the cost of the control equipment. Some benefits of automatic control are:

- Increased productivity and the resulting lower product unit cost.
- Higher product uniformity, which results in less wastage.
- Greater safety for operating personnel.
- The freeing of human beings from dull and menial tasks.
- The control of processes too fast and too complex for human beings to follow.
- A lower labor cost, which is reflected in a lower product cost.

The principle of automatic control, the use of the feedback of a measurement to actuate a controlling mechanism, is quite simple. Diagrammatically, the fundamental closed loop of action and reaction representative of most automatic control systems is shown in Figure 1-1.

The control system can be seen as a series of elements combined to produce a useful result with little or no human intervention. Many years ago, process equipment was designed using steady-state design concepts. Upon completion of the design of the process equipment, the control equipment was added. Painfully, it became apparent that process equipment design with such limited concepts was in many instances literally uncontrollable. Thus evolved the modern concept that process equipment and control equipment interact intimately to form an automatic control system, as shown in Figure 1-1, and must be designed together as a system for best operation.

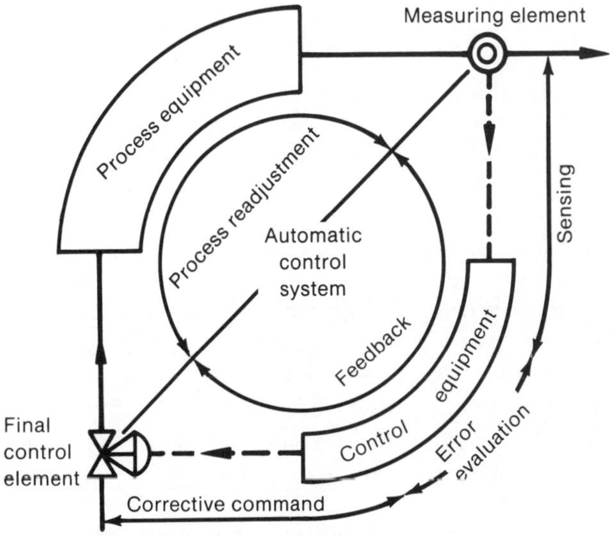

Figure 1-1. Elements of Controlled System

1-2 COURSE CONTENT

The design of an as yet unbuilt automatic control system, or the understanding of the difficulties of a poorly operating system, requires an analysis of each element in the system and the assembly of the model of each element into an analogous system. The model includes the *dynamics* of each element. Dynamics is the science of forces and covers cases where no motion results (statics) and where motion does result (kinetics). The concept of analogy is widely used, and much of the analogy is based on the familiar electrical concepts of resistance and capacitance. Some of the systems that can be made analogous to the electrical system are as follows:

Driving Force	Resulting Action
Net force	Acceleration
Voltage difference	Current flow
Temperature difference	Heat flow
Pressure difference	Fluid flow
Magnetic dipole	Magnetic flux
Composition difference	Chemical reaction

Today's automatic control systems are actuated electrically, pneumatically, electronically, and/or digitally. Essentially no mention is made of these various alternatives in this text, because well applied instrument components and a well designed system are independent of actuation. Each type of actuation has advantages and disadvantages: Electrical systems have a very quick response but sometimes are so noisy that large capacitors must be added for stability; pneumatic transmission is slower, but the dynamics of the tubing stabilizes the control system; digital systems are very accurate but are limited by the number of bits that can be carried and by quantization. Generally, the choice of actuation is economic and has little to do with analysis.

The dynamics of processes and instruments can be determined either mathematically or experimentally. Generally, processes will be analyzed mathematically, because experimental testing is costly in lost production and can be dangerous. Instruments, on the other hand, tend to be small and portable and are easily taken to a convenient place for testing. With this in mind, the early sections that follow emphasize the mathematical modeling of processes and process equipment. A later section is devoted to experimental testing and largely emphasizes the testing of process equipment to determine or verify dynamic response. The analysis of temperature sensors is included, and the section on sinusoidal testing describes the frequency response technique for experimentally determining instrument dynamics.

Frequency response techniques are used to analyze the automatic control system as an entity. Frequency response has been largely replaced as a tool for system analysis by computer simulation but still has great value in assessing the interactive relationship of system components to each other and as a fast, simple, pencil-and-paper technique for determining controller tuning parameters. While computer simulation can handle very large and complex automatic control systems and has the great advantage of providing a time domain solution, it does not provide the insight into a single-loop automatic control system as do the frequency response techniques.

The mathematics of automatic control systems are sophisticated and difficult and will not be remembered unless used regularly. However, the mathematical derivations lead to generalizations and guidelines that are readily usable for design and troubleshooting long after the mathematics have drifted away. Sections on mass balance, flow, and temperature control provide guidelines for systems in widespread use.

Finally, sections on higher-level control systems such as feedforward, cascade, and adaptive control are included. Such systems are commonplace because they are so easy to implement (and to revise in the field) with the new microprocessor-based control equipment. State space techniques are included because of their great strength in designing multiple-loop control systems. They provide insight into system design for pairing (deciding which measured variable to use to manipulate each final control element) and for a design to minimize interaction among the control loops.

1-3 PREREQUISITES

Maximum usefulness will be gained from this text if the reader is familiar with mathematics up to calculus with some knowledge of basic physics and chemistry. Partial differential equations, operational calculus, and statistics are touched on in passing. Some training in chemical unit operations or experience in the petrochemical industry will be helpful. Dimensions are a problem, and stress is laid on carrying dimensions with all numerical factors in order to verify the dimensions of the result. Having developed the mathematical relationships, all calculations can be made with a hand-held calculator.

1-4 LEARNING OBJECTIVES

The principal objective of this text is to develop in each reader a sense of judgment about the adequacy of process equipment design, controller choice, tuning parameters, and valve characteristics for a control system that responds well and is stable. Let it be acknowledged that the broad use of the guidelines contained herein will result in applications that are entirely adequate without further analysis for about 95 percent of control systems. A second objective is to recognize control systems that do present difficulty, that do require analysis, and that might well require more sophisticated control configurations.

1-5 DEFINITION OF TERMS

The following terms are used throughout the book. (Definitions have been taken from SAMA Standard PMC 20-2-1970 Process Measurement and Control Terminology, with the permission of the publisher, Process Measurement and Control Section, SAMA, 370 Lexington Avenue, New York, N.Y. 10017.)

attenuation. (1) A decrease in signal magnitude between two points, or between two frequencies. (2) The reciprocal of gain when the gain is less than one.

Bode diagram. A plot of log amplitude ratio and phase angle values on a log frequency base for a transfer function.

calibrate. (1) To ascertain by the use of a standard the locations at which scale or chart graduations of a device should be placed to correspond to a series of values of the quantity that the device is to measure, receive, or transmit. (2) To adjust the output of a device, to bring it to a desired value, within a specified tolerance, for a particular value of the input. (3) To ascertain the error in the output of a device by measuring or comparing against a standard.

control action. Of a controller or a controlling system, the nature of the change of the output effected by the input.

control action, adaptive. Control action whereby automatic means are used to change the type or influence (or both) of control parameters in such a way as to improve the performance of the control system.

control action, cascade. Control action where the output of one controller is the set point for another controller.

control action, derivative (rate). Control action in which the output is proportional to the rate of change of the input.

control action, feedback. Control action in which a measured variable is compared to its desired value to produce an actuating error signal, which is acted upon in such a way as to reduce the magnitude of the error.

control action, feedforward. Control action in which information concerning one or more conditions that can disturb the controlled variable is converted into corrective action to minimize deviations of the controlled variable.

control action, integral (reset). Control action in which the output is proportional to the time integral of the input; i.e., the rate of change of output is proportional to the input.

control action, proportional. Control action in which there is a continuous linear relation between the output and the input.

control mode. A specific type of control action such as proportional, integral, or derivative.

control system. A system in which deliberate guidance or manipulation is used to achieve a prescribed value of a variable.

controller. A device which operates automatically to regulate a controlled variable.

controller, floating. A controller in which the rate of change of the output is a continuous (or at least a piecewise continuous) function of the actuating error signal.

controller, integral (reset) (I). A controller which produces integral control action only.

controller, proportional (P). A controller which produces proportional control action only.

controller, proportional-plus-integral (reset), (PI). A controller which produces proportional-plus-integral (reset) control action.

frequency, undamped (frequency, natural). (1) Of a second-order linear system without damping, the frequency of free oscillation in radians or cycles per unit of time. (2) Of any system whose transfer function contains the quadratic factor $s^2 + 2\zeta\omega_n s + \omega^2_n$, the value ω_n.

where:

s = complex variable
ζ = constant
ω_n = natural frequency in radians per second

(3) Of a closed-loop control system or controlled system, a frequency at which continuous oscillation (hunting) can occur without periodic stimuli.

gain (magnitude ratio). For a linear system or element, the ratio of the magnitude (amplitude) of a steady-state sinusoidal output relative to the causal input; the length of a phasor from the origin to a point of the transfer locus in a complex plane.

gain, loop. The ratio of the change in the return signal to the change in its corresponding error signal at a specified frequency. (Note: The gain of the loop elements is frequently measured by opening the loop, with appropriate terminations. The gain so measured is often called the open-loop gain.)

gain, proportional. The ratio of the change in output due to proportional control action to the change in input.

integral action rate (reset rate). (1) Of proportional-plus-integral or proportional-plus-integral-plus-derivative control action devices; for a step input, the ratio of the initial rate of change of output due to integral control action to the change in steady-state output due to proportional control action. (Note: Integral action rate is often expressed as the number of repeats per minute because it is equal to the number of times per minute the proportional response to a step input is repeated by the initial integral response.) (2) Of integral control action devices; for a step input, the ratio of the initial rate of change of output to the input change.

Laplace transform — unilateral. Of a function $f(t)$, the quantity obtained by performing the operation.

$$f(s) = \int_0^\infty f(t) e^{-st} dt$$

where:

$s = \sigma + j\omega$

$f(s)$ = function of s

s = complex variable

$f(t)$ = function of t

t = time, seconds

σ = real part of the complex variable s

$j = \sqrt{-1}$

ω = frequency, rad/sec

linearity. The closeness to which a curve approximates a straight line.

loop, closed (feedback loop). A signal path which includes a forward path, a feedback path, and a summing point and which forms a closed circuit.

loop, open. A signal path without feedback.

measured variable. The physical quantity, property, or condition which is to be measured.

offset. The steady-state deviation when the set point is fixed.

parameter. A controllable or variable characteristic of a system or device, temporarily regarded as a constant, the respective values of which serve to distinguish the various specific states of a (the) system or device.

phase difference. (1) Between sinusoidal input and output of the same frequency, phase angle of the output minus phase angle of the input. (2) Of two periodic phenomena (e.g., in nonlinear systems), the difference between the phase angles of their two fundamental waveforms.

primary element. The system element that quantitatively converts measured variable energy into a form suitable for measurement.

process. The collective functions performed in and by the equipment in which the variable(s) is (are) to be controlled.

proportional band. The change in input required to produce a full range change in output, due to proportional control action. (Note: It is reciprocally related to proportional gain.)

response, step. The time response of an instrument when subjected to an instantaneous change in input from one steady-state value to another.

response, time. An output, expressed as a function of time, resulting from the application of a specified input under specified operating conditions.

sensitivity. The ratio of a change in output magnitude to the change of input which causes it after the steady state has been reached.

set point (command). An input variable which sets the desired value of the controlled variable.

signal. Information about a variable that can be transmitted.

signal, error. In a closed loop, the signal resulting from subtracting a particular return signal from its corresponding input signal.

steady state. A characteristic of a condition, such as value, rate, periodicity, or amplitude, exhibiting only negligible change over an arbitrary long period of time.

time, dead. The interval of time between initiation of an input change or stimulus and the start of the resulting response.

time, step response. Of a system or an element, the time required for an output to make the change from an initial value to a large specified percentage of the final steady-state value either before or in the absence of overshoot, as a result of a step change to the input. (Note: Usually stated for 90, 95, or 99 percent change. See *time constant* for use of 63.2 percent value.)

time constant. For the output of a first-order (lag or lead) system forced by a step or an impulse, τ is the time required to complete 63.2 percent of the total rise or decay; at any instant during the process, τ is the quotient of the instantaneous rate of change divided into the change to be completed. In higher-order systems, there is a time constant for each of the first-order components of the process. In a Bode diagram, breakpoints occur at $\omega = 1/\tau$.

time constant, derivative action. Of proportional-plus-derivative control action, a parameter whose value is equal to $1/2\pi f_d$ where f_d is the frequency (cycles per unit time) on a Bode diagram of the lowest frequency gain corner resulting from derivative control action.

time constant, integral action. Of proportional-plus-integral control action, a parameter whose value is equal to $1/2\pi f_i$ where f_i is the frequency (cycles per unit time) on a Bode diagram of the highest frequency gain corner resulting from integral control action.

transducer. An element or device which receives information in the form of one physical quantity and converts it to information in the form of the same or some other physical quantity.

transfer function. A mathematical, graphical, or tabular statement of the influence which a system or element has on a signal or action compared at input and at output terminals.

variable, manipulated. A quantity or condition which is varied as a function of the actuating error signal so as to change the value of the directly controlled variable.

STUDENT SUMMARY NOTES and QUESTIONS FOR INSTRUCTOR

SECTION **2**

JUSTIFICATION FOR AUTOMATIC CONTROL

2-1 INTRODUCTION

Automatic control has played a major role in the evolution of our modern way of life by making more goods available for everyone through increased productivity, by increasing leisure time, and by eliminating many tedious or dangerous jobs. Automation can be found in our homes, our cars, and our workplaces. It is so prevalent that it has become an almost invisible part of our surroundings.

The automatic home heating system is a familiar example of automation in our daily lives. It is difficult to recall fairly recent times when fuel (wood or coal) was piled near the furnace to be added at regular intervals as it burned; the adjusting of dampers for more or less heat; the elimination of ashes from the furnace; and the removal of ashes from the premises. The simple adjustment of a thermostat to the desired temperature not only eliminates constant attention to fuel supply but also the manual labor of fuel and ash transportation. Automatic home heating is widely justified by these conveniences.

The thermostat acts as both sensor and controller for the home heating system. It senses the temperature at a representative location in the house and compares the house temperature with the desired temperature, called the set point temperature. The difference between the sensed temperature and the set point temperature is called the controller error. If the house temperature is below the set point temperature, an electrical circuit is completed that starts a flow of fuel to the furnace and then ignites the fuel.

The fire in the furnace combustion chamber supplies heat to the house until the house temperature rises above the set point temperature, at which time the thermostat shuts off the supply of fuel, the fire goes out, and the house begins to cool. It cools until the house temperature decreases past the set point, at which time the heating cycle starts over again.

2-2 HOME HEATING SYSTEM — ON-OFF CONTROL

The familiar automatic home heating system encompasses many of the basic tenets of automatic control. First, it is an on-off type of control system. This designation results from the manipulated variable (fuel flow to the furnace) being either fully on or completely off. The performance of the system in regulating temperature reflects the type of control, wherein house temperature varies above and below the set point temperature but almost never remains at the set point temperature.

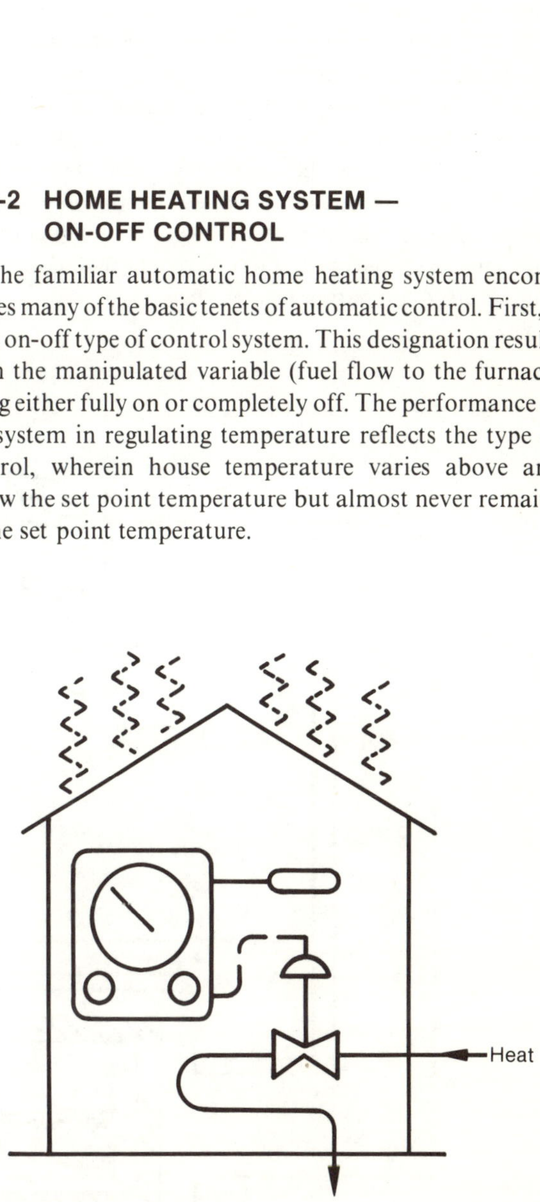

Figure 2-1. Home Heating System[3]
(Courtesy E. I. du Pont de Nemours & Company)

The home heating automatic control system also illustrates the difference between the two basic objectives of feedback control systems, one of which is regulation. To meet this objective, the control system must maintain temperature close to a constant set point with changing external conditions. For example, suppose the temperature outside the house dropped suddenly. House temperature would change its characteristics by decreasing more quickly when the furnace was turned off and increasing more slowly when the furnace was on. The temperature control system must regulate temperature near the set point despite disturbances such as a change in outside temperature. This is the regulatory objective.

The second objective is to follow changes in set point, or to respond as a servomechanism. Should someone in the house feel cold and raise the set point of the thermostat, the control system would be expected to cause the house temperature to follow the set point change. The system would call for the furnace to start and to stay on until the house temperature reached the new set point temperature, at which point the on-off action would again start to maintain the temperature near the new set point.

Lastly, automatic home heating is an example of the type of process that can be well regulated by on-off control. When the furnace is first turned on, the house volume is so large that temperature rises slowly even though heat is being supplied at the maximum furnace capacity. Finally, the temperature passes the set point of the thermostat, which then turns off the furnace. Then, because of the heat contained in the house and the low rate of heat loss, temperature decreases very slowly until it falls far enough to cause the thermostat to actuate the furnace. Thus the characteristic of this on-off control is a temperature that varies between maximum and minimum limits, with a period (elapsed time from one maximum temperature to the next) of many minutes or even hours.

On-off control is less applicable as houses become progressively smaller and less well insulated. Then temperature will rise and fall faster with the furnace being switched on and off more and more frequently. Finally, the furnace would be switched so rapidly as to cause excessive electrical contact wear, and on-off control should be replaced by a continuous type of control.

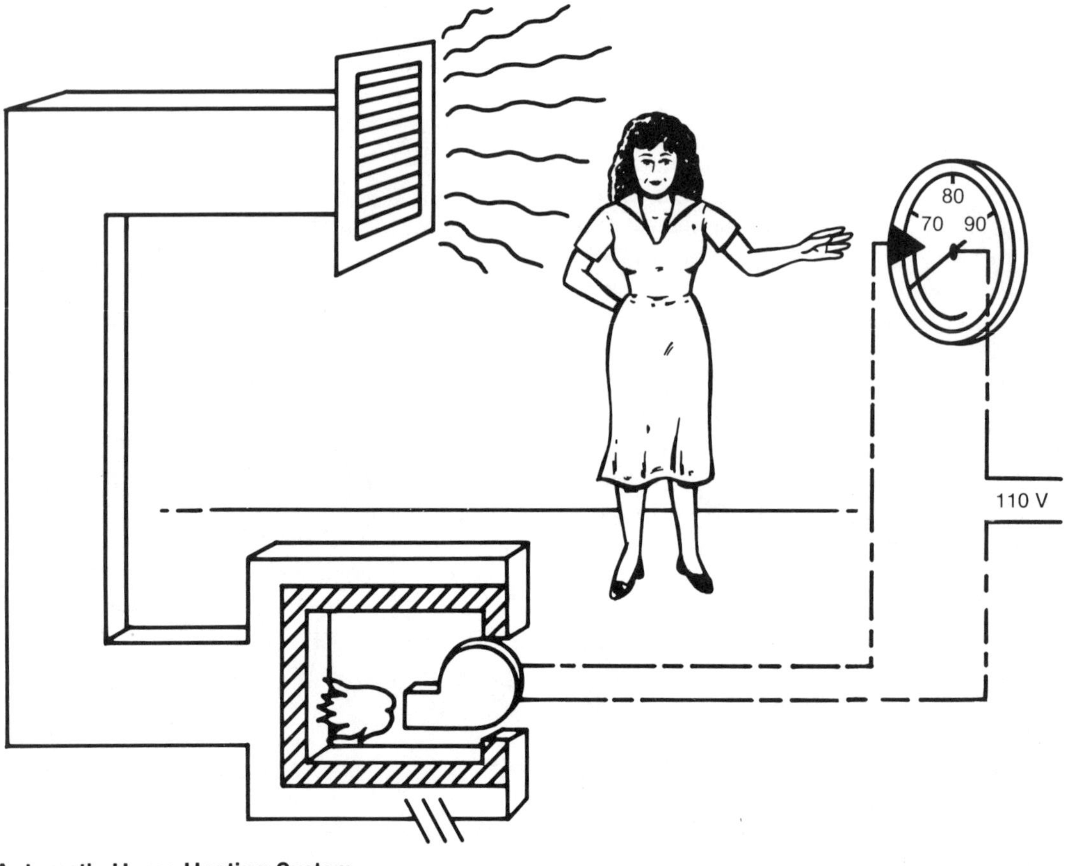

Figure 2-2. Automatic Home Heating System

2-3 HOME HEATING SYSTEM — ON-OFF PLUS DERIVATIVE CONTROL

The home heating system previously described will regulate temperature in such a manner that it will first be a couple of degrees above set point and then a couple of degrees below set point, or it will cycle by plus or minus a couple of degrees. Consider an increasing house temperature with the furnace running. The furnace will shut off as the house temperature crosses the thermostat set point. However, heated air will exist in the furnace and the duct system, and house temperature will continue to increase for a short time after the furnace shuts off. The temperature will increase, reach a maximum, and begin to decrease, at which point it will again cross the set point and start the furnace.

However, the furnace and duct system are cold when the furnace starts, and house temperature will continue to decrease for a short while until enough warm air is available to displace the cooling air in the house. At that point, temperature will reach a minimum and begin increasing, starting the cycle all over again.

The amplitude of the temperature cycle can be reduced by placing a small electrical heating element in the thermostat housing. The heater is turned on by the same switch that starts the furnace. The heater then maintains the temperature sensor inside the thermostat housing at a temperature slightly higher than the house temperature. As the house temperature rises to approach the set point, the sensor temperature is slightly higher, due to the heater, and crosses the set point while the house temperature is still slightly below the set point. The furnace is shut off before the set point is reached, and the overshoot due to residual hot air is reduced.

The thermostat with the heater thus anticipates the reaching of the set point by the house temperature and acts previously (pre-acts) to shut off the furnace. This control action has been called anticipatory or pre-act control, but the modern term is derivative action.

2-4 OPERATING AN AUTOMOBILE — MANUAL CONTROL

A human being can perform well as an automatic controller. Consider the steering of an automobile.[1] The set point is an imaginary line down the center of the driving lane. The eyes act as an error-detecting element by comparing the position of the car with the imaginary line. As the road curves, the imaginary line curves, and the eyes detect the error between the direction of the car and the curving line. The brain represents the controller modes. It decides the amount of correction and transmits this information to the manipulated mechanism, the arms. The arms then reposition the wheels through the steering mechanism. The arms do not necessarily have the strength to move the wheels, but a gain in force between steering wheel and the car wheels is supplied by the gearing or by power steering (another servomechanism).

Figure 2-3. Steering an Automobile

However, the operation of an automobile illustrates some of the weaknesses of the human being as a controller. Continued driving becomes tedious and leads to boredom, which in turn leads to inattention and poor performance. A mechanism, on the other hand, will do precisely the same thing repeatedly for long periods of time.

Another human limitation lies in the ability to discern whether the center of the automobile is over the imaginary line in the center of the driving lane. Precision is of little importance on a modern highway where the driving lane is several feet wider than the automobile. An error of a foot or more can exist with little consequence. But suppose the automobile is to be driven over an old-fashioned narrow bridge. Steering gets progressively more difficult when only a few inches of clearance exist on the sides of the car.

Finally, speed of response is definitely a human limitation. Again, driving is easy when traffic is flowing smoothly. But suddenly the unexpected happens, and an accident occurs before the brakes can be applied or the steering wheel can be turned. Worse yet, speed of response varies with the condition of the driver — age, fatigue, or the effects of medication. Automatic controls will outperform human beings in the areas of repetition, precision, and speed of response whether in the operation of a vehicle or the regulation of an industrial process.

2-5 THE WATER CLOSET — PROPORTIONAL CONTROL

The filling of the common bath room fixture[2] serves to illustrate automatic control in the proportional mode. Water is retained in the tank by a rubber stopper in an opening at the bottom. The stopper is removed by a linkage connected to an external handle. Upon actuation, water runs from the tank very rapidly, and a ball float drops as the water level decreases. The inlet valve plug is lifted from its seat by the linkage that connects it to the ball. Incoming water is admitted as the plug is lifted. The rubber ball reseats itself when the tank is nearly empty. Then the tank begins to refill.

The flow of incoming water gradually decreases as the ball rises. Finally, when the water is at the desired level, the ball closes the valve entirely, and this simple proportional device (the valve opening is proportional to the valve position) has automatically refilled the tank.

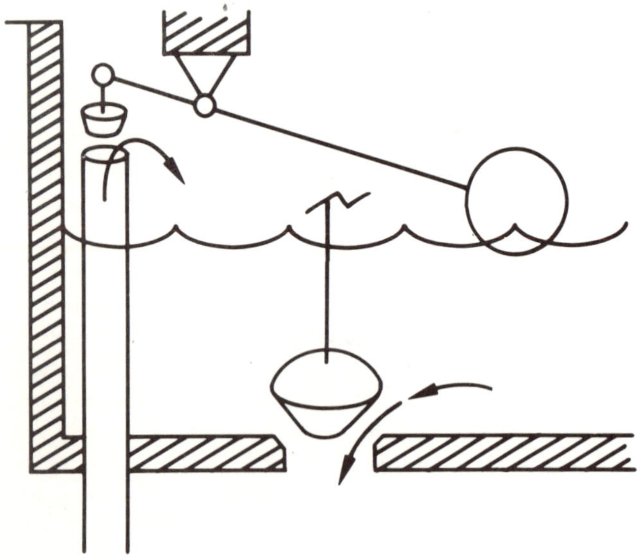

Figure 2-4. Water Closet Level

The water closet float valve illustrates one of the disconcerting characteristics of proportional control. Suppose the rubber stopper developed a small leak. The level would drop and the inlet valve would open until the inlet water flow was equal to the leak rate. At that point the level would remain stationary. Note that it would not return to the desired level because that would close the inlet valve and level would again begin to fall. This characteristic of a proportional control system is called offset, because the controlled variable (level) is offset from the desired condition (the set point) when the system is loaded (by the leak).

2-6 TANK LEVEL — MANUAL CONTROL

Most vessels used in industrial service have liquids flowing into them and out of them continuously. Generally, the tank must take whatever incoming flow it receives, and an exit flow must be manipulated to regulate the level in the tank. The flow can be manipulated by a person adjusting a manual valve. The person would watch the level in a sight glass attached to the tank and would close the valve when level decreases or open it should the level rise.

The difficulty in maintaining tank level increases significantly should the flows be large and the tank small. Under these conditions, a sudden decrease in one of the flows could cause all of the liquid to drain from the tank before the person could close the valve. In this situation, an indication of the incoming flow rate would be an invaluable aid in maintaining level. Then the person regulating the level could watch the flow gage as well as the level. The flow gage would represent feedforward, wherein the regulator (person) would be forewarned of a level change by the change in flow and could begin adjusting the valve to compensate even before level began to change. Then the level indication is a feedback, permitting the regulator to bring the variable of interest to precisely the desired value.

Quite obviously, the technique used by the person acting as a regulator to cause level in the tank to remain stationary is to manipulate the valve until the outflow equals the inflow. However, if the flows are rematched after a change in the inlet flow, the level will not be at the desired set point. Thus an offset will exist, similar to that in proportional-only control. To bring the level back to the set point from a low level condition, the outlet flow must be decreased to a rate less than the incoming flow until the level is at the set point, then be increased to equal the inflow. More generally, this control strategy would continue to close the control valve (in this example) until the level was at the set point. In this sense, it would reset the level to the set point and, appropriately, is called reset control.

2-7 TANK LEVEL — PROPORTIONAL CONTROL

While maintaining tank level by manipulating a valve is a simple task, not only would the person to whom the task is given soon be overcome by tedium, but a simple mechanism can do the job more reliably. The simplest mechanism is a ball float inside the tank, attached to a rod that is pivoted somewhere along its length. The movement of the float up and down with level causes the opposite end of the rod to move up and down by a proportionate amount. If the stem of a control valve is attached to the opposite end of the rod, the movement will open and close the valve. Thus a device has been constructed that operates in the

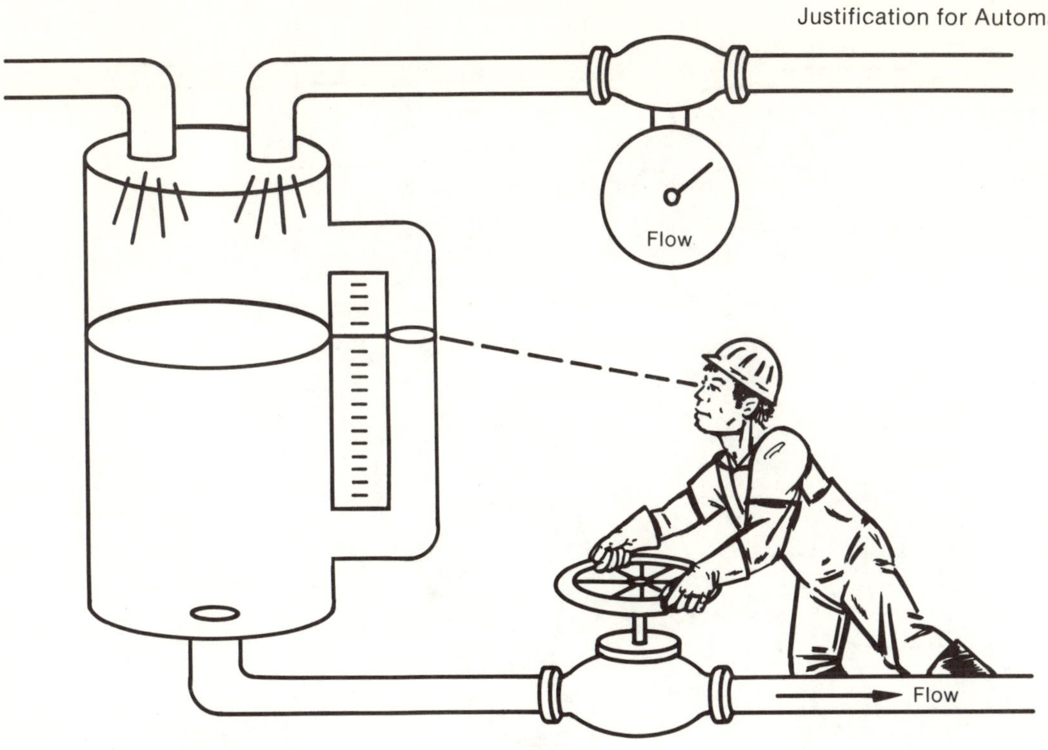

Figure 2-5. Manual Tank Level Control

same proportional mode as the water closet previously described.

While there is little interest in the level in a home water closet, it is quite necessary to visually indicate the level measurement in industrial vessels, since high level indicates a danger of running over, and low level might cause interruption of supply to the next vessel in the process. A pointer can be added to the pivot point of the float rod and be referenced to a scale showing feet, gallons, or percent full of the tank. The pointer can be used as an indicator because the float and rod have sufficient motion to provide a readable scale. This mechanism would be classified as a motion-balance instrument, and this type of instrument will usually provide sufficient movement to actuate an indicating pointer. Linkage movement is almost imperceptible in the newer force-balance instrument designs and does not lend itself to indicator actuation. With force-balance instruments, a separate measurement of the transmitted signal is required as an indication of the measured variable.

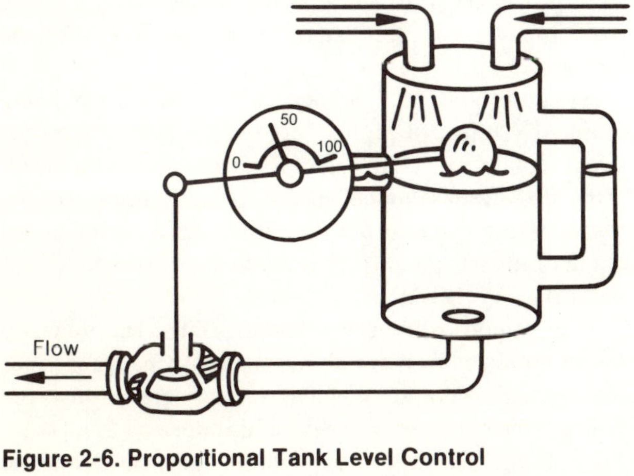

Figure 2-6. Proportional Tank Level Control

2-8 TANK LEVEL — SAMPLED DATA

Since the level in one vessel can be controlled by a person watching the level in a sight glass and manipulating a manual valve to keep the level constant, the level in two tanks could be regulated by manipulating individual valves in the pipeline from each tank. By the same reasoning, level could be controlled in several vessels by manipulating their respective valves.

In this case, the human operator would be required to look at the level in one tank and adjust its valve, look at the level in the next tank and adjust its valve, and so on until the levels in all tanks had been checked and the operator starts the sequence with the first tank. The sequence would be repeated as often as necessary to maintain all the levels within their specifications.

The reason for having a single operator oversee several tanks is obviously economic. He would be unable to devote

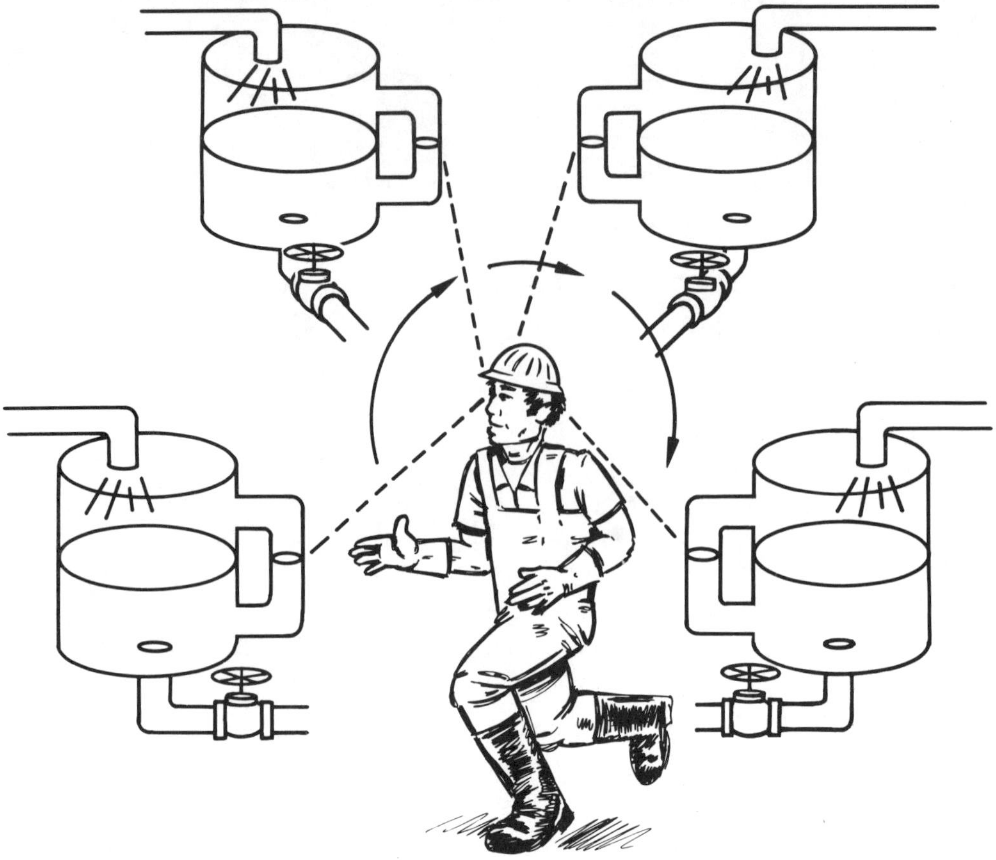

Figure 2-7. Tank Level Control, Multiple Tanks

his full attention to watching one tank level gage, but would look at the levels in sequence. In a sense, he would obtain the information (data) from which to decide how to adjust the valves by sampling the continuously available level information from each tank. By this description, the technique of having one (human) controller regulate several variables is called sample data. The economic advantage remains with the application of an automatic controller, since the sampled data system requires the purchase, installation, and maintenance of less control equipment.

The ability to regulate, or the regulation efficiency, deteriorates as more variables are sampled for regulation by the same controller. In the case of the manual level control of tanks, the operator can spend only one half of his time on each tank if he is responsible for two tanks, and one might expect level to vary twice as much in a given tank before the operator is able to return and adjust the valve. Likewise, progressive deterioration in regulation would be expected as three tanks are served, then four, and so on.

The economic benefits of sampled data control must therefore be balanced against the deterioration in regulation performance.

2-9 A SHOWER BATH — DEAD TIME

A common manual temperature control is found in the shower bath in the home or gym, where hot and cold faucets are available to the user. Both faucets are opened, and the temperature of the resulting mixed stream is sensed by the skin, which signals through the nervous system that the water is hot or cold. This information reaches the brain (the controller), which directs the manipulated variable (position of the faucet handles) to move to make the temperature comfortable.

Suppose, however, that the construction of the building required 120 feet of piping between the point where the streams of hot and cold water mixed and the shower head. If the mixed water flow velocity is 2 ft/sec, some 60 seconds would elapse between the time that a mixture of a new temperature left the mixing point and the time it reached the surface of the skin.

Now, suppose the water felt a little chilly. The hot water faucet is nudged open. Nothing apparently happens, since the lower temperature water is still coming through the pipe. So the hot water valve is nudged open a little more... still nothing... a little more... still nothing... again... and the

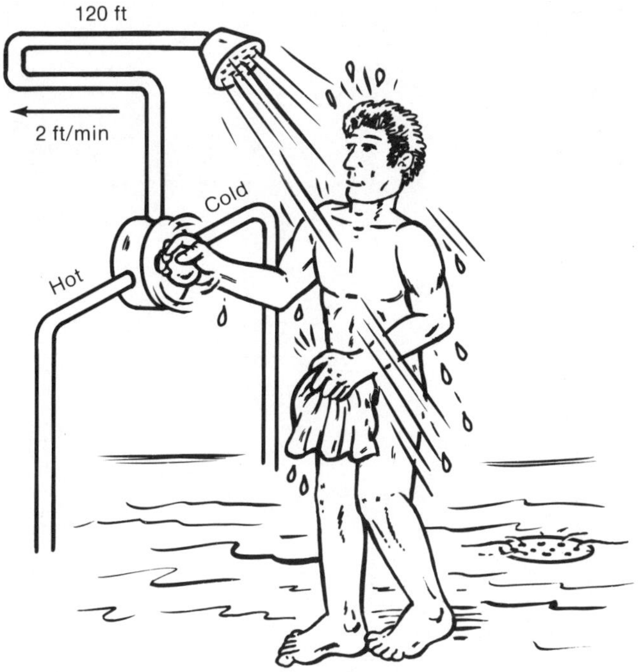

Figure 2-8. Mixing of Hot and Cold Water

water begins to get warmer because the 60 seconds during which the water is in the pipe has elapsed. During this 60-second time period, the controlled variable (temperature) has not changed or has been dead (unmoving). Thus the 60 seconds (in this example) is aptly named dead time.

But back to the story. The water begins to get warmer... and warmer... and the hot water faucet is cut back... but the water is now hot... the hot water faucet is closed frantically... it is scalding... the hot water is shut off... and the water begins to cool. Obviously, the water will be freezing cold in another minute, much to the irritation of the bather.

The bather's experience is remarkably similar to that of an automatic controller as it attempts to regulate a process that includes a significant dead time. In this case, the controller would compute an error between the set point temperature and the temperature of the flowing liquid. It would simultaneously begin to open the hot water valve and close the cold water valve. Like the bather, it would not be able to sense the result of its action until the dead time had elapsed. Thus it would continue to open the hot water valve and close the cold until a period longer than the dead time had passed, at which point the process fluid would be too hot. Then, like the bather, the controller would reverse itself, and a period of oscillating temperature would ensue until the set point temperature was finally attained.

The automatic controller, however, has a major advantage over the human bather — infinite patience. While the human wants the water at the correct temperature right now, the controller can be desensitized and made to respond very slowly, thus, in a sense, patiently waiting for the dead time to pass before it moves the valves too far.

2-10 SYSTEM VARIABLES — CAUSE AND EFFECT

A. H. McKinney[3] has pointed out that physical systems follow the laws of cause and effect. Careful distinction must be made between these two happenings. The conventions used in plotting graphs are helpful. The abscissa, or *x*-axis, is traditionally the cause, or "what I did". The ordinate, or the *y*-axis, is the effect, or "what happened".

A cause, then, is a source of material or energy flow that can produce a measurable change in one or more of the effects being observed. Synonyms for a cause are independent variable, external source, activation, forcing function, manipulated variable, and disturbance variable. In the home heating system previously described, the disturbance variable is the outdoor temperature; in the water closet, pushing the handle; in the level tank, a change in inflow.

An effect is a property or distinguishing feature of a system, that is subject to measurement, that may have significance or interest, and that responds to changes in one or more of the causes. Synonyms for effect are property, dependent variable, result, meter reading, measured variable, and controlled variable. The effect in the home heating system was quite obviously temperature and in the level tank, the level itself.

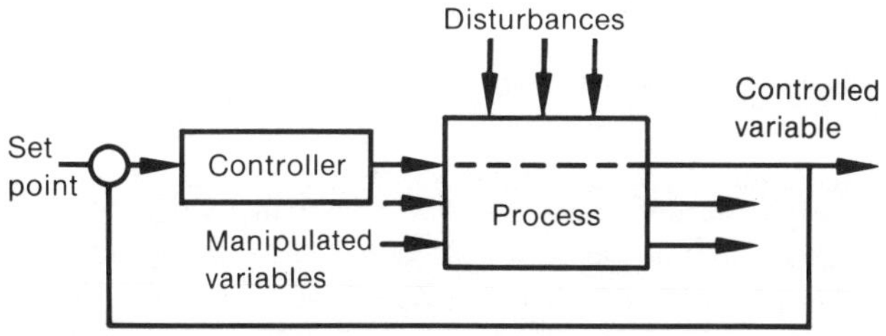

Figure 2-9. Feedback Control System[14]

Definitions become more difficult when a variable that qualifies as a cause is also an effect. Consider the level tank shown in Figure 2-5 with the feedforward flowmeter. The flow through the flowmeter is a cause that changes the level (an effect). However, the flow is measured and therefore is an effect that can be used in the feedforward system. Thus, the definition of a variable depends on its function in the control system.

The discussion of cause and effect indicates that every physical system has disturbance variables, manipulated variables, measured variables, and controlled variables. While the home heating system and the level control system are

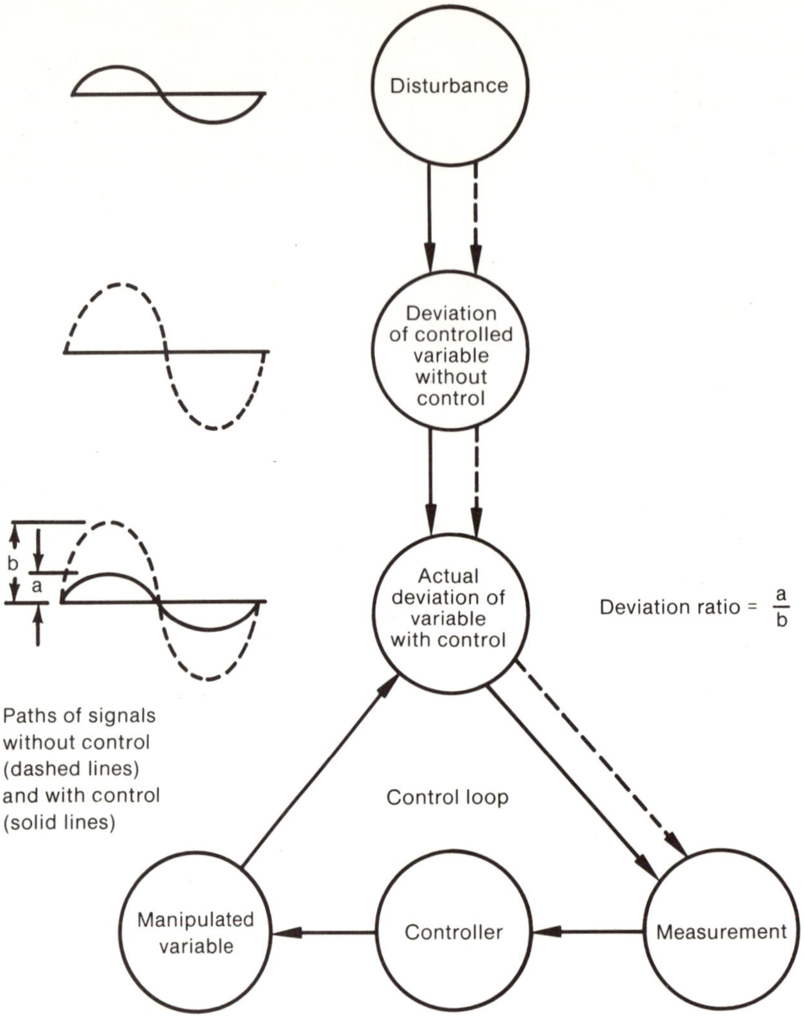

Figure 2-10. Control System Schematic[4]

vastly different in appearance, they can be described in similar terms. This realization has led to diagramming techniques that can be used to describe any control system. The diagrams consist of blocks where some action takes place and lines between the blocks that represent information being transmitted. Figure 2-10 shows a plant control system schematic that is applicable to any physical system. A disturbance is generated in a block and transmitted to another block where a deviation or error is computed; the error is transmitted and measured; the measurement goes to a controller block; the controller output goes to a manipulated variable block; and the manipulated variable is transmitted to the error block to reduce the error. The input to each block is a cause, and the output is an effect.

2-11 BLOCK DIAGRAMS

Nourse[5] has shown a diagram of a proportional closed-loop control system such as those already discussed. The controlled variable is measured and transduced to an analog signal (pneumatic, hydraulic, or electric); the measured status is compared to the desired status (set point), and an error is generated; proportional control action is taken; the action is transmitted through a logical switching point (manual-automatic station) and to an actuator, which changes the measured variable. This action causes a change in the controlled variable. The limited capabilities of the controller require human intelligence as a backup.

In order for the human operator to optimize the system on-line, he is furnished with set point and manual reset knobs and automatic-manual switching to permit him to take over entirely. Consider, for example, the on-off home heating system discussed earlier. The operator (home-owner) is given a set point with which to optimize the heating system. The system can be optimized for maximum comfort, for minimum temperature variation, or for maximum economy. For maximum comfort, a high set point

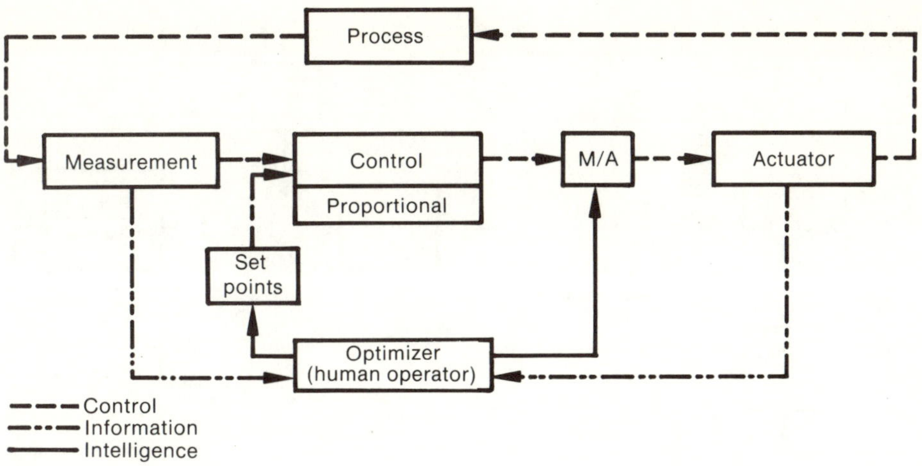

Figure 2-11. Simple Proportional Control[5]
(Courtesy *Instruments and Control Systems)*

would be chosen. To minimize the temperature variation due to the on-off control action, the set point could be lowered slightly while the furnace is running and raised slightly while it is off. For maximum economy, the set point would be lowered when the house is empty or during sleeping hours (a strategy that produces discomfort during the period when the set point is raised and the house is heating). Thus it is apparent that different optimization strategies are required by different optimization objectives. Generally, optimum performance and optimum economy are mutually exclusive. Consider only the "jackrabbit" automobile driver who "floors" either the accelerator or the brake. Barring accidents, a faster (optimum) average speed is attained but at a significant cost in gasoline and tires.

2-12 BLOCK DIAGRAM — SELF-CONTAINED PRESSURE REGULATOR

The pressure regulator provides proportional regulation of pressure within a single device. The operator is given only a set point adjustment, which consists of a screw in the metal casing. Human intervention is very difficult but is usually not required. The set point represents the desired status of the pressure in the previous discussion, and set point screw position is converted into a force by a spring within the regulator.

The controlled variable (pressure) is converted to a force by a diaphragm that opposes the movement of the spring. The diaphragm turns the pressure into a force, which opposes the force of the spring. The opposition of spring force and diaphragm force is a comparison of set point and controlled variable, and diaphragm movement is the generated error. But the valve stem is connected to the diaphragm, and diaphragm movement is also the actuator. Therefore, diaphragm movement causes valve movement, which adds more or less fluid to raise or lower the pressure to satisfy the set point-controlled variable comparison.

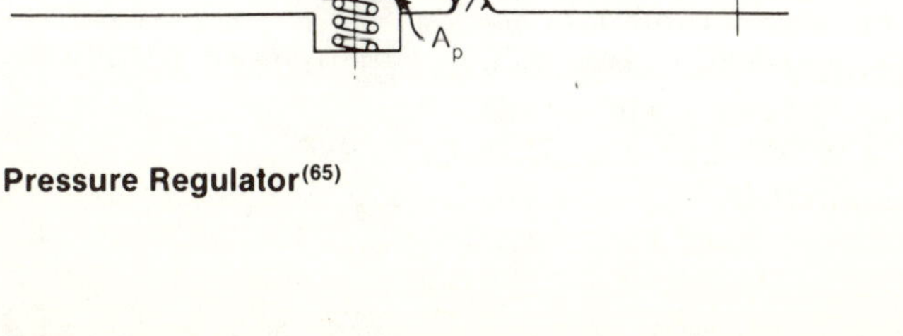

Figure 2-12A. Self-Contained Pressure Regulator[65]

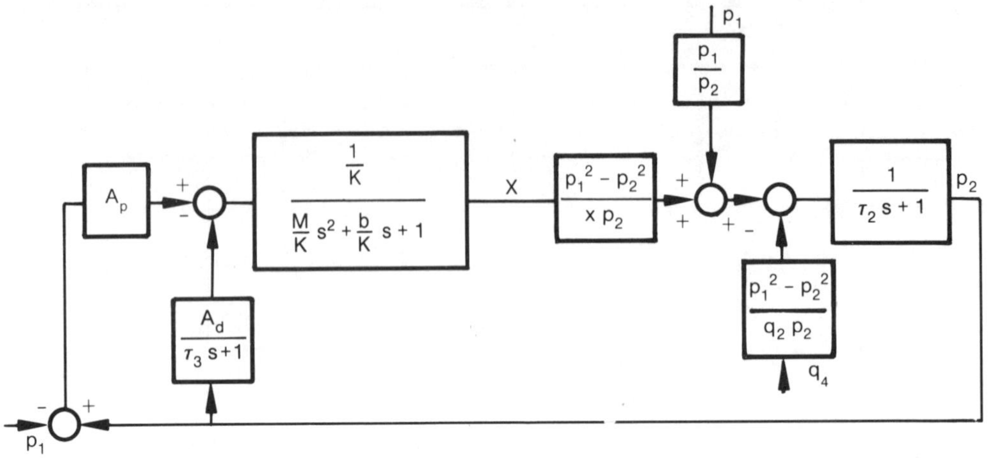

Figure 2-12B. Block Diagram of Regulator Shown in Figure 2-12A[65]

Similar block diagrams can be drawn for the level and temperature processes previously discussed.

2-13 CONTROL DYNAMICS

The importance of time in automatic control has already been mentioned. The on-off home heating system works well because the volume of the house causes temperature to take a long time to change. Manual control of level in a tank is easy when the tank is large (has a long residence time) but is very difficult when residence time is small. Dead time and sampling time have also been discussed, and the deterioration of controllability with increasing values of either one has been illustrated. The selection of instruments for the control of a process depends on the time-varying nature of the process and the speed of response of the instruments. Controllability is assessed by considering the dynamics of the system made up of the process and the instruments.

The inherent implication of the foregoing examples is that the commonly used descriptions of process behavior include timing in a changing situation. The terms "dead time," "sampling time," and "time constant" all suggest that effects in one part of the system will occur at some later time after the effects in another part of the system. By putting numbers on the "times" in the system, it becomes possible to describe system performance in numbers rather than in relative descriptions. The most familiar description of time response is in terms of the response of a measuring system or a process variable to a "step" disturbance. Basic response curves are shown in Figure 2-13.

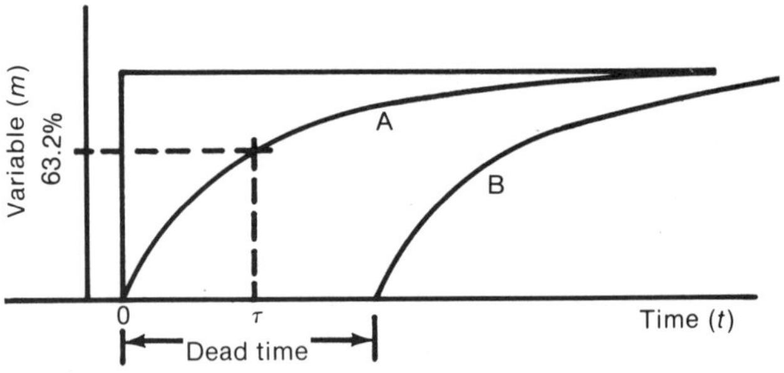

Figure 2-13. Basic Response Curves

The response curves shown above are obviously made up of an infinite number of time values. The "time constant" (τ) is a single number that has been chosen to typify response and is used universally. Curve A illustrates the time constant. Curve B shows the same time constant in combination with the dead time already described.

Mathematically, many natural phenomena such as the foregoing are described by the equation of exponential growth. This equation describes processes where the rate of change of the variable (m) is proportional to the quantity (m) itself.[6] If m is the amount of the quantity at any time t, the fact that m changes at a rate proportional to itself can be expressed in the form

$$\frac{\Delta m}{\Delta t} = km$$

Since delta (Δ) represents a difference, the left-hand side of this equation can be stated as a change in m with a change in time. The k is a proportionality constant which, as will be shown in later sections, can be calculated from a knowledge of the physical elements and thus can be calculated even before the process is built.

The foregoing difference equation (a "difference" in m with a "difference" in t) is in the form of finite differences. That is, two values of m are subtracted to obtain the difference in m. The limiting concepts of calculus [7] are used to convert the differences into an instantaneous derivative:

$$\frac{dm}{dt} = -km$$

which is still stated as the change in m with a change in time.

2-14 MATHEMATICAL MODELING

The emphasis on "timing" and "changing" in describing a control system indicates the dynamic or nonsteady-state basis for the design and adjustment of processes and their associated control systems. Quite obviously, a process that operates at one steady-state condition can be adjusted once to that condition, and no controls are required. Consider the home furnace previously described. If the outdoor ambient temperature were to remain always the same and it was desired that the indoor temperature remain constant, a furnace could be designed to provide these conditions, and no temperature control system would be required.

Control systems are thus designed on a non-steady-state basis. Controls for processes not yet built are conceived and designed using "difference" or differential equations[6] such as the one shown above. However, a process and its associated control system will be described, or modeled, by a number of algebraic and differential equations that permit calculation of the changes in the variables in the process. The combined equations are called a mathematical model of the process and control configuration.

Williams[8] has pointed out that mathematical models have the following benefits in addition to the design of the control system:

- Process performance can be improved with a continuous dynamic, as opposed to steady-state, method of operation.
- The amount of off-specification material from a given process can be determined, and oversizing the process can be avoided.
- Process dynamics information can supplement steady-state investigations in elucidating the nature of the process, which is helpful in the design of the process itself.

Goff[14] reviews the techniques for the development of dynamic process models. These models fall into the following two broad categories:

- *Theoretical models.* Manipulated and controlled variables, as well as defined and undefined disturbances, must be described in detail. Each step in the process must be defined so that physical and chemical laws may be applied to develop a complete system of equations interrelating all of the process variables. Advantages of the theoretical model include adaptability to other similar processes, ability to include highly specific phenomena, and the availability of variables in the model not presently measurable or manipulatable in the actual process. Conversely, disadvantages include the numerous simplifications and assumptions required in model development, the difficulty of solving the equations, and the uncertainty of studying only one part of a large interacting process.
- *Empirical models.* A series of tests are designed and carried out in an operating plant, and the data is analyzed to yield the dynamic process responses that comprise the empirical model. These responses are then approximated by analytical expressions that make up the model. Advantages of the empirical model are (1) avoidance of the lengthy and expensive study required to develop the theoretical model and to solve the equations, and (2) the consideration of only important variables. Disadvantages include (1) the inability to develop the model for a process not already in operation, (2) restriction to only those variables that are measurable and manipulatable, and (3) a limited scope of application.

Thomas and Tong[9] state that feedback control is used principally to compensate for uncertainty in the process, in disturbances, and in the controls themselves. Thus a mathematical model is developed from the following conditions:

- Equations completely describing the process are available.
- The environment in which the process is to operate can also be completely described by equations that will result in minimum control being required. Minimum control has the following advantages: (1) Physical systems are such that a control loop added to improve one phase of system performance usually degrades another aspect of system performance, and (2) additional control implies additional cost.

2-15 EFFECTS OF MEASUREMENT ACCURACY AND SPEED

The performance of the control system depends on both steady-state and dynamic errors of the instruments in the control loop.[10] Consider the home heating system. If the temperature-sensing device is inaccurate, the furnace (the process) and control system will never be able to bring the house temperature to the desired temperature; likewise, the shower bath with the long mixing pipe where the measurement is delayed so long after the process (the mixing nozzle) that the control system has great difficulty in bringing the temperature to the desired set point.

Instrument inaccuracy and response time are responsible for system errors the controller cannot suppress but which can be minimized by selecting the right combination of accuracy and speed in the measuring instrument. A controller is considered optimum when its performance minimizes the magnitude of the deviations of the controlled variable from the set point. VanderGrinten[10,13] has introduced the factor r as a measure of control efficiency: for $r = 0$, system completely uncontrollable; for $r = 1$, optimum control of system.

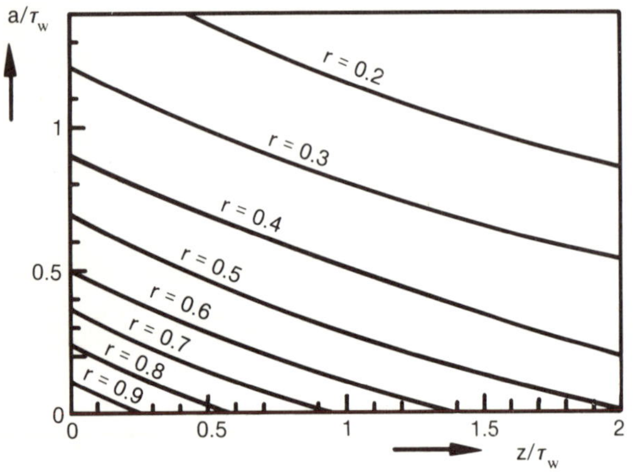

Figure 2-14. Variation of *r* as Joint Function of Dead Time *a* and Sampling Time *z*[10]

The factor r can be calculated as follows:

Dead Time Effect

$$r_d = e^{-a/\tau_w}$$

Sampling Effect

$$r_z = e^{-z/2\tau_w}$$

Accuracy Effect

$$r_a = 1 - \frac{\sigma_m}{\sigma_w} \quad \sqrt{\tau_T/\tau}$$

Combined Inefficiency

$$r = (r_d)\,(r_z)\,(r_a)$$

where τ_w = disturbance time constant

τ_T = measurement time constant

The relationship of actual to optimum deviations in the measured variable can be stated in terms of statistical variances:[11]

$$\sigma_e^2 = \sigma_w^2\,(1 - r^2)$$

where σ_e^2 = variance of actual measurement

σ_w^2 = variance of disturbance

The variance is a statistical measure of the spread of data, which permits comparison of quantities of a widely differing nature.

The fundamental causes of the difference between actual and optimal control are:

- Dead time (a). The combination of instrument and process dead times cause the correction to be applied too late. Most analyzers (chromatographs, automatic titrators) introduce dead time into industrial control systems.
- Sampling time (z). If this time is too long, fast disturbances between two sampling instants may go unnoticed.
- Measuring instrument inaccuracy. The controller cannot make the distinction between measurement deviations and instrument inaccuracy.

These factors can all be interpreted as a shortage of information, and combine to make up the control efficiency r. Besides these fundamental causes, nonlinearities and poor controller tuning can further impair performance. A rule of thumb says that in systems where $r < \frac{1}{2}$, a commercially available feedback controller will not significantly reduce the effect of disturbances.

2-16 CONTROL SYSTEM DESIGN

The control system must compensate for the ever present process disturbances in order to minimize their undesirable effects and must therefore be matched to the process.[14]

The process is identified by the process variables, or properties, listed in Table 2-1.

The manipulated variables are those valve positions, damper positions, motor speeds, and so on that can be arbitrarily controlled over operating ranges. Controlled variables, on the other hand are the pressures, temperatures, flow rates, product compositions, etc., that the control system attempts to maintain at set point values.

Load variables or disturbances can be classified as measurable and controllable, which, in principle, can be controlled and eliminated at the source; measurable and uncontrollable, such as ambient temperature; and either unmeasurable or their measurement is not economically feasible. The number of disturbances affecting a process may be one or many: Their magnitude and speed of occurrence (τ_w) must be determined to ascertain control system performance.

Table 2-1
Process Properties(15)

Term	Description	Example
Controlled variable	That process output that is to be kept at a particular value.	Solids content of the product leaving an evaporator
Load variable	Any input to the process that disturbs the controlled variable	Feed rate change to an evaporator
Manipulated variable	That process input deliberately changed to offset or counter a change in a load variable	Steam flow to an evaporator changed due to feed rate change
Capacity	The ability to store material or energy	Tank
Dead time, a	The time it takes for any change in the output of a process to result from a given change in the process input	Transportation time

(Courtesy *Instrumentation Technology*)

The typical control system is one in which the controlled variable is compared to the set point; the resulting error is operated on by the controller modes to move the manipulated variable so as to return the controlled variable to the set point. The control modes are as follows:

- Proportional band = $\frac{100}{\text{Gain}}$
- Reset time. The time for the controller to return a constant offset in the controlled variable to the set point.
- Reset rate. 1/reset time in repeats per minute

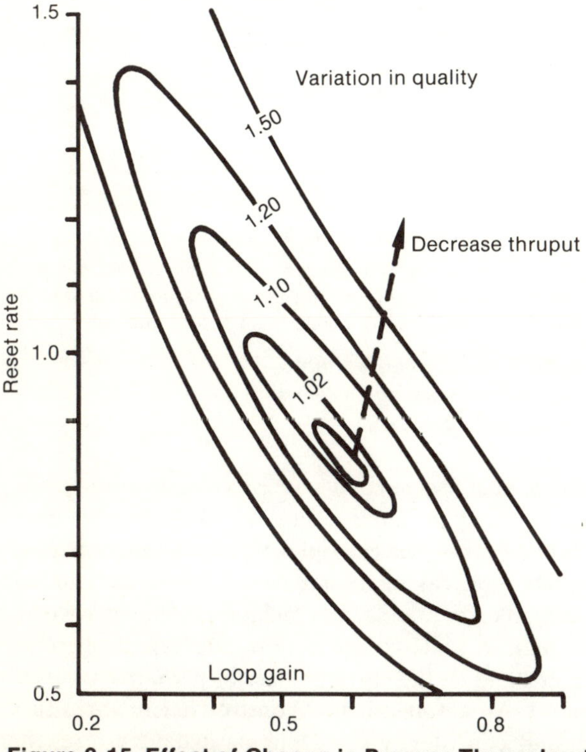

Figure 2-15. Effect of Change in Process Throughput on Controller Tuning(14)
(Courtesy *Ceramic Age*)

Proportional band is a term that is falling into disuse, and will be used sparingly. The proportional band and the reset rate are adjustable on the controller, and can be "tuned" for good system performance, thus matching it dynamically to the process.

The quotient of the reset time and controller gain (τ_i/K_c) is a measurement of the effectiveness of the control system. The variance of the controller variable (σ_e^2) will be reduced as this quotient becomes smaller. As an example, the quotient for a flow control system might be 2.5, while for a pH control system it could be 2000.

The quantification of the process and the instruments into gains (K) and time constants (τ) for mathematical modeling provides a means for assessing the cost versus performance of the instrument components.[16,17]

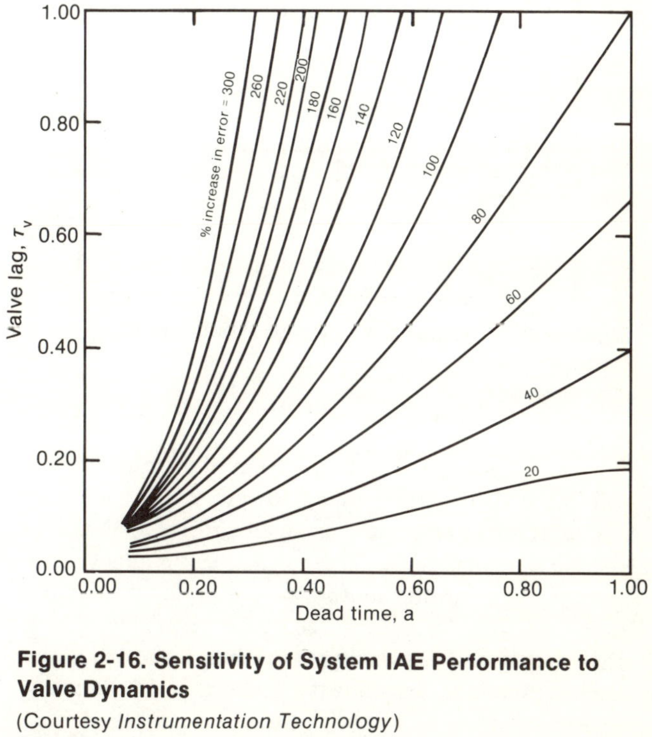

Figure 2-16. Sensitivity of System IAE Performance to Valve Dynamics
(Courtesy *Instrumentation Technology*)

The integrated absolute error (IAE) is a measure of the cumulative deviation from set point of a controlled variable responding to a disturbance. Thus it serves as a measure of control system performance. Figure 2-16 shows the effect of increasing control valve lag on performance with various (normalized) system dead times. As might be expected, performance is very sensitive to valve lag at small dead times (where valve lag predominates) but is significantly less sensitive at larger dead time values.

The control valve is an example of where the trade-off between cost and performance can be made, since control valves can be positioned with pneumatic, hydraulic, and electrical actuators. Further, pneumatic actuator performance can be improved by the addition of valve positioners and booster relays. Buckley[17] found a 600 percent improvement in dynamic performance of a valve by adding a booster relay at a 5 percent increase in cost; and a 2000 percent improvement with an electrohydraulic positioner at a cost increase of 100 percent over the pneumatic actuator alone.

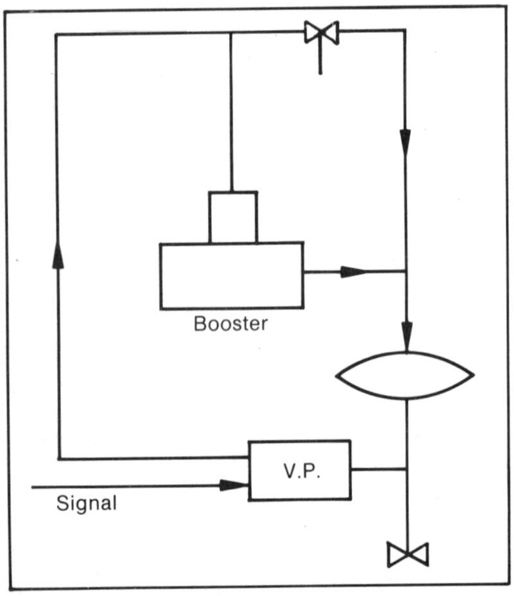

Figure 2-17. Hookup of Valve with Positioner plus Booster[17]
(Courtesy ISA)

Analysis also permits the selection of instruments that will cause the control system to ignore extraneous or unimportant signals, such as small process fluctuations and measurement noise. These signals cause minor variations in the controller output or the input to the valve actuator. Some actuators, such as a reversible electric drive, wear rapidly as the signal to them is reversed. Wear can be reduced dramatically if unimportant signals can be ignored.

Badavas(18,19) has shown how an information processor can be inserted into the control loop and error signals accepted or rejected on the basis of a prescribed doctrine. The processor passes the error signal to the controller when the control quality criterion indicates that control quality will deteriorate if the signal is lost. Further, the processor passes the signal along directly when the error exceeds a preset value, such as in the case of a sudden upset.

2-17 JUSTIFICATION OF CONTROL

In the final analysis, automatic control must be economically justified on one or more of the following bases:

- Elimination of tedius or dangerous jobs
- Additional production with the same investment
- Greater production within quality specifications

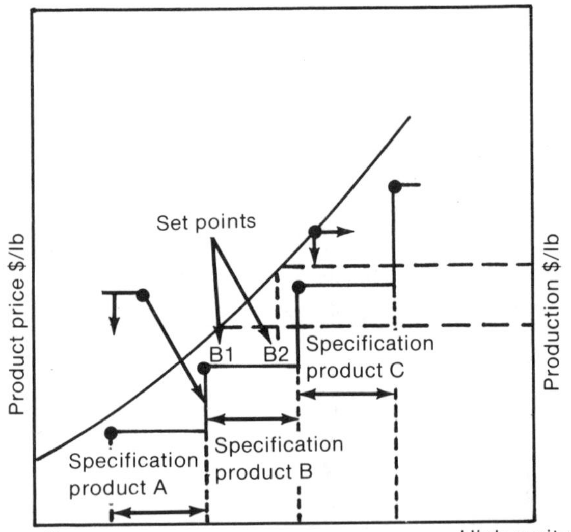

The economic potential of improved control (feedforward in this case) is evident in the amount of product that must be sold at lower prices as a result of fluctuations in output quality[15]

Figure 2-18. Controlled Variable
(Courtesy *Instrumentation Technology*)

The manual firing of a furnace, steering an automobile, and maintaining the liquid level in a tank — all mentioned earlier in the text — are examples of tedious and sometimes dangerous jobs. Taking the liquid level control as an industrial example, the manual manipulation of the valve would be so tedious that the operator would rapidly become distracted and let the tank run over or perhaps quit from boredom. An automatic level control system works tirelessly and consistently for 24 hours a day. Worse, the human operator's salary would be such that the operation would be economically prohibitive.

Silva[20] investigated the economic justification for improvement of the process for the catalytic reforming of naptha by (1) improving existing pneumatic instrumentation, (2) adding process analyzers, or (3) adding analyzers and digital supervisory control. Justification includes cost estimates of gross savings per day with each instrument system, installation costs, operating costs, and manpower expenditures. From these figures, a return on investment for each alternative can be determined. The results show that additional analyzers with digital supervisory control provide the largest profit potential at a high return.

Justification also can be based on process optimization,[21] where optimization refers to identifying and enforcing controller set points that correspond to maximum profit or minimum costs. Optimization requires that a significant measure of process performance — profit, production rate, yield, or cost — be expressed as a mathematical relationship of some key process variable. Good regulation is closely related to optimization, since it permits operation closer to some limiting condition, i.e., flooding of a distillation column or surge in a centrifugal compressor. Benefits from optimization depend on the extent of prior poor control, the ability to calculate and track the optimum operating point, and the degree to which limiting conditions interface with operating at the optimum.

Finally, justification of control can be determined by an analysis of market price for quality specifications.[15] The analysis includes market price fluctuations as a result of control quality and production cost as a function of control system specification. Thus, if regulation is poor and the quality variable spends considerable time away from the set point, the set point must be set for a much higher quality than required so as not to violate the lower limit on quality specification. Better control (costing more) permits operation nearer the lower limit, thereby reducing production costs.

Figure 2-18 shows market price as a function of product specification (set point) and production cost (right hand scale) as a function of control specification. The product is fictitious with low, intermediate, and high purity levels. Poor control necessitates setting the set point higher than required (say, at B2) so as not to violate the lower limit of product B specification. As a result, some product C is made but sold at the product B press level. Improved control permits set point operation closer to the lower limit (say at B1), with lower production costs (right hand scale) but without violating the lower specification limit.

The sections that follow will describe techniques for the determination of the performance of control systems.

Problems

1. Match the following automatic control systems with the appropriate mode of control:

(a) Home heating	1. Proportional only
(b) Industrial liquid tank	2. On-Off
(c) Water closet	3. Proportional reset

2. What is the manipulated variable in each of the following control systems?
   (a) Water closet:

(b) Home heating:

(c) Industrial liquid tank:

3. Describe:
(a) Regulation

(b) Servomechanism

(c) Manual

(d) Automatic

4. Concerning the water closet, is a leak in the stopper (or plug) a cause or an effect?

5. What is the measuring sensor in a self-contained pressure regulator?

6. Calculate the control efficiency (r_d) for a process described dynamically by
(a) Dead time = 1 minute; time constant = 10 minutes
(b) Dead time = 10 minutes; time constant = 1 minute

STUDENT SUMMARY NOTES
and QUESTIONS FOR INSTRUCTOR

SECTION 3
ELEMENTS OF PROCESS DYNAMICS

3-1 COMMENTARY

Some years ago, processes were designed using steady-state concepts; after they were completed, instruments were added that permitted an operator to run the process. The process could be a heating system, a chemical plant, or a locomotive. Correlations for traditional engineering design at equilibrium were expressed and solved largely by arithmetic and algebra.

Gradually, improved performance was demanded of the processes. In chemical plants, throughput was increased, drastically decreasing the time available to detect and correct a process upset that resulted in a poor quality product. Equilibrium became the exception rather than the rule. With the new demands, it became necessary to consider the process and the instruments as a system that had to be designed as a unit. It also became necessary to consider periods of operation when equilibrium had not been reached. These periods are referred to as transients.

The term "dynamic analysis" is used to describe a method for quantitatively analyzing the time-varying response of a system that has been subjected to a disturbance. Mathematical descriptions serve as the basis of dynamic analysis. All material and energy transfers that occur in the system of interest must be described in detail. This can become a rather formidable task, since the description must be in terms of differential equations to model the system under transient (time-varying) conditions. When sufficient expressions to describe any practical system are correlated into a number of differential equations, it is economically impractical to solve them manually by the methods of classical calculus. Results are attained by resorting to operational calculus or to computer simulation.

The terms "systems engineering" and "dynamic analysis" are often used interchangeably. The terms are tending to diverge with the passage of time, however, and the sets of information in each field might be described as shown in Figure 3-1.

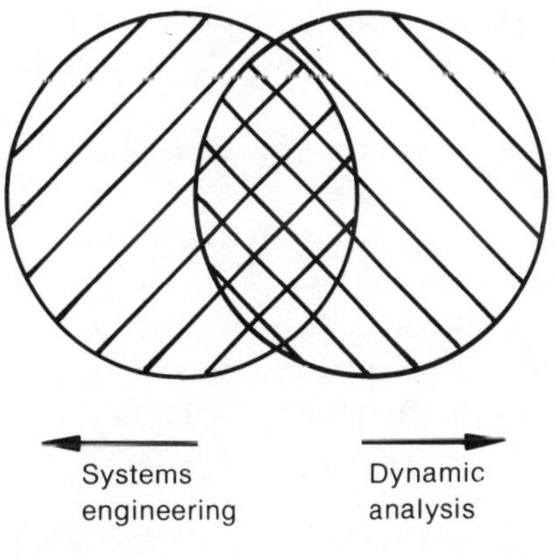

Figure 3-1. Overlap

The portion of systems engineering not included in dynamic analysis has to do with instrument accuracy, serviceability, and applicability, while the unique features of dynamic analysis are the development and solution of the multiple differential equations that describe the system. The following questions can be answered by systems engineering for a petrochemical process:

(1) Can the system be controlled?

(2) What is the effect of dead time or instrument response?

(3) What controller modes should be used?

(4) How should the controller be tuned?

(5) What control valve flow characteristic should be used?

(6) Can the instrument system meet the accuracy specifications?

These questions are not easily answered. Today, however, the tolerance for empirical and seat-of-the-pants solutions is diminishing. The increasing economic squeeze and the increasing size of processing plants demand that new installations be designed for control while still on the drawing boards. The postmortem investigation for the purpose of correcting difficulties while delaying start-up or operating on manual control is too costly. Rapid technological change demands higher engineering performance; and the questions that demand answers do become more difficult.

In addition, dynamic analysis is directed toward answering these questions:

(1) How much improvement in control can be expected by substituting 3/8-inch pneumatic tubing for 1/4-inch?

(2) What vessel volumes are required?

(3) Will the control system be stable?

(4) What is the magnitude of variations in the controlled variable?

(5) Is feedforward required (and feasible) to meet project specifications?

(6) Will the control systems interact?

(7) Which of several alternative systems will perform best?

The utility of dynamic analysis can be illustrated by these questions that were asked about a recycle centrifugal compressor:

(1) Would resonance occur in the recycle line around the compressor?

(2) What should be the size of the surge tank on the suction side of the compressor to permit control of inlet pressure to between +10 inches of water and −1 inch of water?

Some 10 man-hours of study resulted in the recommendation that the compressor be powered by a variable speed drive and in the elimination of the surge tank and its associated instruments because inlet pressure was shown to then be extremely stable. In this case, a few man-hours of mathematical analysis eliminated the use of $10,000 worth of instruments and a tank.

Systems engineering, then, requires an understanding of the process as well as the instruments. A process, in the science of automatic control, is an operation or series of operations on a fluid or on solid materials during which the materials are placed in a more useful state. A unit process in a chemical plant will usually involve a change in chemical state. A mechanical process, such as those in the metal-working industries, usually produces a change in physical state.

Many external and internal conditions affect the performance of a process. These conditions are expressed in terms of process variables such as weight, temperature, flow, composition, boiling point, and so on. The process may be controlled by measuring a variable that represents the state of the desired end point and automatically adjusts one of the other variables of the process.[22]

3-2 INTRODUCTION

College, professional, and short courses provide considerable detail on the fundamentals of mathematics and instrument component operation. In fact, they have provided far more detail than is needed by the practicing instrument engineer or control system engineer who is not specializing in the study of dynamics.

Accordingly, the objective of this text is to demonstrate some simple methods whereby one can, in general, determine whether a potential control problem exists by combining some basic calculations of time constants and dead times with a few rules of thumb and with experience.

For example, suppose that gas pressure in a header must be controlled to ± 1 psi in spite of step changes in downstream demand. A quick calculation might show that this step change in demand would cause header pressure to change at the rate of 0.1 psi per second. From prior knowledge of response rates of pressure transmitters and control valves, it might be concluded that the control loop would be adequate to handle the upset. On the other hand, if the calculation showed that pressure would change at the rate of 10 psi per second, there would probably be a control problem since the control loop would now have only 0.1 second in which to exert full corrective action.

Thus, by combining some simple calculations with experience and judgment, one can in many cases decide whether or not a control problem exists. To make such calculations readily, however, requires practice — enough so that they become as familiar as orifice sizing calculations. To this end, questions and problems throughout and at the end of each section illustrate the principles of that section. If the principles are understood, the problems can be solved or the questions answered in a few minutes.

Such practice will serve as the basis for rapid calculations as to the severity of the potential control problem.

3-3 THE BASIC CONCEPTS

Generally, process or equipment dynamics can be defined in terms of the two fundamental concepts: the unsteady-state energy or material balance, and the driving-force principle. The unsteady-state equation can be simply stated:

Input – output = Rate of accumulation

The unsteady-state equation is well illustrated by the liquid level in a tank as shown in Figure 3-2.

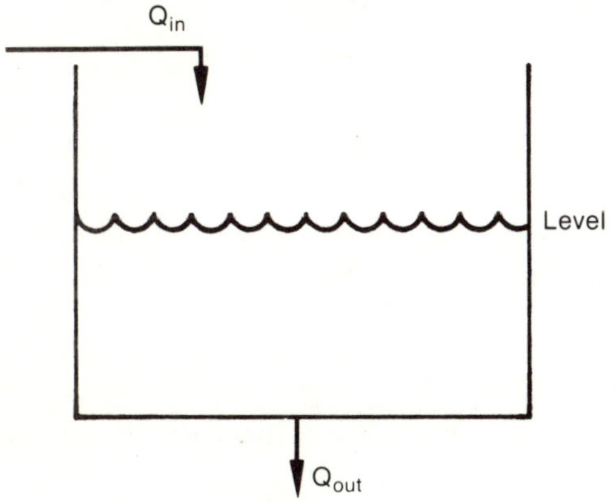

Figure 3-2. Liquid Level System

The rate of accumulation in the tank is observed by the change in level and is the result of a mismatch between input and output.

The driving-force equation is equally simple:

$$\text{Rate of transfer} = \frac{\text{Driving force}}{\text{Resistance}}$$

Ohm's Law is a classical example of the driving-force equation:

$$\text{Current flow} = \frac{\text{Voltage}}{\text{Resistance}}$$

These are the concepts that will be dealt with in this text. Note that they apply equally well to either material or energy transfer. It is necessary only to properly define the driving force and the nature of the accumulation and resistance terms to apply them to problems that involve electrical circuitry, heat transfer, fluid flow, mass transfer, etc.

Before considering in detail the application of these concepts to physical systems, familiarity with two other topics will be necessary.

3-4 LAPLACE NOTATION

For purposes of this text, very little need be known about the Laplace transform method of solving differential equations. It is worthwhile to be familiar with a few simple transforms, however, since much of the literature presents frequency response information in terms of the Laplace notation.

Without concern for the mathematical derivations of the transforms, s will be defined as the Laplace operator and it can be stated:

$$s = \frac{d}{dt}$$

Note that the units of s must be reciprocal time (time^{-1}). Reference to extensive tables of transforms[23] permits the listing of some common conversions between the time (t) and complex (s) domains:

$f(t)$	$f(s)$
$\frac{d}{dt}$	s
$\frac{d^2}{dt^2}$	s^2
$\int dt$	$1/s$
$e^{-t/a}$	$\frac{a}{as+1}$

The first derivative in the time domain is denoted by s, the second by s^2, etc. Similarly, since integration is the inverse of differentiation, integration is denoted by $1/s$. The exponential decay term $e^{-t/a}$ transforms to the first-order lag.

The following examples will indicate the use of the Laplace transform:

$$\frac{dX}{dt}(t) = sX(s)$$

$$A\int Q(t)dt = \frac{AQ(s)}{s}$$

$$Be^{-t/c}(t) = \frac{Bc}{cs+1}$$

$$A\frac{d^2x}{dt^2}(t) + B(t) = As^2x(s) + B(s)$$

Frequently, the parenthetical t and s are omitted to simplify the notation. Note that each term in an equation is transformed separately; the same is true of the inverse transform; i.e., the return from the complex domain to the time domain. Note also that the differential equations in the time domain have become algebraic equations in the complex domain and can generally be manipulated like any other algebraic equations; that is, they can be added, subtracted, factored, etc.

3-5 TRANSFER FUNCTIONS

A transfer function defines the dynamic relationship between output and input, expressed in Laplace notation. It is considered as having a steady-state gain term (K) and a time-varying or dynamic term (G). In block diagram form, it is expressed as:

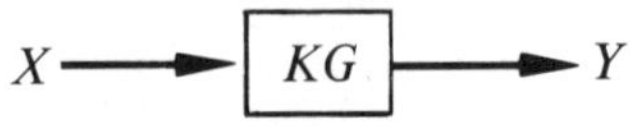

Symbolically, the block diagram states that X multiplied by KG equals Y, or

$$\frac{Y}{X} = KG \tag{3-1}$$

Example

Consider a typical heat transfer equation:

$$Q - \frac{T}{R} = C\frac{dT}{dt}$$

where Q = heat flow rate

T = temperature

R = resistance to heat transfer

C = heat capacitance

t = time

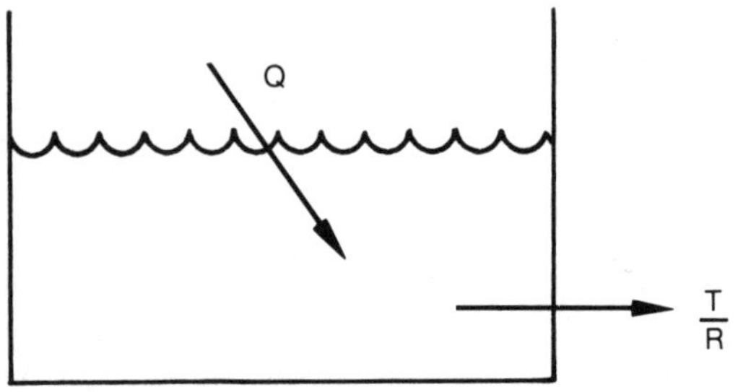

Figure 3-3. Generalized Heat Exchanger

Transforming to the Laplace notation (omitting the s in parenthesis),

$$Q - \frac{T}{R} = CsT$$

From this expression, the transfer function defining the response of T to a change in Q can be derived:

$$RQ - T = RCsT$$

$$RQ = RCsT + T$$

$$RQ = (RCs + 1)T$$

$$\frac{T}{Q} = \frac{R}{RCs + 1} \tag{3-2}$$

Figure 3-4 shows this in block diagram notation.

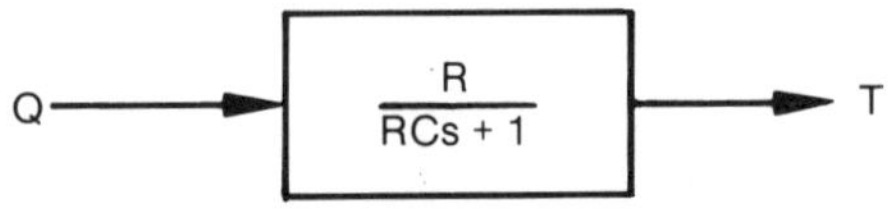

Figure 3-4. Block Diagram

Two useful concepts will now be considered. For purposes of this text, s has been defined as:

$$s = \frac{d}{dt}$$

This relationship indicates that if $dt \to 0, s \to \infty$; and if $dt \to \infty, s \to 0$. Further, if the time at the start of the transient (t_o) is considered to be zero, then $dt \to t \to 0$ results in $s \to \infty$, and $dt \to t \to \infty$ results in $s \to 0$. These concepts lead to two very important theorems concerning Laplace transforms.[109]

Final Value Theorem. The final value theorem gives the final value, if it exists, of the time function in terms of the behavior of the Laplace transform as s tends to zero, as follows:

$$\lim_{t \to \infty} g(t) = \lim_{s \to 0} s\,G(s)$$

Initial Value Theorem. The initial value theorem states that

$$\lim_{t \to 0} g(t) = \lim_{s \to \infty} sG(s)$$

These theorems are subject to mathematical constraints and are not applicable in some special circumstances. However, in general they provide a convenient method for evaluating a transfer function at time = 0 and at time = ∞, when the new steady-state condition has been reached.

The Laplace transform of a sudden change in a variable, called a step change, is $1/s$. The theorems when used with a step change simplify to:

$$\text{Lim} \frac{s\,G(s)}{s} = \text{Lim}\; G(s)$$

Applying the theorems to the transfer function shown in Figure 3-4, the initial value of that relationship is

$$\lim_{s \to \infty} \frac{R}{RCs + 1} = \frac{R}{RC\infty + 1} = 0$$

Thus, at time equals zero, the change in T resulting from a change in Q is zero. The final value of the transfer function in responding to a step change in Q is:

$$\lim_{s \to 0} \frac{R}{RCs + 1} = \frac{R}{RC(0) + 1} = R$$

Then, after a step change in Q, the final steady-state change in T will be

$$\Delta T = R \Delta Q \qquad (3\text{-}3)$$

The final value theorem permits determination of the final value that the output will assume after a step change to the input.

Problems

1. Write the Laplace transforms for the following:

(a) $q_1 - q_2 = C\dfrac{dL}{dt}$

(c) $A\dfrac{d^2x}{dt^2} + B\dfrac{dx}{dt} = 0$

(b) $E = \dfrac{1}{C}\displaystyle\int I dt$

(d) $\left(\dfrac{dx}{dt}\right)^2$

2. Derive the transfer function p/w_1 by combining the following:

$$w_1 - w_2 = C\frac{dp}{dt}$$

$$w_2 = \frac{p}{R}$$

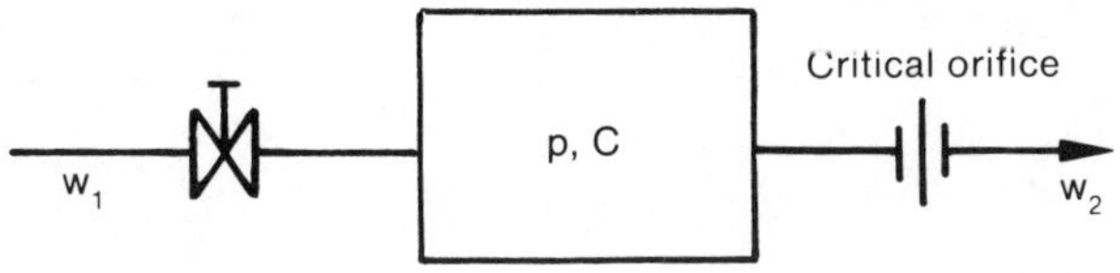

3. The equation that relates the output M of a proportional-plus-reset controller to the input error signal e is:

$$M = K_c e + \frac{Kc}{\tau_i} \int e\,dt$$

where K_c = controller gain

τ_i = reset time

Develop the transfer function M/e.

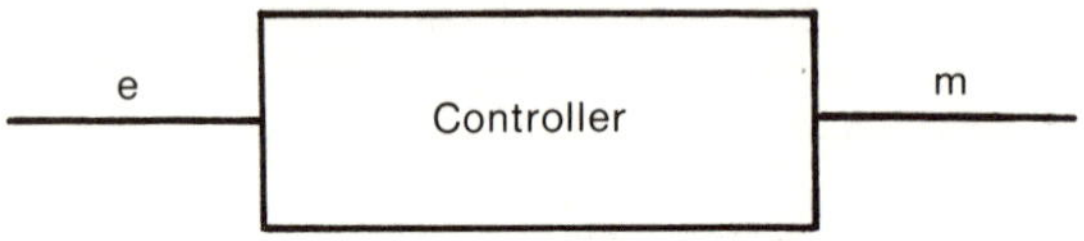

4. The equation relating control valve stem position x with topworks pressure p is:

$$M \frac{d^2x}{dt^2} + B \frac{dx}{dt} + Kx = pA$$

where M = mass of moving parts

B = damping factor (viscous friction)

K = spring constant

A = effective diaphragm area

Derive the transfer function x/p.

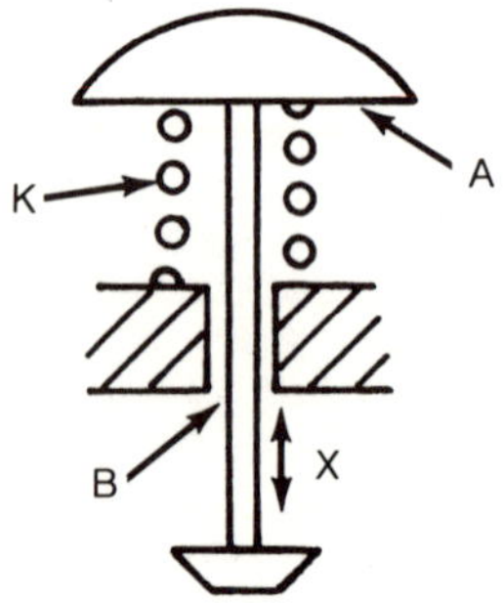

3-6 CAPACITANCE AND THE UNSTEADY-STATE EQUATION

Section 3-3 described the basic unsteady-state equation as:

Input − output = Rate of accumulation

If input does not equal output, which would be the case in the unsteady state, there must be an accumulation or depletion of energy or material on the right-hand side of the equation. This implies that the system under consideration has capacitance—the ability to store energy or material. If the system does not have capacitance, it can neither accumulate nor supply energy or material; then the rate of accumulation is zero, and output must equal input. By definition,

$$\text{Capacitance} = \frac{\text{Quantity stored}}{\text{Potential}}$$

Thus, for a given capacitance, the quantity stored can be increased by increasing the potential. Similarly, if the potential is reduced, the quantity stored is reduced; i.e., energy or material has flowed out of the system.

The actual value of capacitance is generally defined by the dimensions or physical properties of the system. Capacitance will be considered as related to the unsteady-state equation for a variety of specific systems in the rest of this section.

3-6-1 Electrical System

Consider a simple electrical circuit as shown in Figure 3-5.

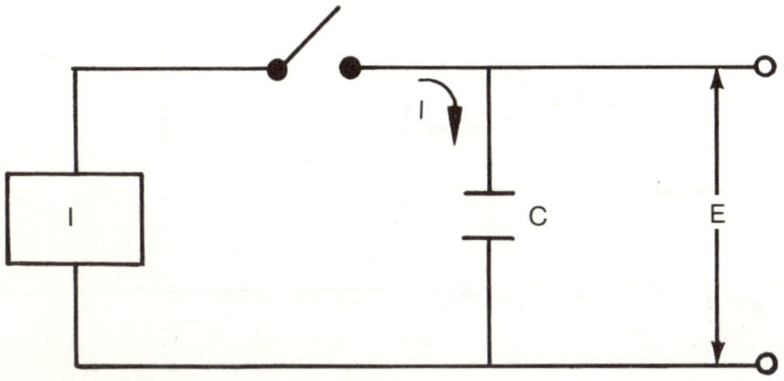

Figure 3-5. Electrical Circuit

Assume a constant current source I available to charge capacitor C from an initial value of zero to some voltage E when the switch is closed. Assume that there is no other current flow; i.e., that the output flow = 0. From the unsteady-state energy balance equation,

Input − output = Rate of accumulation

the following can be written:

$$I - O = \frac{dq}{dt}$$

where I = current flow

q = electrical charge

Interest here, however, is in the relationship between voltage E and current I. Defining, then,

$$C = \frac{q}{E} = \frac{\text{Quantity stored}}{\text{Potential}}$$

or

$$q = CE$$

Then, differentiating,

$$\frac{dq}{dt} = C\frac{dE}{dt}$$

and

$$I = C\frac{dE}{dt} \tag{3-4}$$

First check the dimensional consistency of Equation 3-4. In popular units,

I = amperes

C = farads

E = volts

t = seconds

Then, substituting these dimensions, it can be seen that

$$\text{amperes} = \text{farads} \times \frac{\text{volts}}{\text{second}}$$

This does not appear dimensionally correct. If, however, the more basic electrical units are used,

I = current in coulombs/second = amperes

C = capacitance in coulombs/volt = farads

then,

$$\frac{\text{coulombs}}{\text{second}} = \frac{\text{coulombs}}{\text{volt}} \times \frac{\text{volts}}{\text{second}}$$

Thus, the equation is dimensionally consistent.

The capacitance term can be considered as a factor that converts from the units of the flow terms on the left (in this case, current flow) to the units of rate of change of the potential term on the right (in this case, volts/second).

Since interest was in the relationship between voltage E and current I, find the transfer function E/I from Equation 3-4 above. Transforming gives

$$I = CsE$$

whence,

$$\frac{E}{I} = \frac{1}{Cs} = \left(\frac{1}{C}\right)\left(\frac{1}{s}\right) \tag{3-5}$$

and the block diagram becomes

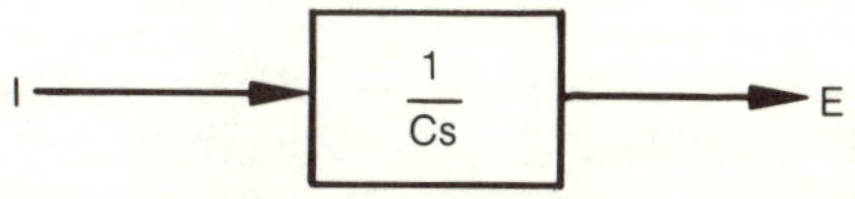

From the inspection of Equation 3-5 or the block diagram, several conclusions can be drawn:

(1) From the Laplace transforms, it is known that $1/s$ represents pure integration; as long as I has a finite value, E will continue to change, since $I/s = \int I dt$.

(2) The rate of change of E will vary inversely with C. If C is large, E will change slowly.

(3) Applying the final value theorem as $t \rightarrow \infty$, $s \rightarrow 0$, it can be seen that eventually $E = I/O = \infty$. Thus, in principle, as long as current can be forced into the capacitor, the voltage will continue to rise. There are, of course, practical physical limits to the system; at some value of E, either the capacitor will break down or the current generator will be unable to maintain constant current. Nevertheless, the concept is valid.

Appendix A shows the mathematical development of the dynamic relationship in an electrical circuit using the formal solution of the differential equations. That circuit is shown as going from one voltage level to another. The great simplicity introduced by the Laplacian operator notation and by assuming that a design or set point datum is defined as zero is striking when Appendix A is compared to the foregoing development.

3-6-2 Thermal Systems

Consider a tank of well mixed liquid (i.e., uniform temperature throughout) as shown in Figure 3-6.

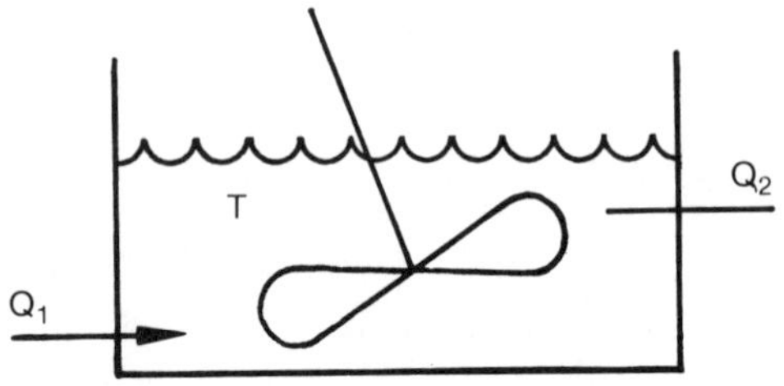

Figure 3-6. Heat Transfer Process

Assume a heat input of Q_1 and a heat loss of Q_2. From the unsteady-state heat balance,

$$Q_1 - Q_2 = C\frac{dT}{dt} \qquad (3\text{-}6)$$

Let us define Q in PCU/hour (where PCU is the amount of heat necessary to raise 1 lb of water 1°C), T in °C, and t in hours. By inspection, the units of C must be:

$$\frac{\text{PCU}}{\text{hour}} - \frac{\text{PCU}}{\text{hour}} = \frac{\text{PCU}}{^\circ\text{C}} \times \frac{^\circ\text{C}}{\text{hour}}$$

Intuitively, it would be expected that the heat capacitance C would depend on the amount of liquid in the tank and on the ability of the liquid to store heat; i.e., its specific heat.

Define M = weight of tank contents in lbs and c = specific heat of liquid in PCU/lbs °C. Then,

$$Mc = (\text{lbs})\left(\frac{\text{PCU}}{\text{lbs }^\circ\text{C}}\right) = \frac{\text{PCU}}{^\circ\text{C}} = C$$

Substituting $Mc = C$ in Equation 3-6,

$$Q_1 - Q_2 = Mc\frac{dT}{dt}$$

or, in Laplace notation,

$$Q_1 - Q_2 = McsT$$

The transfer function for T as a function of $Q_1 - Q_2$ becomes

$$\frac{T}{Q_1 - Q_2} = \frac{1}{Mcs} \qquad (3\text{-}7)$$

and the block diagram is as shown in Figure 3-7. Note the necessity for the sign convention at the summation point (Σ) in the block digram; the diagram must be consistent with the equation, otherwise T may be increasing as the heat loss Q_2 increases.

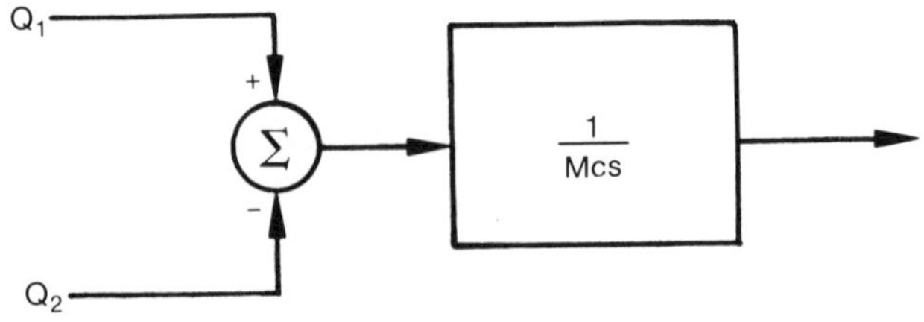

Figure 3-7. Block Diagram—Heat Transfer

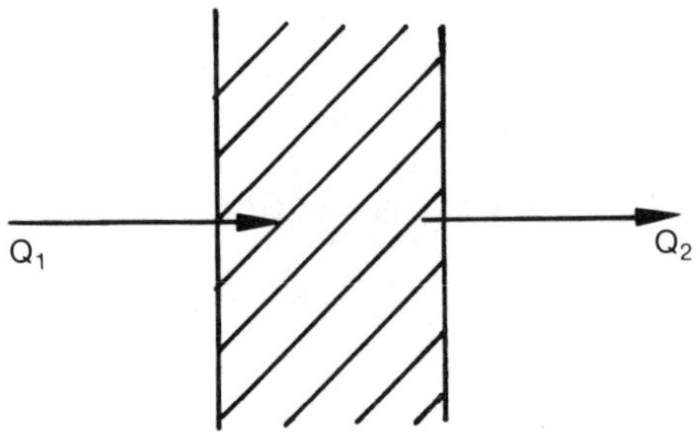

Figure 3-8. Heat Exhanger Tube Wall

From inspection of Equation 3-7 or the block diagram of Figure 3-7, the same conclusions can be drawn as those drawn for the electrical system:

(1) As described, the system represents pure integration (i.e., contains $1/s$). In practice, Q_2, the heat loss, can be expected to vary with T, but Q_2 has not yet been defined as a function of T.

(2) The rate of change of T for any given values of Q_1 and Q_2 will vary inversely with the capacitance; i.e., with the quantity of liquid in the tank and with the specific heat of the liquid.

(3) Applying the final value theorem, it can again be seen that as $t \to \infty$ and $s \to 0$, $T \to \pm\infty$ depending on the relative values of Q_1 and Q_2. This response is typical of pure integrating systems and is valid within the physical limits of the system.

The above analysis could be equally well applied to the wall of a heat exchanger tube if the wall were thin enough to be considered at uniform temperature from inside to outside. In most cases this is a valid assumption, at least for a first approximation. In the case of a solid, of course, M and c would refer to the weight and specific heat of the solid.

Problems

1. A controller delivers a constant output of 1 milliamp dc. What will be the rate of voltage change if the controller delivers into:

   (a) A 1-microfarad condenser?

   (b) A 100-microfarad condenser?

   (c) A 1-farad condenser?

2. Equation 3-7 described a tank heating system. With $Q_1 = 20{,}000$ PCU/hour and $Q_2 = 5000$ PCU/hour, what will be the rate of change of temperature if the tank contains:

   (a) 1000 lbs water with specific heat = 1.0 PCU/lbs °C?

   (b) 1000 lbs kerosene with specific heat = 0.5 PCU/lbs °C?

   (c) 500 lbs water with specific heat = 1.0 PCU/lbs °C?

3. A heat balance around the tank above, with 1000 lbs water contained, showed a temperature rise of 20° C/hour with a constant heat input of 30,000 PCU/hour. What is the heat loss?

4. Suppose, in Equation 3-7, heat loss Q_2 is defined as being proportional to T; that is, $Q_2 = KT$. What is the transfer function for T/Q_1?

3-6-3 Incompressible Fluid Systems

Consider a tank partially filled with liquid as shown in Figure 3-9. Application of the unsteady-state material balance equation gives

$$w_1 - w_2 = \frac{dM}{dt} \tag{3-8}$$

where w = flow in lbs/second

M = holdup in lbs

t = time in seconds

Figure 3-9. Liquid Level Process

Assume the need to determine the influence of changes in flow rate on level L rather than on weight M. By definition,

$$M = \rho V = \rho AL$$

where ρ = density in lbs/ft³

V = volume in ft³

A = cross-sectional tank area in ft²

L = liquid level in feet

Assuming that A and ρ are constant, differentiation gives

$$\frac{dM}{dt} = \rho A \frac{dL}{dt}$$

which can be substituted into Equation 3-8 to yield

$$w_1 - w_2 = \rho A \frac{dL}{dt} \tag{3-9}$$

By comparison with previous unsteady-state equations, it can be recognized that

$$C = \rho A$$

and that in this case C must have units of lbs/ft.

If w had been defined as gallons/minute, L as liquid level in inches, and t as time in minutes, C would have to have units of gallons/inch. Thus, while C represents a capacitance, its dimensions will depend on the units used in the balance of the equation.

Transforming and rearranging Equation 3-9, the transfer function turns out to be

$$\frac{L}{w_1 - w_2} = \frac{1}{\rho As} \tag{3-10}$$

Again it can be seen that this system will integrate the difference between w_1 and w_2, at a rate inversely proportional to the liquid density and the cross-sectional area of the tank, and that by the final value theorem, L will eventually approach ∞ (in this case, the top or bottom of the vessel).

3-6-4 Compressible Fluid Systems

Consider the following compressible fluid system shown in Figure 3-10. If w_1 exceeds w_2, pressure p in the volume V will increase, raising the density ρ and the amount of material stored M. Unsteady-state material balance will allow writing

$$w_1 - w_2 = \frac{dM}{dt} \tag{3-11}$$

where w = flow rate in lbs/second

M = weight of compressible fluid in the vessel in lbs

t = time in seconds

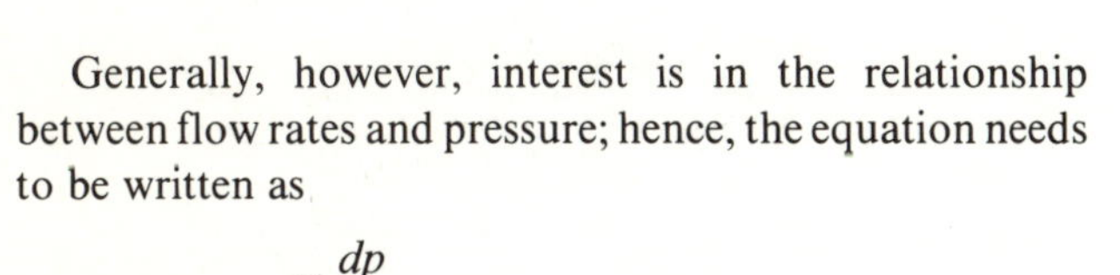

Figure 3-10. Gas Pressure Process

Generally, however, interest is in the relationship between flow rates and pressure; hence, the equation needs to be written as

$$w_1 - w_2 = C\frac{dp}{dt} \tag{3-12}$$

There are several methods of evaluating C, depending on the nature of the fluid and the thermodynamic relationships that apply. If the fluid can be considered as an ideal gas (i.e., it is far above its boiling temperature), C may be defined in terms of the gas law, which states

$$pV = \frac{M}{m}\bar{R}T$$

where p = pressure

V = volume of container

M = weight of fluid

m = molecular weight

$\bar{R}$ = gas constant

T = absolute temperature

Rearranging,

$$\frac{mVp}{\bar{R}T} = M$$

and, for constant values of V and T, differentiation yields

$$\frac{mV}{\bar{R}T}\frac{dp}{dt} = \frac{dM}{dt}$$

If Equations 3-11 and 3-12 are equated:

$$C\frac{dp}{dt} = \frac{dM}{dt}$$

Therefore,

$$C = \frac{mV}{\bar{R}T} \tag{3-13}$$

Strictly speaking, this capacitance term applies only to isothermal processes. The more general equation is

$$C = \frac{mV}{n\bar{R}T} \tag{3-14}$$

where n is the polytropic exponent. For isothermal processes, $n = 1$. For adiabatic processes, $n = c_p/c_v$ = ratio of specific heats at constant pressure and volume; values of n range from 1.1 for steam through 1.4 for air to 1.67 for monatomic gases. Generally, however, gas systems tend to be isothermal; that is, they lose the heat of compression readily, except where compression or expansion is the primary objective. Thus, Equation 3-13 can be used in most instances (or Equation 3-14 with $n = 1$).

Combining Equations 3-12 and 3-14 gives

$$w_1 - w_2 = \frac{mV}{n\bar{R}T}\frac{dp}{dt} \tag{3-15}$$

The absolute value of the gas constant $\bar{R}$ will depend on the units used in defining the other variables in the equation. One convenient set of engineering units is:

w in lbs/sec

m in lbs/mol

V in ft^3

p in lbs/in.^{2} absolute

n is dimensionless

T in °K = 273 ± °C

t in seconds

$\overline{R} = 19.3$ lbs-ft^3/in.^{2} mol °K

The units of $\overline{R}$ are unusual only because they contain the conversion between square feet and square inches. If these dimensions are eliminated, $\overline{R}$ would have a different value but would have the more familiar units of ft lb/°K. Note that in the above units

$$C = \frac{\text{lbs}}{\text{mol}} \,\bigg|\, \frac{\text{ft}^3}{} \,\bigg|\, \frac{\text{in.}^2 \text{ mol °K}}{\text{lbs ft}^3} \,\bigg|\, \frac{}{\text{°K}} = \text{in.}^2$$

The above dimensions for capacitance seem a bit strange. But

$$C = \frac{\text{Quantity stored}}{\text{Potential}} = \frac{\text{in.}^3}{\text{in.}} = \text{in.}^2$$

Problems

1. The tank described by Equations 3-8 through 3-9 has a cross-sectional area of 16 ft^2. Inflow exceeds outflow by 1000 lbs/min. What is the rate of change of level in ft/min if:

   (a) Liquid specific gravity = 1.0 (i.e., $\rho = 62.5$ lbs/ft^3)?

   (b) Liquid specific gravity = 0.70?

   (c) Liquid specific gravity = 2.95?

2. A 10 ft^3 pressure vessel contains an ideal gas at 27°C under isothermal conditions. What is its capacitance if:

   (a) The gas is hydrogen (mol. wt. = 2)?

   (b) The gas is air (mol. wt. = 29)?

   (c) The gas is radon (mol. wt. = 222)?

3. In the above problem, what happens to the capacitance if the storage temperature is 327°C instead of 27°C?

4. Air at 127°C flows through a pipe at 1.0 lb/sec. If the inflow remains constant while the outflow is reduced stepwise by 10 percent, what is the required volume of a surge tank that will limit the rate of pressure rise to 0.5 lb/in.^{2} sec? (Neglect pipe capacitance.)

3-6-5 Other Compressible Fluid Systems

The previous subsections discussed the evaluation of the capacitance term for incompressible fluid systems and for systems containing ideal compressible fluids. Frequently, however, one must deal with compressible fluids that are not ideal and deviate substantially from the Ideal Gas Law. Vapors (especially those near their boiling points) and compressible liquids (such as high polymers) fall into the latter category. For these non-ideal fluids, it is necessary to evaluate capacitance from experimental or empirical data.

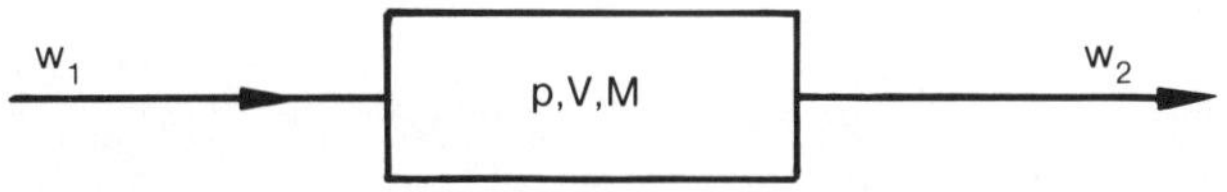

Figure 3-11. Vapor Pressure Process

Consider the system shown in Figure 3-11. By unsteady-state material balance, combining Equations 3-11 and 3-12,

$$w_1 - w_2 = \frac{dM}{dt} = C\frac{dp}{dt}$$

It can be stated that

$$M = V\rho$$

where ρ is the gas density. Differentiating,

$$\frac{dM}{dt} = V\frac{d\rho}{dt}$$

Combining the last term in combination with the equations above, since both equal dM/dt,

$$C\frac{dp}{dt} = V\frac{d\rho}{dt}$$

Solving for C,

$$C = V\frac{d\rho}{dt}\frac{dt}{dp} = V\frac{d\rho}{dp}$$

If the following are defined:

V = volume in ft^3

ρ = density in lbs/ft^3

p = pressure in lbs/in.^{2}

then

$$C = \frac{\text{ft}^3}{} \left| \frac{\text{lbs}}{\text{ft}^3} \right| \frac{\text{in.}^2}{\text{lbs}} = \text{in.}^2$$

i.e., the same dimensional units obtained for C when the Ideal Gas Law was used. It is now necessary to evaluate $d\rho/dp$ in order to evaluate C. Depending upon the nature of the fluid, two alternatives are available.

If the fluid is a common vapor such as steam, Dowtherm®, sulfur dioxide, etc., for which tables of thermodynamic properties are available, $d\rho/dp$ (or more specifically, $\Delta\rho/\Delta p$) may be evaluated by interpolating in the appropriate table of thermodynamic properties in the pressure and temperature region of interest. The average of several values might be used to cover the required pressure and temperature range.

From the steam tables, for example, $\Delta\rho/\Delta p$ might be evaluated in the range of 300 to 350 psia and 500 to 560° F as follows:

p	T	$1/\rho$	ρ	$\Delta\rho$	$\Delta\rho_{ave}$	Δp	$\frac{\Delta\rho}{\Delta p}$
300	500	1.77	0.565				
				0.107			
350	500	1.49	0.672				
					0.102	50	0.00204
300	560	1.92	0.521				
				0.097			
350	560	1.62	0.618				

The foregoing is actually a numerical evaluation of the total derivative

$$\Delta\rho = \left.\frac{\delta\rho}{\delta T}\right|_p \Delta T + \left.\frac{\delta\rho}{\delta p}\right|_T \Delta p$$

The assumption of constant temperature for each calculation of $\Delta \rho$ is required. Thus interchanging the rows of the above tabulation will result in an incorrect calculation.

Note that the units of $\Delta \rho / \Delta p$ are in.^{2}/ft^3, since the steam tables express p in lbs/in.^{2} and specific volume (i.e., $1/\rho$) in ft^3/lbs. When $\Delta \rho / \Delta p$ is multiplied by V in ft^3, however, C will have the units of in.^{2} as before.

The effect of compressibility or bulk modulus of liquids, especially high polymers, should not be overlooked as an aid to controllability, especially when the main function of the control loop is to limit pressure surges on load changes to rather wide tolerances.

Consider a polymer flowing at 180 lbs/hour through a mixer of 0.6 ft^3 volume. The polymer is supplied by a gear pump and is withdrawn by 10 spinning pumps (gear pumps), any one of which can be shut down stepwise. A ballast position will bleed polymer to waste in order to maintain pressure in the spinning pump manifold at 500 lbs/in.^{2}. Laboratory tests show that the fractional change in polymer volume is 7.5×10^{-6} per lbs/in.^{2}, at an average polymer density of 60 lbs/ft^3. Can pressure be controlled within 150 lbs/in.^{2} of the set point when one spinning pump is suddenly shut down?

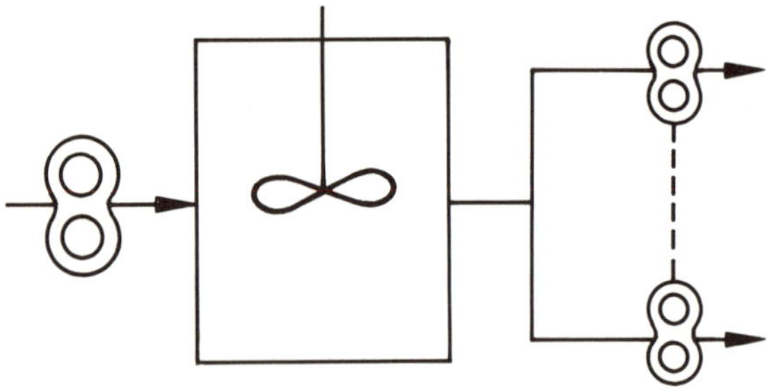

Figure 3-12. Polymer System

Consider the system shown in Figure 3-12. Assume that the system contains a constant weight (M) of polymer. Then,

$$M = \rho V$$

$$M = \text{constant}$$

$$dM = 0 = \rho dV + V d\rho$$

$$d\rho = -\frac{\rho}{V} dv$$

$$\frac{d\rho}{dp} = -\frac{\rho}{V}\frac{dV}{dp}$$

$$V\frac{d\rho}{dp} = (V)\left(-\frac{\rho}{V}\frac{dV}{dp}\right)$$

But

$$\frac{1}{V}\frac{dV}{dp} = (7.5)(10^{-6})$$

and

$$C = V\frac{d\rho}{dp} = V(-\rho)\left(\frac{1}{V}\frac{dV}{dp}\right)$$

$$= -(0.6\ \text{ft}^3)(60\ \text{lb/ft}^3)(7.5)(10^{-6})\ \text{in.}^2/\text{lb}$$

$$= (2.7)(10^{-4})\ \text{in.}^2$$

Then

$$\Delta W = C\frac{dp}{dt}$$

or

$$\frac{dp}{dt} = \frac{\Delta W}{C}$$

$$\frac{dp}{dt} = \left(\frac{180 \times 10\ \text{percent}}{360}\ \frac{\text{lbs}}{\text{sec}}\right)\left(\frac{1}{2.7 \times 10^{-4}\ \text{in.}^2}\right)$$

$$= 18.5\ \frac{\text{lbs}}{\text{in.}^2\ \text{sec}}$$

While this is a fast rate of pressure rise, the tolerance is 150 lbs/in.^{2} Therefore, the control loop has $150/18.5 = 8$ seconds to make its correction. With typical transmitter response of 0.5 second and typical valve response of 0.1 second, control within tolerance appears entirely feasible. Note that if the fluid had been water ($d\rho/dp = 3 \times 10^{-8}$), pressure rise would have been about 10,000 times faster, and control within tolerance would have been impossible.

Problems

1. A refrigeration machine contains Freon® (F-12) in a 5 ft^3 expansion tank. What is the average capacitance of the tank if the following data are known:

P (lbs/in.^{2} abs)	T (° F)	$1/\rho$ ft^3/lb
20	50	2.203
30	50	1.448
20	60	2.250
20	60	1.480

2. Steam at 300 to 350 lbs/in.^{2} and 500 to 560° F (same conditions as given in the example in this section) flows at 1.0 lb/sec. If the inflow remains constant while the outflow is reduced stepwise by 10 percent, what is the required volume of a surge tank that will limit the rate of pressure rise to 0.5 lb/in.^{2} sec? (Neglect pipe capacitance.)

3. In the last example given in 3-6-5 concerning rate of rise of polymer pressure, assume that there are n number of polymer pumps withdrawing from the volume, and that $0.1n$ of the pumps were shut down simultaneously. Does it matter whether n = 10, 50, or 100 if throughput remains 180 lb/hr? On what do you base your conclusion?

4. In the polymer pressure control example, what would be the rate of pressure decrease if two spinning pumps were started up simultaneously?

3-6-6 Composition Systems

Consider the system shown in Figure 3-13.

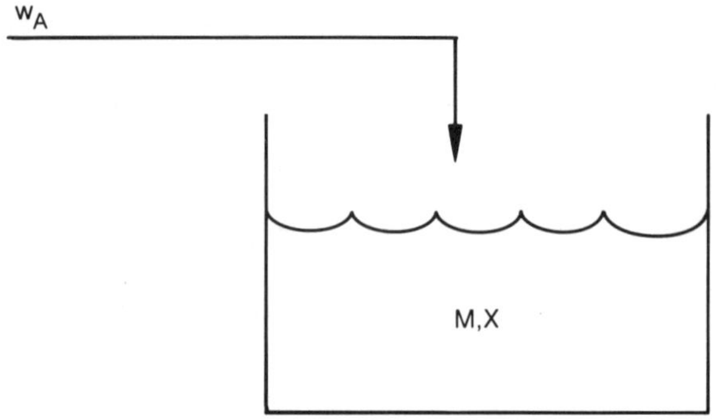

Figure 3-13. Composition Process

Assume that a stream of pure component A flows at w lbs/sec into a tank containing M lbs of liquid. If M_A is defined as the weight of component A present in the tank at any time, the unsteady-state material balance on component A can be written as

$$w_A - 0 = \frac{dM_A}{dt} \tag{3-16}$$

If X_A is the weight fraction of component A in the liquid and M is the total weight of liquid, then

$$M_A = MX_A \tag{3-17}$$

and, if M is essentially constant (i.e., w_A is very small compared to M and hence can change M only very slowly), then

$$\frac{dM_A}{dt} = \frac{MdX_A}{dt} \tag{3-18}$$

and, substituting in Equation 3-16,

$$W_A = \frac{MdX_A}{dt}$$

By comparison with previous equations, it can be seen that the capacitance term C is identical to the weight term M. If X had been defined as volume fraction and w as volumetric flow rate, such as gallons/minute, C and M would have had units of gallons. In a gas system, w would probably have been expressed in ft^3/second, X in volume fraction, and C and M in ft^3.

Suppose that the influent stream was not the pure component w_A but a flow W containing the weight fraction X_{A1} of component A, as shown in Figure 3-14.

The material balance on component A would become

$$wX_{A1} = \frac{MdX_{A2}}{dt}$$

where X_{A2} is the weight fraction of A in the tank contents, and the subscripts 1 and 2 distinguish between the influent stream and the tank contents.

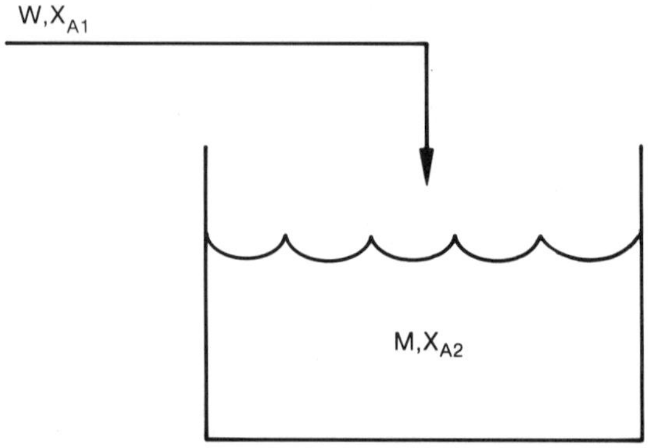

Figure 3-14. Batch Composition Process

Consider the system shown in Figure 3-15, in which the feed stream w contains the weight fractions X_{A1} and X_{B1} of components A and B.

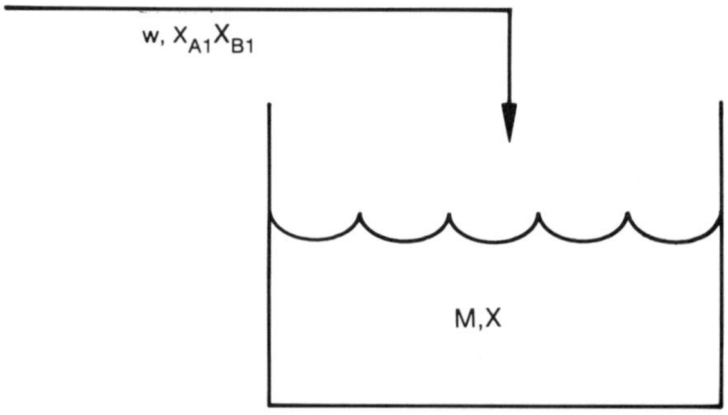

Figure 3-15. Multicomponent Batch Composition Process

It is evident that the amount of each component (A and B) in the tank at any time will be the product of its weight fraction (X_{A2} or X_{B2}) and the the total weight of the liquid in the tank. Thus, a material balance on each component would give:

$$wX_{A1} = M\frac{dX_{A2}}{dt}$$

$$wX_{B1} = M\frac{dX_{B2}}{dt}$$

This approach could be extended to any number of components in the feed stream or to any number of feed streams containing any number of components.

It should be noted that in no case was the starting composition of the tank contents specified. This was not necessary, since the rate of change of composition would be the same, regardless of the initial composition.

3-6-7 Mass Transfer Systems

Mass transfer is concerned with the movement of material from one region to another by means of diffusion or convection. The regions may be two zones in the same medium (a liquid with a concentration gradient, for example) or the regions may represent two different phases (such as mass transfer between liquid and vapor). In any case, the nature of the capacitance term is similar to that for composition systems, except that the mass transfer capacitance term may have a greater variety of dimensional units applied to it.

Consider a plate in a distillation column as shown in Figure 3-16. An unsteady-state material balance around the tray can be written in the conventional manner to relate the tray composition in component X (i.e., X_o) and the capacitance for this component in terms of input and output flows:

$$L_iX_i - L_oX_o + V_oY_o - V_iY_i = C\frac{dX_o}{dt}$$

where L = liquid flow rate in mols/hour

V = vapor flow rate in mols/hour

X = mol fraction of the component in the liquid

Y = mol fraction of the component in the vapor

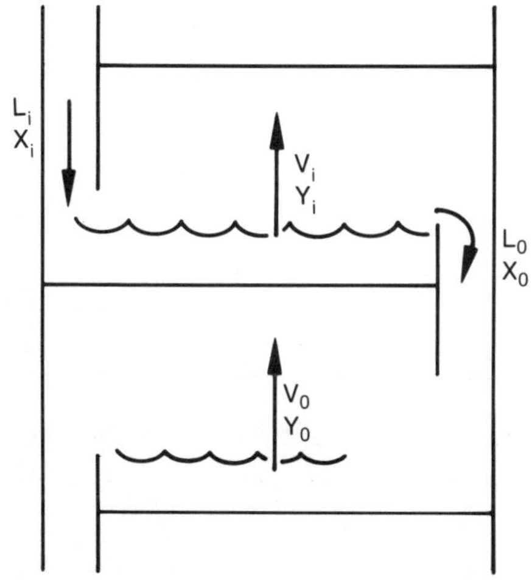

Figure 3-16. Distillation Column Tray

Since the terms on the left-hand side of the equation have units of

$$\left(\frac{\text{mols}}{\text{hour}}\right) \text{(mol fraction)}$$

and since dX/dt has units of mol fraction/hour, C must have units of mols and must in fact represent the total mols of liquid retained on the tray. Then, if M is also the number of mols of liquid on the tray,

$$C = M$$

and

$$L_iX_i - L_oX_o + V_oY_o - V_iY_i = M\frac{dX_o}{dt}$$

If L and V had been expressed in lbs/hour and X and Y in weight fraction, then M would have had the units of lbs.

Frequently in adsorption work the concentration of adsorbate (i.e., adsorbed material) in a solid adsorbent is expressed as lbs adsorbate/ft^3 adsorbent. Consider the system shown in Figure 3-17. Let X be the weight fraction of adsorbable material A (adsorbate) in a stream flowing at w lbs/hour. Let Y be the concentration of adsorbate A in lbs adsorbate/ft^3 adsorbent. By material balance

$$wX = C\frac{dY}{dt}$$

By inspection of dimensional units,

$$\left(\frac{\text{lbs}}{\text{hour}}\right) (\text{wt. fraction } A) = (C) \left(\frac{\text{lbs } A}{\text{hr ft}^3 \text{ adsorbent}}\right)$$

Hence,

$$C = \text{ft}^3 \text{ adsorbent}$$

Thus, the dimensional units of C will be dependent on the units of the input and accumulation terms.

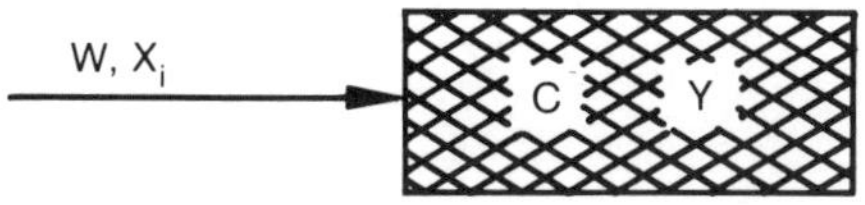

Figure 3-17. Adsorption Process

3-6-8 Summary

The concept of capacitance and the unsteady-state equation might be summarized in a stepwise fashion as follows:

(1) The unsteady-state equation balances inflow, outflow, and rate of accumulation, all in the same set of units. For example,

$$\frac{\text{lbs}}{\text{hour}} \text{ in } - \frac{\text{lbs}}{\text{hour}} \text{ out } = \frac{\text{lbs}}{\text{hour}} \text{ accumulation}$$

(2) Since, generally, it is necessary to express the rate of accumulation in some units other than those of the input and output, a capacitance term that relates the units of flow

to the desired units of accumulation can be inserted. For example,

$$\frac{\text{lbs}}{\text{hour}}\text{ in } - \frac{\text{lbs}}{\text{hour}}\text{ out } = C\left(\frac{\text{lbs}}{\text{in.}^2\text{ hour}}\right)$$

(3) It is then necessary to evaluate C in terms of the physical properties of the system or material or the appropriate thermodynamic relationships. It is wise always to check the equations for dimensional consistency; failure to do so will produce some unusual results.

Problems

1. An acid adjusting tank containing 10,000 lbs of weak sulfuric acid (1 percent by weight) is supplied by two streams. One, flowing at 10 lbs/hour, contains 120 percent by weight of sulfuric acid. The other, flowing at 600 lbs/hour, is a waste acid stream containing 50 percent by weight of sulfuric acid. What is the rate of change of sulfuric acid concentration in the tank? Use the simplifying assumption of Equations 3-17 and 3-18. (Note: Sulfuric acid exists at 120 percent concentration as oleum.)

   (a) Can the simplifying assumptions of Equations 3-17 and 3-18 still be used if flow of 120 percent sulfuric acid = 1000 lbs/hr, flow of waste acid = 6000 lbs/hr, and initial acid concentration = 50 percent?

2. A tank contains 1000 pounds of weak nitric acid. At what rate must 30 percent nitric acid be added to raise the concentration in the tank at the rate of 1.0 percent per minute?

3. A stream of gas containing 20 percent by weight of acetylene and 80 percent inerts (unabsorbable) is bubbled through a tank containing 1000 pounds of DMF. Total inlet gas flow is 1000 lbs/hr. If absorption efficiency is 50 percent, what is the rate of change of acetylene concentration in DMF?

4. A sieve tray in a distillation column holds 60 pounds of liquid. If the liquid inflow and outflow through the tray are 2000 lbs/hour and the vapor inflow and outflow are 1800 lbs/hour, at what rate does tray composition X_o change if:

   (a) $X_i = 0.85$?

   (b) $X_o = 0.90$?

   (c) $Y_o = 0.80$?

   (d) $Y_i = 0.75$?

3-7 RESISTANCE AND THE DRIVING-FORCE EQUATION

In Section 3-3, brief mention was made of the concept of resistance and the driving-force equation, which was defined as

$$\text{Rate of transfer} = \frac{\text{Driving force}}{\text{Resistance}}$$

This is a fundamental concept: to transfer energy or material, a driving force is needed to overcome a resistance. It opposes the concept of perpetual motion in which mass or energy could be transported with no expenditure of energy, i.e., no driving force. The driving-force equation will be applied to a variety of systems in this and the next few subsections.

3-7-1 Electrical System

Application of the driving-force concept is best illustrated by the familiar Ohm's Law equation:

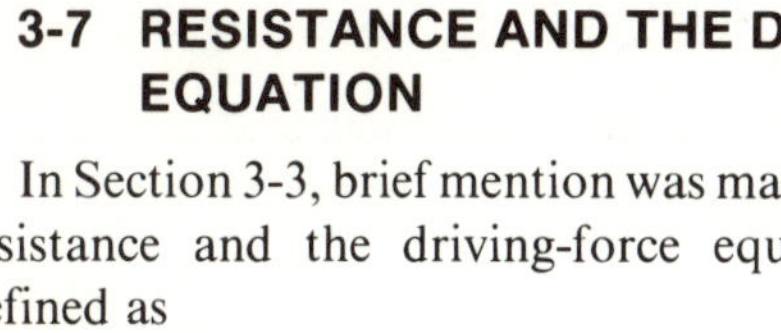

$$I = \frac{E_1 - E_2}{R}$$

In this instance, the driving force is the potential difference $E_1 - E_2$ and the rate of transfer is the current flow I. The resistance term is of course the electrical resistance R. The concept is so familiar that no further explanation is required.

A word on the dimensional units of R, however, is worthwhile, because R will assume unusual dimensional units in some systems. Yet, like the capacitance term in the unsteady-state equation, R must lead to a dimensionally correct equation.

In popular units, I is expressed in amperes, E in volts, and R in ohms. In more fundamental units, I is expressed in coulombs/second. Then, the dimensions of R become

$$R = \frac{E}{I} = \frac{\text{volts}}{\text{coulombs/second}} = \frac{\text{volt seconds}}{\text{coulomb}}$$

It will be found that in any system R will have units of time in its dimensions. This becomes evident when one considers that the rate of transfer term must contain units of time, whereas the driving-force term refers to conditions at a given instant of time and therefore does not contain the units of time; the time dimension must then be provided by the resistance term R.

3-7-2 Thermal Systems

There are three fundamental mechanisms for heat transfer: conduction, convection, and radiation. Each of these mechanisms will be considered as it relates to the driving-force equation,

$$\text{Rate of transfer} = \frac{\text{Driving force}}{\text{Resistance}}$$

or, rearranging to define resistance,

$$\text{Resistance} = \frac{\text{Driving force}}{\text{Rate of transfer}}$$

- *Conduction*

The equation describing the flow of heat by conduction is

$$Q = \frac{k}{b} A \; (T_1 - T_2) \tag{3-19}$$

where Q = heat flow, PCU/hour

k = thermal conductivity, PCU ft/hour ft^2 °C

b = length of conductive path, ft

A = area for heat transfer, ft^2

T = temperature, °C

In terms of the driving-force equation,

$$Q = \frac{T_1 - T_2}{R} \tag{3-20}$$

Comparison of Equations 3-19 and 3-20 therefore defines R as

$$R = \frac{b}{kA} \; \frac{\text{hr °C}}{\text{PCU}}$$

The resistance to the conduction of heat increases as the length of the conductive path increases and decreases with increasing conductivity or heat transfer area. Note also that R contains the units of time, as did the electrical resistance term.

- *Convection*

When heat is transferred from one location to another by physical movement of a fluid, the heat transfer mechanism is defined as convective. The equation describing convective heat transfer is

$$Q = h_c A \; (T_1 - T_2)$$

where h_c = film coefficient, PCU/hour ft^2 °C

Again, by comparison with the driving-force equation,

$$Q = \frac{T_1 - T_2}{R}$$

it is evident that

$$R = \frac{1}{h_c A} \; \frac{\text{hr °C}}{\text{PCU}} \tag{3-21}$$

The film coefficient h may be obtained from heat exchanger design specifications, or it may be evaluated by the appropriate equations in the literature.[24] The film coefficient is quite dependent on the velocity of the fluid flowing past the film. For fluids flowing inside tubes, the correction is

$$h_{c2} = h_{c1} \left(\frac{v_2}{v_1}\right)^{0.8}$$

where v is the velocity. For fluids flowing across tube banks, the correction is

$$h_{c2} = h_{c1} \left(\frac{v_2}{v_1}\right)^{0.6}$$

The variation of the heat transfer film coefficient (h) with fluid flowing velocity is extremely important in the application of temperature-measuring sensors.

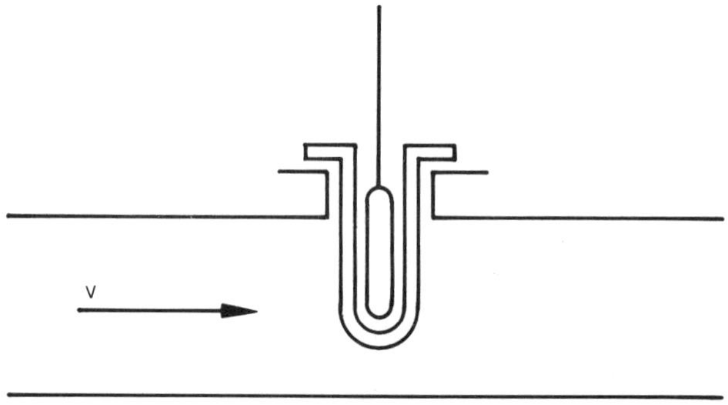

Figure 3-18. Temperature Sensor

Consider the sensor in a thermowell as shown in Figure 3-18. The difference between flowing temperature and sensor temperature and the ability of the sensor to track changes in flowing temperature both depend on a low heat transfer resistance between flowing fluid and sensor. But resistance is inversely proportional to film coefficient, Equation 3-21, and film coefficient is a function of velocity. Thus low velocity results in a high resistance and a poor measurement. See Appendix B for a detailed analysis of the thermal sensor.

Because the film coefficient depends on physical properties of the fluid (specific heat, thermal conductivity, and fluid viscosity in the film), it will also vary somewhat with temperature, although this variation can normally be neglected in preliminary calculations.

- *Radiation*

Radiant heat transfer involves the transfer of heat by electromagnetic radiation from one body to another not in contact with it. For purposes of this text, the rate of radiant heat transfer will be described as

$$Q = h_r A \; (T_1 - T_2) \tag{3-22}$$

where h_r = radiant coefficient, PCU/hr ft^2 °C, and from which, by analogy to the previous driving-force equations,

$$R = \frac{1}{h_r A} \quad \frac{\text{hr °C}}{\text{PCU}} \qquad (3\text{-}23)$$

If the emissivities of the radiating bodies are 0.9 or greater, h_r may be evaluated from graphs of h_r versus the temperature difference between the hot and cold bodies.[24]

Sometimes it is necessary to evaluate R from the basic equation for radiant heat transfer:

$$Q = \varepsilon A \, (10^{-8}) \, (T_1^4 - T_2^4)$$

where Q = heat flow, PCU/hr

ε = emissivity

A = area, ft^2

T = temperature, °K

The temperature difference term can be factored to yield

$$Q = \varepsilon A \, (10^{-8}) \, (T_1^2 + T_2^2) \, (T_1 + T_2) \, (T_1 - T_2) \qquad (3\text{-}24)$$

Comparison of Equations 3-22 and 3-24 shows that

$$h_r A = \varepsilon A \, (10^{-8}) \, (T_1^2 + T_2^2) \, (T_1 + T_2)$$

If T_1 is not more than twice T_2 (on an absolute temperature basis), a satisfactory approximation using the average temperature T_{ave} is

$$h_r A = 4 \varepsilon A \, (10^{-8}) \, T^3_{ave}$$

Then, comparising Equations 3-23 and 3-25,

$$R = \frac{10^{-8}}{4 \varepsilon A \, T^3_{ave}} \quad \frac{\text{hr °K}}{\text{PCU}}$$

- *Combined Coefficients*

The heat loss from a bare hot pipe results from the combined heat transfer due to convection and radiation and is described by

$$Q = (h_c + h_r) \, A \, (T_1 - T_2)$$

The resistance term thus is

$$R = \frac{1}{(h_c + h_r) \, A} \quad \frac{\text{hr °C}}{\text{PCU}}$$

The individual coefficients h_c and h_r are evaluated as described above.

- *Overall Coefficient*

Consider the case of heat transfer from a fluid inside an insulated pipe to atmospheric air. The heat passes serially through the inside film, the pipe wall, the insulation, and the outside film. It is sometimes convenient to consolidate the various coefficients into an overall coefficient such that, if T_f is the fluid temperature and T_a is the ambient air temperature, the heat transfer rate can be expressed as

$$Q = UA \, (T_f - T_a)$$

where U = overall coefficient, PCU/hr ft^2 °C.

By analogy with the driving-force equation, it can be stated that

$$R_o = \frac{1}{UA} \quad \frac{\text{hr °C}}{\text{PCU}}$$

It should be recognized that the area must be an average area based in some manner on the inside pipe area and the outside area of the insulation, such as an area based on the log mean average of inside and outside diameters.

Since the heat flows through the resistance in series, the resistance can be added thus:

$$R_o = R_f + R_p + R_i + R_a$$

where the subscripts refer respectively to the overall resistance, fluid film resistance, pipe resistance, insulation resistance, and outside (air) film resistance. Substituting the appropriate coefficients and areas,

$$\frac{1}{UA_{ave}} = \frac{1}{h_f A_f} + \frac{b_p}{k_p A_p} + \frac{b_i}{k_i A_i} + \frac{1}{(h_{ca} + h_{ra}) \, A_a}$$

which is the familiar form of the overall coefficient. Note that, when the proper areas are applied to the calculation of the individual resistances, it is not necessary to determine an "average" area for R_o.

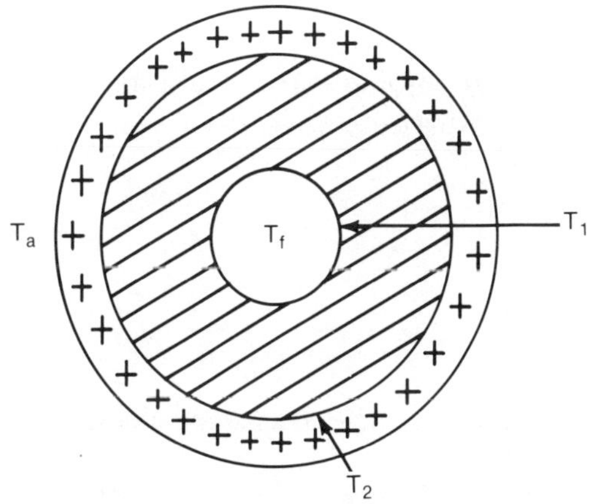

Figure 3-19. Insulated Pipe

Problems

1. Heat transfer rate through a 1.2-inch thick by 1-foot square plate is 3600 PCU/hour. What is the temperature gradient (driving force) if the place is
   (a) Steel (k = 30 PCU ft/hr ft^2 °C)?

   (b) 316 S/S (k = 8.8)?

   (c) Copper (k = 220)?

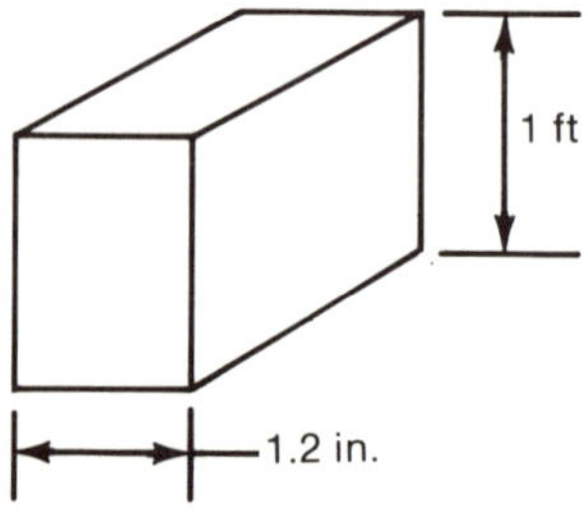

2. In the above example, what are the corresponding values of resistance?

3. In convective heat transfer, what is the relationship between the resistance R and the velocity V of a fluid flowing inside of a pipe?

4. What is the value of R in a radiant heat transfer system of $Q = 9.8$, A = 10 ft^2, T_1 = 600° K and T_2 = 400° K?

3-7-3 Fluid Flow Systems

In a fluid flow system, the fluid may be incompressible or compressible, and the flow regime may be streamline (laminar), turbulent, or critical. It is necessary therefore to select the proper combination of fluid property and flow regime to evaluate the resistance term. There is also the added complication of isothermal versus nonisothermal flow, since fluid properties will change with temperature. For purposes of this text, however, a satisfactory approximation in the case of nonisothermal flow would be the evaluation of fluid properties at the average of inlet and exit temperatures.

To determine the flow regime, it is necessary to evaluate the Reynolds number. The convenient form of the equation is

$$\text{Re} = \frac{4w}{\pi D \mu}$$

where w = flow rate, lbs/sec

D = pipe diameter, ft

μ = viscosity, lbs/ft sec = $\frac{\text{centipoise}}{1488}$

Flow in circular pipes is always streamline for Reynolds numbers less than 2000 and is generally turbulent for Reynolds numbers above 3000. In the critical range of Reynolds numbers of 2000 to 3000, the flow pressure drop relationship is poorly defined because the flow regime is in transition between streamline and turbulent conditions. In commercial installations, however, the Reynolds number usually lies clearly to one side or the other of the transitional region except in the occasional case where a facility is being operated at a small fraction of its design capacity.

- *Streamline Flow — Incompressible Fluids*

The equation describing the isothermal flow of an incompressible fluid in a circular duct can be written as

$$p_1 - p_2 = \frac{128 \mu l w}{144 \pi g \rho D^4} \qquad (3\text{-}26)$$

where p = pressure, lbs/in.^{2}

μ = viscosity, lbs/ft sec = $\frac{\text{centipoise}}{1488}$

l = pipe length, ft

w = flow rate, lbs/sec

g = gravity, 32.2 ft/sec^2

ρ = fluid density, lbs/ft^3

D = inside pipe diameter, ft

144 = conversion factor, in.^{2}/ft^2

Recasting Equation 3-26 into the form of the driving-force equation, which defines resistance as driving force over rate of transfer, gives

$$R = \frac{p_1 - p_2}{w} = \frac{128 \mu l}{144 \pi g \rho D^4} \quad \frac{\text{sec}}{\text{in.}^2}$$

Note again that R contains the units of time. If design or manufacturer's information is available relating pressure drop to streamline flow rate for a given piece of equipment, R can be evaluated directly from the driving-force equation:

$$R = \left. \frac{p_1 - p_2}{w} \right|_{ss} \quad \frac{\text{sec}}{\text{in.}^2}$$

where the $\big|_{ss}$ signifies that R is evaluated from some set of steady-state conditions. The calculated value of R will then apply over the entire range in which the flow remains streamline.

- *Streamline Flow — Compressible Fluids*

In the case of compressible liquids such as high polymers, the change in density even at large values of $p_1 - p_2$ is negligible; accordingly, the above equations can be safely applied to compressible liquids.

The foregoing equations can also be applied to the streamline flow of gases, if the density is based on the average of the upstream and downstream pressures. It should be recognized that it is difficult to maintain a gas in streamline flow at very appreciable pressure drops, except in very small capillaries. Thus, the pressure drop (hence, change in gas density) will generally be small in commercial systems involving the streamline flow of gases.

Note that resistances in series can be added directly. For example, for two resistances in series, the following could be written:

$$w = \frac{p_1 - p_2}{R_1}$$

and

$$w = \frac{p_2 - p_3}{R_2}$$

Rearranging,

$$wR_1 = p_1 - p_2$$
$$wR_2 = p_2 - p_3$$

and adding,

$$w(R_1 + R_2) = p_1 - p_3$$

Whence,

$$R_T = R_1 + R_1 = \frac{p_1 - p_3}{w}$$

Series hydraulic resistances are thus analogous to series electrical resistances.

- *Turbulent Flow — Incompressible Fluids*

The equation relating pressure drop to the isothermal turbulent flow of an incompressible fluid is

$$p_1 - p_2 = \frac{32\, flw^2}{144\, \pi^2\, g\rho D^5} \qquad (3\text{-}27)$$

Dimensional units are the same as in Equation 3-26, except that a new variable, the friction factor f, has been added. The friction factor, which is dimensionless, may be evaluated from graphs of f versus Reynolds number, or it may be calculated from the expression

$$f = \frac{0.04}{(\mathrm{Re})^{0.16}}$$

In considering Equation 3-27 in terms of the driving-force equation, two complications become evident: the relationship between pressure drop and flow is nonlinear (square law); and the friction factor, being a function of Reynolds number (which varies with flow), is not a constant. Thus, a value of resistance R derived from Equation 3-27 will be constant only over a limited range of flows and pressure drops.

The nonlinearity of the relationship between pressure drop and flow is emphasized in the graph of the relationship shown in Figure 3-20.

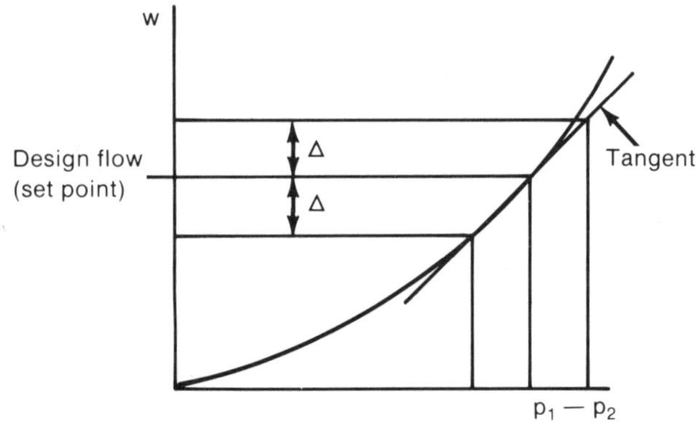

Figure 3-20. Square Law Nonlinearity

The nonlinearity of this common relationship cannot be accommodated by linear mathematics. But there is the advantage of dealing with a system that is being controlled to a set point. Thus the equations must be valid for only small deviations (Δs) around the set point. A linear relationship is then obtained by drawing a tangent to the nonlinear curve at the set point and agreeing that the equation of the tangential straight line represents the function for only, say, ±10 percent of the span. The driving-force equation for the straight line can be determined either graphically or mathematically. The total derivative is used for the latter linearization technique.

$$\Delta w = \left.\frac{\delta w}{\delta p_1}\right|_{ss} \Delta p_1 + \left.\frac{\delta w}{\delta p_2}\right|_{ss} \Delta p_2 \qquad (3\text{-}28)$$

The partial derivatives, when evaluated at set point conditions, can be considered as gain terms; that is, they define the change in flow in terms of small changes in upstream or downstream pressure based on steady-state conditions. The Δs are sometimes stated in the total derivative to emphasize the small variations but are generally omitted for ease in writing the equations.

For convenience in taking partial derivatives, Equation 3-27 can be rewritten as:

$$w^2 = K\,(p_1 - p_2) \qquad (3\text{-}29)$$

and the partial derivative with respect to p_1, with p_2 constant,

$$2w\delta w = K\,\delta p_1 \qquad (3\text{-}30)$$

Substituting for K from Equation 3-29,

$$\frac{\delta w}{\delta p_1} = \frac{1}{2w}\left.\left(\frac{w^2}{p_1 - p_2}\right)\right|_{ss} = \left.\frac{w}{2(p_1 - p_2)}\right|_{ss}$$

Since, in Equations 3-28 the only difference between the p_1 and p_2 terms is the sign, it can readily be written that

$$\frac{\delta w}{\delta p_2} = \left.\frac{-w}{2(p_1 - p_2)}\right|_{ss}$$

Substituting Equations 3-29 and 3-30 into 3-28, and factoring out the common steady-state coefficient,

$$W = \left.\frac{w}{2(p_1 - p_2)}\right|_{ss} (p_1 - p_2) = \frac{1}{R}\,(p_1 - p_2) \qquad (3\text{-}31)$$

Now, the driving-force equation defines R as

$$R = \frac{p_1 - p_2}{w} \qquad (3\text{-}32)$$

Comparison of Equations 3-31 and 3-32 then defines R for the turbulent flow equation as

$$R = \left.\frac{2(p_1 - p_2)}{w}\right|_{ss} \quad \frac{\text{sec}}{\text{in.}^2} \qquad (3\text{-}33)$$

There is now a linear (or linearized) relationship for flow, pressure drop, and resistance. If flow pressure drop relationships are known for a given piece of equipment, R may be evaluated directly from that information. Otherwise, it is necessary to return to the basic turbulent flow equation (Equation 3-27), evaluate pressure drop for the desired flow rate, and substitute the results into Equation 3-33 to evaluate R.

It is important to realize that the "linearized" version of R is only applicable to perturbation calculations (that is, for calculating changes in pressure drop due to changes in flow, or vice versa). It cannot be used for calculating absolute values of flow or pressure drop. In other words, it must be used in the following manner:

$$\Delta w = \frac{\Delta(p_1 - p_2)}{R}$$

The computed Δ values can then be added algebraically to the steady-state values used in calculating R.

Problems

1. A polymer having a viscosity of 30,000 cP and a density of 60 lbs/ft^3 flows through a 0.01-ft diameter capillary of 1-ft length. Assuming streamline flow:

   (a) Calculate R.

   (b) Calculate $p_1 - p_2$ if flow rate is 1 lb/hour.

2. A pressure impulse line 0.24-inch ID by 20 feet long connects the gas space of a pressure vessel to a pressure transmitter. Assuming that flow would be streamline for moderately fast pressure changes, calculate R if the fluid is air (density = 0.24 lbs/ft^3, viscosity = 1.5×10^{-5} lbs/ft sec).

3. In the above problem, what would be the value of R if the fluid had been helium under the same conditions ($\rho = 0.033$ lbs/ft^3, $\mu = 1.5 \times 10^{-5}$ lbs/ft sec)?

4. Dowtherm A® condensate flows through 100 feet of 2-inch pipe (ID = 2.067 inches) at 20,000 lbs/hour under turbulent flow conditions. Calculate the linearized value of R if $\rho = 53.1$ lbs/ft^3 and $f = 0.0053$.

• *Turbulent Flow — Compressible Fluids*

Compressible liquids, such as high polymers, can be treated as incompressible fluids since the change in density, even at high pressure drops, is too small to affect significantly the flow-pressure drop relationship.

For gases, however, rigorous treatment is more complex because density changes with pressure. If the flow-pressure drop relationship is known (from experimental data, manufacturer's information, etc.), the resistance may be calculated directly in the manner developed for incompressible fluids, since density changes would necessarily have been included in the data. Then,

$$R = \left.\frac{2(p_1 - p_2)}{w}\right|_{ss} \tag{3-34}$$

More generally, however, only the flow and upstream pressure and temperature are known, and it is necessary to determine the downstream pressure in order to evaluate R. As a first approximation, the pressure drop can be calculated using the equation for the turbulent flow of incompressible fluids:

$$p_1 - p_2 = \frac{32\, flw^2}{144\, \pi^2\, g\rho D^5} \tag{3-35}$$

with the density calculated from upstream conditions. If the pressure drop is less than 10 percent of the upstream absolute pressure, this approach is satisfactory for purposes of this text, and R can be evaluated by substituting the flow and pressure drop calculated in Equation 3-35 into Equation 3-34.

If the pressure drop calculated by Equation 3-35 is 10 to 50 percent of the upstream pressure, a more accurate relationship between flow and pressure drop may be calculated from the equation:

$$p_1^2 - p_2^2 = \frac{64\, fl\bar{R}T_1\, w^2}{144\, \pi^2\, gmD^5} \tag{3-36}$$

Equation 3-36 is developed by substituting for ρ in Equation 3-35 an average value of density based on p_1, p_2, and the Ideal Gas Law:

$$\rho_{ave} = \frac{m}{\bar{R}T_1}\,\frac{p_1 + p_2}{2}$$

where m = molecular weight in lbs/mol

$\bar{R}$ = gas constant = 19.3 lbs/ft^3/in.^{2} °K mol

The value of friction factor f used in Equation 3-35 can be used in Equation 3-36. The resistance term R can then be evaluated by substituting the appropriate values of p_1, p_2 and w in Equation 3-34.

• *Critical Flow Systems*

If the downstream pressure is less than about 53 percent of the upstream pressure, the fluid attains sonic velocity, which is also the velocity of propagation of a pressure wave in the fluid. The flow rate therefore becomes unaffected by downstream pressure and is dependent only on upstream conditions. The critical pressure ratio p_2/p_1 is a function of the ratio of specific heats (c_p/c_V); hence, it varies with the nature of the gas molecule. Taking a representative value of c_p/c_V = 1.4, the value for diatomic gases, flow, and pressure drop can be related by the equation

$$w = 8.32 \left(\frac{m}{T_1}\right)^{1/2} D^2 p_1$$

where w = flow rate, lbs/sec

m = molecular weight, lbs/mol

T_1 = upstream temperature, °K

D = restriction diameter, ft

p_1 = upstream pressure, lbs/in.^{2} absolute

The constant 8.32 contains units to make the equation dimensionally consistent. The equation is within about ± 10 percent for c_p/c_V ratios between 1.1 and 1.67. Note that the driving force is now the upstream pressure rather than the pressure difference. The resistance term thus becomes

$$R = \left.\frac{(T_1/m)^{1/2}}{8.32\, D^2}\right|_{ss} \quad \frac{\text{sec}}{\text{in.}^2}$$

• *Control Valves*

The equations used for sizing control valves incorporate the physical properties of the fluid, the upstream and downstream pressures, and the required flow rate. Once the valve has been sized, the valve flow formulas can be simplified to

$$w = KX\,(p_1 - p_2)^{1/2} \tag{3-37}$$

in which X represents the fractional valve capacity and in which K can be evaluated in terms of the flow obtained through the specified valve at the specified pressure drop with X = 1 (valve full open). Note that X does not represent stem position (fractional lift) but fractional capacity; thus, for nonlinear valves, X must be determined from the valve characteristic curve relating fraction of full capacity to fraction of lift, in the region of interest.

Since the flow-pressure drop relationship in Equation 3-37 is nonlinear, the equation must be linearized before evaluating R. Treating the fractional valve capacity X as constant, the linearized equation becomes

$$w = \left.\frac{w}{2(p_1 - p_2)}\right|_{ss} (p_1 - p_2)$$

and

$$R = \left.\frac{2(p_1 - p_2)}{w}\right|_{ss} \quad \frac{\text{sec}}{\text{in.}^2}$$

Note that K and X have disappeared from the linearized equations. They have not been forgotten, however; to obtain the steady-state values required for evaluating R, it is necessary to solve Equation 3-37 using actual values of K and X. Thus the resistance R will vary as a function of the fractional capacity used.

Consideration of the total derivative will show flow w to be a function of pressures p_1 and p_2, and valve stem position x. The relationship of x to w can be linear or nonlinear, depending on the valve flow characteristic and the magnitude of the pressure drop in series with the control valve. These relationships are considered in greater detail in Section 5, Control Valves.

3-7-4 Mass Transfer Systems

The driving force for mass transfer by diffusion or convection between two locations in the same medium or between two phases is the concentration gradient. The driving-force equation can thus be applied to mass transfer systems in the same manner that it is applied to other systems of material or energy transfer.

Treatment of mass transfer systems is complicated by two factors: concentrations may be expressed in a variety of ways (lbs mol/ft^3, partial pressure, vapor phase mol fraction, etc.), and evaluation of the resistance term is often limited by lack of data on diffusion coefficients. Accordingly, only one representative mass transfer system will be considered in this text.

- *Counterdiffusion of Two Gases*

The diffusional mixing of two gases at constant pressure may be described by

$$w_A = \frac{DpA}{\bar{R}Tb_f}(Y_A - Y_{Ai}) \tag{3-38}$$

where w_A = rate of transfer of component A, mol/hr

D = diffusion coefficient, ft^2/hr

p = total pressure, lbs/in.^{2} absolute

A = diffusional area, ft^2

$\bar{R}$ = gas constant 19.3 lbs ft^3/in.^{2} mol °K

T = temperature on which D is based, °K

b_f = film thickness, ft

Y_A = mol fraction of A in the bulk gas

Y_{Ai} = mol fraction of A in the gas at the interface

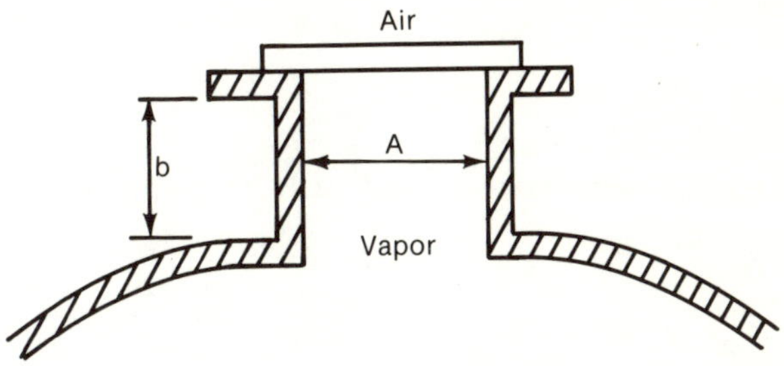

Figure 3-21. Vapor Diffusion Process

By comparison with the basic driving-force equation, it can be seen that

$$R = \frac{\bar{R}Tbf}{DpA} \quad \frac{\text{hr}}{\text{mol}}$$

Equations of the form of Equation 3-38 also apply to other methods of mass transfer, such as the absorption of a soluble gas through a film of insoluble gas, adsorption of a gas into the pores of an adsorbent, evaporation of a liquid through a gas film, etc. The terms that comprise resistance R will differ from one system to another as a function of the physical character of the system and the method of expressing the driving force.

3-7-5 Summary

The concept of resistance and the driving-force equation can be summarized as follows:

(1) To transfer mass or energy, a driving force is needed to overcome a resistance to transfer. These three factors are related by an equation of the general form

$$\text{Rate of transfer} = \frac{\text{Driving force}}{\text{Resistance}}$$

(2) The resistance term will always contain the units of time; the actual dimensions of R will depend on the units describing the rate of transfer and the driving force.

(3) Evaluation of the resistance term from the physical parameters of the system is often difficult because of lack of reliable data on some of the factors that comprise the resistance term.

(4) it is always wise to check the transfer equations for dimensional consistency.

Problems

1. Air flows through 100 feet of 0.25-ft ID pipe at 1800 lbs/hr. Taking $p_1 = 45$ psia, $f = 0.0063$, and $\rho = 0.15$ lbs/ft^3, calculate R.

2. Helium (molecular weight = 4) at 127°C bleeds through a 0.001-ft diameter orifice at critical pressure drop:

   (a) Calculate R.

   (b) Calculate w in lbs/hour if $p_1 = 45$ psia.

3. A control valve has been sized to pass 5,400 lbs/hour of steam (when fully open) at a 25-psi pressure drop:

   (a) What is the valve capacity at $X = 0.5$ and $(p_1 - p_2) = 36$ psi?

   (b) What is the value of R under these conditions?

4. A vessel contains pure ammonia vapor at 27°C and atmospheric pressure. A manhole cover of 5-square-feet area is suddenly removed. What is the resistance to mass transfer if the diffusion coefficient for ammonia in air (or vice versa) is 0.915 ft^2/hour? Assume that the film thickness equals the nozzle height of 1 foot.

SECTION **4**

SELF-REGULATING PROCESS SYSTEMS

4-1 INTRODUCTION

Analysis of processes is made much easier through the development of the analogous elements, resistance and capacitance, as discussed in Section 3. These are the "building blocks" of systems and almost all industrial processes can be described by arrangements of numbers of these elements in various parallel and series configurations.

Some processes are described by resistance alone and are designated proportional elements. A liquid flow measurement is an example of a proportional element. Other processes are described by capacitance alone and are called capacitance elements. A liquid level tank is an example of a capacitance element. Many processes are described by the product of resistance and capacitance (RC), a term designated as the time constant (τ), which always has the dimension of time. Processes described by time constants are called self-regulating processes.

4-2 NONSELF-REGULATING SYSTEMS

Consider a generalized system of the type shown in Figure 4-1, which can be described by the unsteady-state equation

$$A - B = C\frac{dD}{dt}$$

$$w_1 - w_2 = C\frac{dL}{dt}$$

where A = inflow of energy or material
B = outflow of energy or material
C = capacitance
D = accumulation variable
t = time

Rearranging,

$$\frac{dD}{dt} = \frac{A - B}{C} \qquad (4\text{-}1)$$

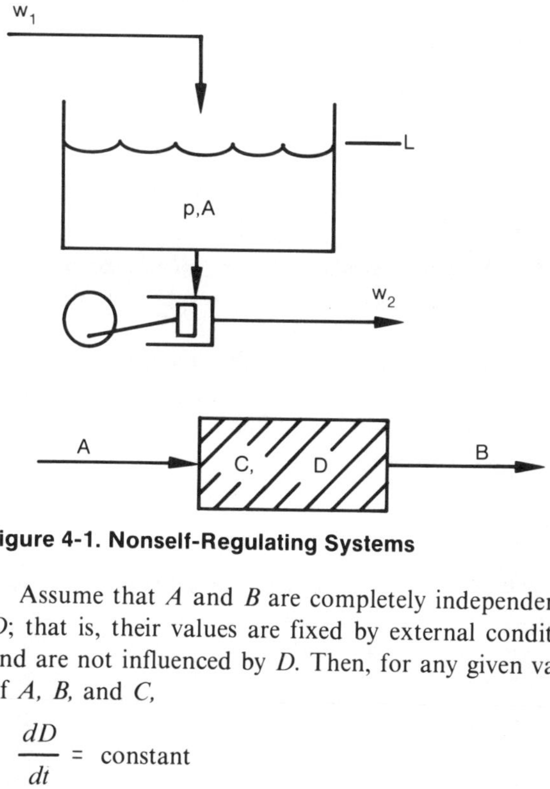

Figure 4-1. Nonself-Regulating Systems

Assume that A and B are completely independent of D; that is, their values are fixed by external conditions and are not influenced by D. Then, for any given values of A, B, and C,

$$\frac{dD}{dt} = \text{constant}$$

and D will change at a constant rate. To put it another way, a plot of D versus t will be a straight line of slope $(A-B)/C$.

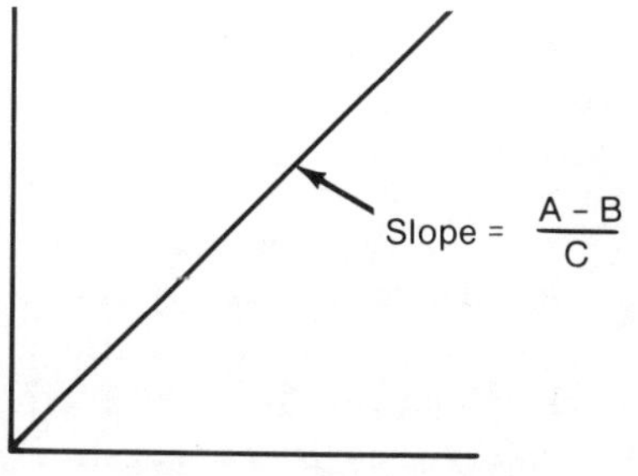

Figure 4-2. Graph of Integrating Process

Thus, a nonself-regulating system may be defined as one in which the accumulation variable exerts negligible influence on the input and output terms. An example would be a storage tank in which the inflow and outflow rates were established by positive displacement pumps. Since the pump capacities are independent of level, any difference between inflow and outflow rates would remain constant, and level would change at a constant rate (limited, of course, by the top and bottom of the tank.)

Nonself-regulating systems are pure integrating systems; they integrate any error between input and output into a constant rate of change of the accumulation variable. Transforming Equation 4-1 into the Laplace notation,

$$sD = \frac{A - B}{C}$$

or

$$D = \frac{A - B}{Cs}$$

The $1/s$ term is recognized as the Laplace notation for integration. Applying the final value theorem, as $t \to \infty$ and $s \to 0$, then $D \to \infty$, a response typical for integrating (nonself-regulating) systems.

Note that if a control loop were provided to manipulate A or B as a means of holding D constant, the system would still be nonself-regulating; it would merely be controlled.

4-3 SELF REGULATING SYSTEMS

Now consider another generalized system of the type shown in Figure 4-3, wherein:

$$A - KD = C\frac{dD}{dt} \qquad Q_1 - \frac{T}{R} = C\frac{dT}{dt} \tag{4-2}$$

and K is a proportionality constant. Rearranging,

$$\frac{dD}{dt} = \frac{A - KD}{C} \tag{4-3}$$

If, initially,

$$KD = B$$

Equation 4-3 would duplicate Equation 4-1 (which described the nonself-regulating system) and the initial relationship of D and t would be a straight line of slope $(A-B)/C$. In Equation 4-3, however, the value of $(A-KD)/C$ changes as D changes. If D is increasing, $(A-KD)/C$ is decreasing. At some point, KD will equal A, and dD/dt will become zero; the slope of D versus t will become zero. The system thus reaches some upper (or lower) limit, defined by the value that D must reach to make $A - KD = 0$ or $\frac{A}{K} = D$.

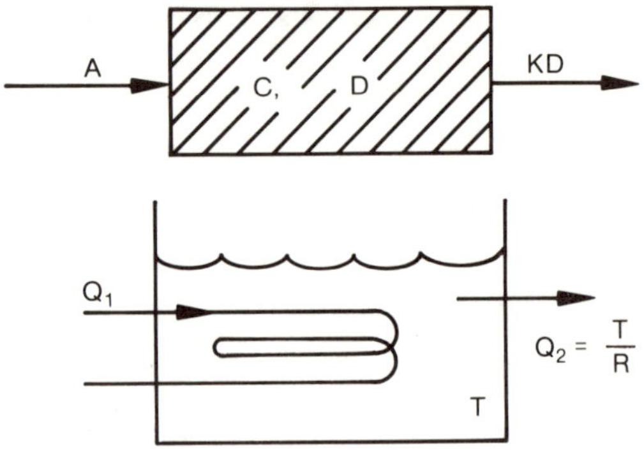

Figure 4-3. Self-Regulating Systems

A self-regulating system is thus defined as one in which the accumulation variable acts on the inflow or outflow to force them into balance. Referring again to Equation 4-3, the following conclusions about self-regulating systems can be drawn:

(1) The initial slope of D versus t will be the same as that of a nonself-regulating system.

(2) At any time, the slope of D versus t will vary directly with the instantaneous value of $(A - KD)$ and inversely with C.

(3) The ultimate value of D must equal A/K, at which time the slope of D versus t becomes zero.

Equation 4-2 can be written in the Laplace notation and rearranged as follows:

$$A - KD = CsD$$

$$A = CsD + KD$$

$$\frac{A}{K} = \frac{C}{K}sD + D$$

$$\frac{A}{K} = \left(\frac{C}{K}s + 1\right)D$$

The transfer function D/A then becomes

$$\frac{D}{A} = \frac{1/K}{\frac{C}{K}s + 1}$$

or, defining the time constant τ as

$$\tau = \frac{C}{K}$$

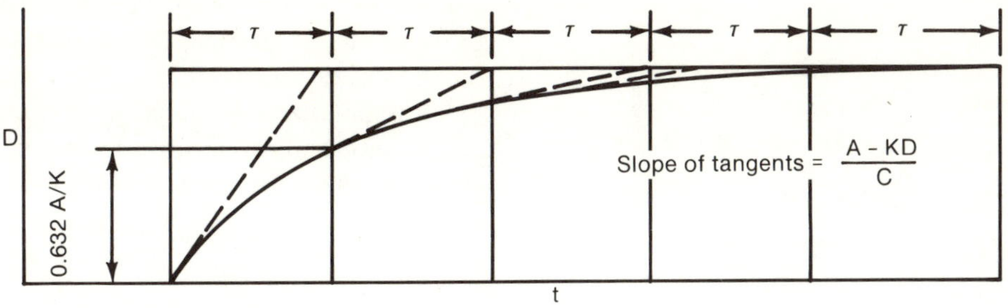

Figure 4-4. First-Order Response

gives

$$\frac{D}{A} = \frac{1/K}{\tau s + 1} \tag{4-4}$$

There are two features of this transfer function worthy of mention. First, the denominator is dimensionless: the time constant has units of time, whereas s has dimensions of reciprocal time (i.e., 1/time). Second, the numerator $(1/K)$ must therefore have the same dimensions as the left side of the equation (D / A).

If A is now considered as a step change in the input quantity and Equation 4-4 is converted back to the time domain, the resulting equation is:

$$D = \frac{A}{K}\left(1 - e^{-t/\tau}\right) \tag{4-5}$$

Equation 4-5 can be evaluated at three convenient points:

$t = 0$	$e^{-0} = 1.0$	$1 - 1.0 = 0$	$D = 0$
$t = \tau$	$e^{-1} = 0.368$	$1 - 0.368 = 0.632$	$D = 0.632\ A/K$
$t = \infty$	$e^{-\infty} = 0$	$1 - 0 = 1$	$D = A/K$

Thus, at the end of one time constant ($t = \tau$), D will have progressed 63.2 percent of the way to its ultimate ($t = \infty$) steady-state value of A/K. In fact, during any time constant (i.e., during any interval equal to the time constant), D will move 63.2 percent of the way from its value at the start of the "time constant" period to its ultimate value. Graphically, the response is as shown in Figure 4-4.

Note that the application of the final value theorem ($s \rightarrow 0$) to Equation 4-4 would have given a final value of $D = A/K$ also.

The value of the time constant τ thus uniquely defines the reponse of a self-regulating system. Since the self-regulating system is defined by a first-order differential equation (no derivatives higher than first order), the characteristic response is termed a first-order lag.

One can think validly of a self-regulating system as one with built-in negative feedback. Or, in fact, one could consider it as a nonself-regulating system in which negative feedback is provided by an external proportional controller. The controller gain would then correspond to the proportionality constant K. The larger the proportionality constant K, the smaller will be the ultimate value of $D = A/K$. Similarly, the higher the controller gain, the smaller will be the offset D for a given change in A. Thus, the self-regulating system can be redrawn in the form shown in Figure 4-5 to demonstrate the negative feedback concept. In fact, a high gain amplifier diagram for a first-order lag is directly comparable (see Figure 4-6).

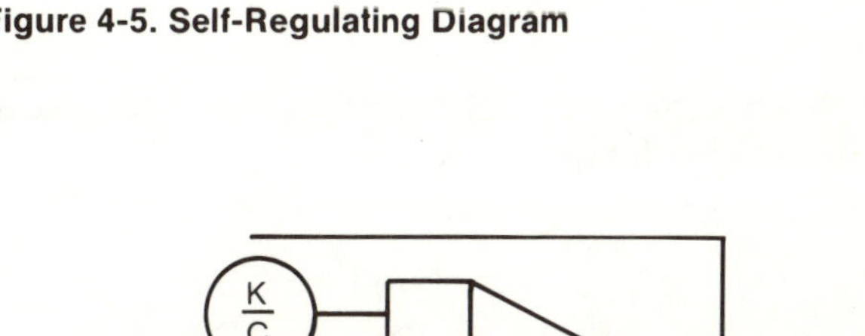

Figure 4-5. Self-Regulating Diagram

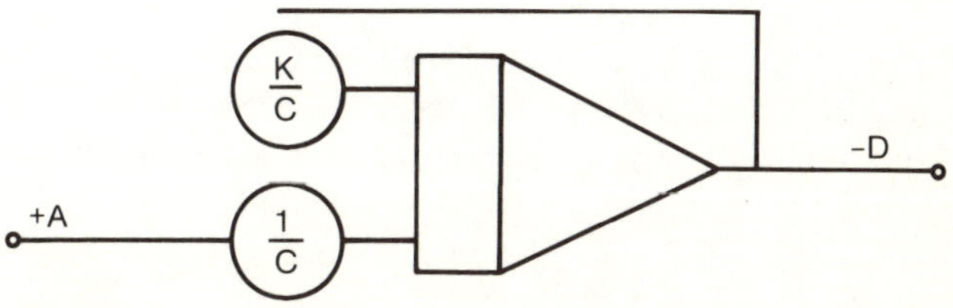

Figure 4-6. Amplifier with Feedback

The diagram is mechanized from the rearranged form of Equation 4-2:

$$\frac{dD}{dt} = \frac{A}{C} - \frac{K}{C} D$$

or, integrating,

$$D = \int_0^t \left(\frac{A}{C} - \frac{K}{C} D\right) dt$$

The triangular symbol in Figure 4-6 represents an integrator; the circles are coefficient potentiometers by which the constants are inserted. The integrator inherently inverts the signs of the inputs.

To summarize, a self-regulating system can be defined as one with built-in negative feedback; it is a first-order system in which the response is characterized by the time constant τ.

Problems

1. Nitrogen is supplied from a 100-psig source through a restricting orifice to the vapor space of a vessel that has a self-contained backpressure regulator set at 30-psig. Is this a self-regulating system? Justify your answer.

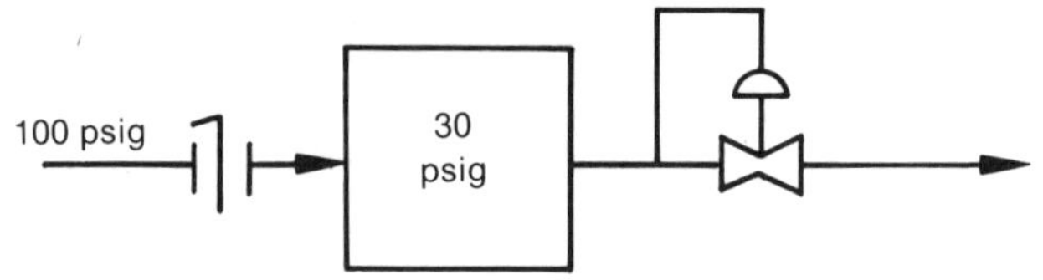

2. A screw pump, in which slip (internal backflow) is proportional to the discharge pressure, delivers polymer to a group of spinning pumps. Is this system self-regulating? Justify your answer.

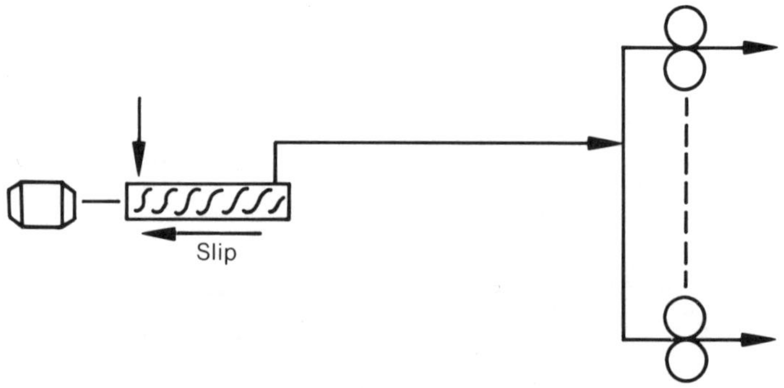

3. Referring to the response curve for a first-order system, what is the slope of the second tangent to the response curve (i.e., the tangent for the second time constant interval)?

4. The response curve for a first-order lag shows that the response is 63.2 percent complete (or 36.8 percent incomplete) after the first time constant. What is the percent incompletion after
   (a) Three time constant intervals?

   (b) Five time constant intervals?

4-4 FIRST-ORDER SYSTEMS

In Section 4-3, a first-order system was defined as one in which the accumulation variable can exert an influence on the inflow or outflow rates so as to bring them into balance. It has, in essence, a built-in negative feedback. It is characterized by its time constant, which is defined as the time required for the system to reach 63.2 percent of its ultimate value. Now specific types of first-order systems will be considered.

4-4-1 Electrical Systems

Consider the system shown in Figure 4-7
where E = electrical potential, volts
i = current flow, coulombs/sec
R = resistance, volt sec/coulomb
C = capacitance, coulomb/volt
and the only current flow i is that required to charge the capacitor C. From the driving-force equation, it can be written that

$$i = \frac{E_1 - E_2}{R} \qquad (4\text{-}6)$$

and, from the unsteady-state energy balance,

$$i = C\frac{dE_2}{dt}$$

or, in Laplace notation,

$$i = CsE_2 \qquad (4\text{-}7)$$

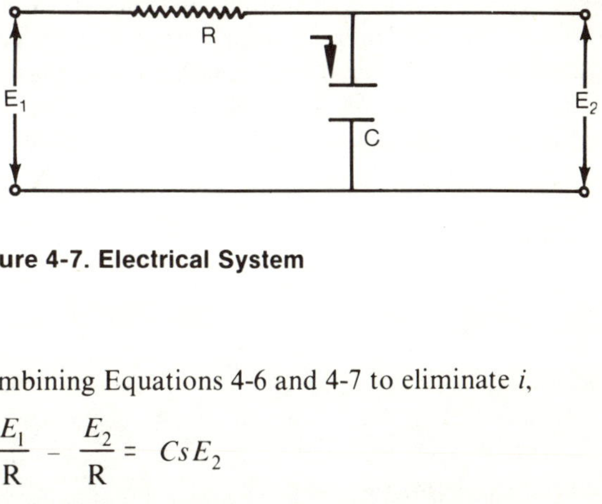

Figure 4-7. Electrical System

Combining Equations 4-6 and 4-7 to eliminate i,

$$\frac{E_1}{R} - \frac{E_2}{R} = CsE_2$$

Rearranging to obtain the transfer function E_2/E_1,

$$E_1 - E_2 = RCsE_2$$

$$E_1 = RCsE_2 + E_2$$

$$\frac{E_2}{E_1} = \frac{1}{RCs + 1} \qquad (4\text{-}8)$$

Let the time constant $\tau = RC$; then,

$$\frac{E_2}{E_1} = \frac{1}{\tau s + 1} \qquad (4\text{-}9)$$

Checking the units of τ, it can be seen that

$$\tau = RC = \frac{(\text{volt sec})}{(\text{coulomb})}\ \frac{(\text{coulomb})}{(\text{volt})} = \text{sec}$$

The time constant always has units of time only. Note that if the conventional units of ohms and farads are used, the time constant will have units of seconds also, since the dimensions used in Equation 4-8 are those of resistance and capacitance in ohms and farads respectively. The product of resistance in the more typical units of megohms and capacitance in microfarads also result in units of seconds.

If a step change is made in E_1, the response would be expected to be as shown in Figure 4-8.

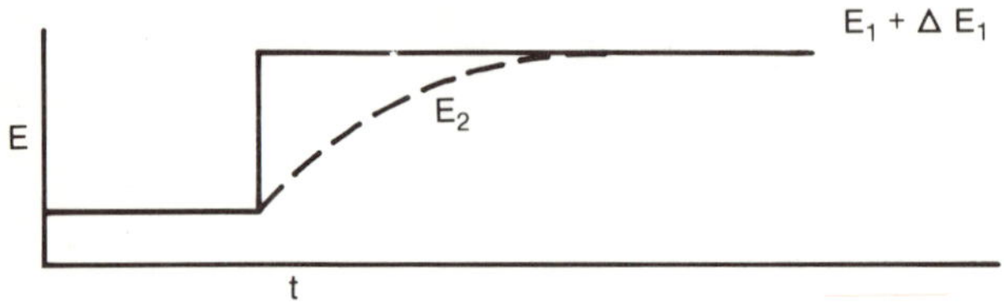

Figure 4-8. Response of Self-Regulating System

The response should be about 95 percent complete after three time constants and 99 percent complete after five time constants.

Appendix A is an analysis of an electrical circuit where voltage has a finite value at time equals zero and where the analysis is carried out by the formal solution of the differential equation. By comparison, the simplicity of the foregoing solution using operational notation is striking.

Note that application of the final value theorem to Equation 4-9 would have also shown that eventually ΔE_2 would equal ΔE_1.

4-4-2 Thermal Systems

- *Conduction*

Consider a bar of metal, insulated on the sides, through which heat flows longitudinally. Consider two sections of the bar, each of length *b*, the central temperatures of which are T_1 and T_2 (see Figure 4-9). By unsteady-state heat balance, it can be written that

$$Q_1 - Q_2 = CsT_1 \tag{4-10}$$

and, from the driving-force equation,

$$Q_2 = \frac{T_1 - T_2}{R} \tag{4-11}$$

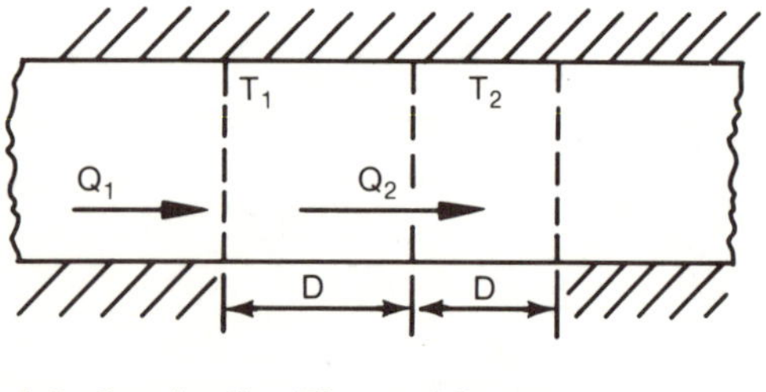

Figure 4-9. Conductive Thermal System

Combining Equations 4-10 and 4-11, the transfer function for T_1 can be found:

$$Q_1 - \frac{T_1}{R} + \frac{T_2}{R} = CsT_1$$

$$RQ_1 + T_2 = RCsT_1 + T_1$$

$$RQ_1 + T_2 = (RCs + 1)\ T_1$$

$$T_1 = \frac{RQ_1 + T_2}{RCs + 1} \tag{4-12}$$

This is the first-order lag. In block diagram form, it would appear as shown in Figure 4-10.

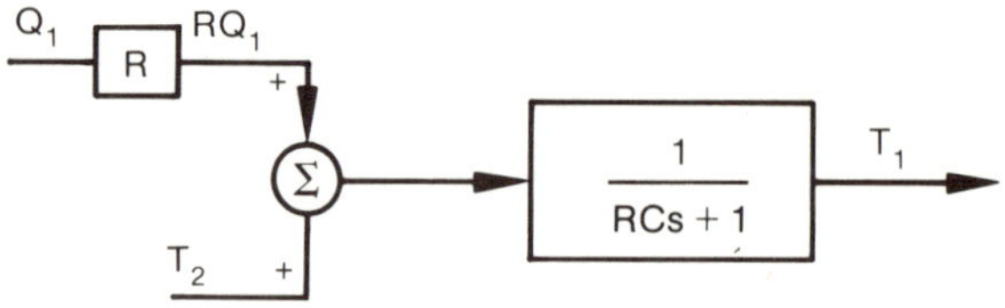

Figure 4-10. Block Diagram, Conductive Thermal System

Section 3 stated that the capacitance term is comprised of the weight of metal and the specific heat:

$$C = M_1\, c_1 \quad \frac{\text{PCU}}{^\circ\text{C}}$$

Section 3 also stated that the resistance term depends on the thickness or length of the conduction path, the thermal conductivity, and the heat transfer area:

$$R = \frac{b}{K_1 A_1} \qquad \frac{\text{hr}^\circ\text{C}}{\text{PCU}}$$

Then,

$$\tau = RC = \frac{M_1 c_1 b}{k_1 A_1} \quad \text{hr} \tag{4-13}$$

Substituting Equation 4-13 into 4-12,

$$T_1 = \frac{RQ_1 + T_2}{\dfrac{M_1 c_1 b}{K_1 A_1} s + 1} \tag{4-14}$$

It is evident that response will be slower (or time constant larger) as the weight *M*, the specific heat *c*, and the path length *b* are increased. The response will be faster as the thermal conductivity *k* and the area *A* are increased. From the final value theorem ($s \rightarrow 0$), it can be seen that eventually ΔT_1 will equal ΔT_2 or $R\Delta Q_1$, depending on which input was step-changed.

Note that the transfer function refers only to deviations around some steady-state value. The new steady-state value would be the algebraic sum of the original steady-state value and the deviation. This is typical of linearized analyses of control problems.

• *Convection*

Consider a thermal element immersed in a flowing fluid thus (see Figure 4-11). By an unsteady-state heat balance, the bulb temperature (neglecting heat losses by conduction to the air outside the pipe) can be defined as

$$Q_b = CsT_b \tag{4-15}$$

From the driving-force equation, it is known that

$$Q_b = \frac{T_f - T_b}{R} \tag{4-16}$$

where the subscripts refer to fluid and bulb, respectively.

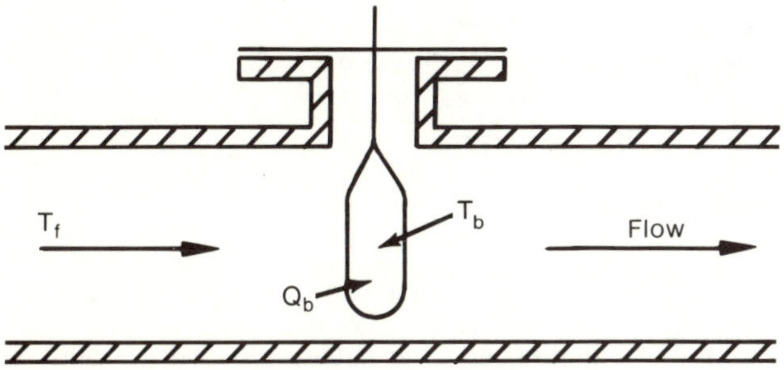

Figure 4-11. Thermal Sensor Response

Combining Equations 4-15 and 4-16 to eliminate Q and arranging the result in the form of the transfer function T_b/T_f gives

$$\frac{T_f}{R} - \frac{T_b}{R} = CsT_b$$

$$T_f = RCsT_b + T_b$$

$$T_f = (RCs + 1)\ T_b$$

$$\frac{T_b}{T_f} = \frac{1}{RCs + 1}$$

In Section 3-6 it was stated that the heat capacitance of a thermal system depended on the weight and specific heat of the material involved, i.e.,

$$C = M_b c_b \ \frac{\text{PCU}}{°\text{C}}$$

Section 3-6-5 also defined the resistance to convective heat transfer as being dependent on the film coefficient and the heat transfer area, i.e.,

$$R = \frac{1}{h_b A_b} \ \frac{\text{hr °C}}{\text{PCU}}$$

Consolidating these values into the time constant,

$$\tau = RC = \frac{M_b c_b}{h_b A_b} \ \text{hrs} \tag{4-17}$$

Defining the variables as:

M = weight, lbs

c = specific heat, PCU/lbs°C

h = film coefficient, PCU/hr ft^2 °C

A = area, ft^2

the units of π become

$$(\text{lbs}) \frac{(\text{PCU})}{(\text{lbs °C})} \ \frac{(\text{hr ft}^2\ °\text{C})}{(\ \text{PCU}\ )} \ \frac{(1)}{(\text{ft}^2)} = \text{hrs}$$

Substituting Equation 4-17 into the transfer function,

$$\frac{T_b}{T_f} = \frac{1}{\dfrac{M_b c_b}{h_b A_b} s + 1} \tag{4-18}$$

Again, the time constant increases as the heat capacitance increases and decreases as the resistance decreases (i.e., as the film coefficient or the area increases) and that ultimately (at $t = \infty$ and $s = 0$), ΔT_b will equal ΔT_f. The film coefficient (h_b) varies as the (velocity)$^{0.6}$ for flow normal to the temperature bulb. Equation 4-18 is discussed in detail as a part of temperature sensor dynamics in Appendix B.

• *Radiation*

Consider a hollow cylinder suspended in an evacuated radiant heating furnace as shown in Figure 4-12, where the heater is at temperature T_1 and the average temperature of the cylinder is T_2. The cylinder has an outside area A and a heat capacitance C.

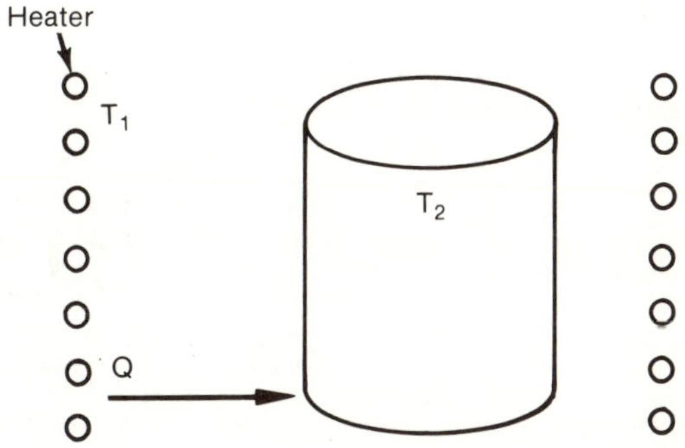

Figure 4-12. Thermal Radiation System

The unsteady-state heat balance gives

$$Q = CsT_2 \tag{4-19}$$

and the driving-force equation gives

$$Q = \frac{T_1 - T_2}{R} \tag{4-20}$$

Combining Equations 4-19 and 4-20 to obtain the transfer function T_2/T_1,

$$\frac{T_1 - T_2}{R} = CsT_2$$

$$T_1 = RCsT_2 + T_2$$

$$T_1 = (RCs + 1)\ T_2$$

$$\frac{T_2}{T_1} = \frac{1}{RCs + 1} \qquad (4\text{-}21)$$

By now, it is known that

$$C = Mc\ \frac{\text{PCU}}{°\text{C}} \quad \text{or} \quad \frac{\text{PCU}}{°\text{K}}$$

that is, weight times specific heat. From Section 3-6-1, Equation 3-23,

$$R = \frac{1}{h_r A} = \frac{10^8}{4\,\epsilon\, A T^3_{\text{ave}}}\ \frac{\text{hr}\ °\text{C}}{\text{PCU}}$$

whence,

$$\tau = RC = \frac{10^8\, Mc}{4\,\epsilon\, A T^3_{\text{ave}}}\ \text{hrs} \qquad (4\text{-}22)$$

Thus, increasing the emissivity or increasing the heater temperature (which will increase T_{ave}) will reduce the resistance and hence the time constant. Combining Equations 4-21 and 4-22,

$$\frac{T_2}{T_1} = \frac{1}{\dfrac{10^8\, Mc}{4\,\epsilon\, A T^3_{\text{ave}}}\, s + 1} \qquad (4\text{-}23)$$

- *Combined and Overall Coefficients*

Combined or overall coefficients can be used in calculating time constants in the same fashion as film coefficients. It is merely necessary to use the appropriate expression for evaluating the resistance R. For example, the resistance term for the heat loss due to convection and radiation was developed in Section 3-6-1 as

$$R = \frac{1}{(h_c + h_r)A}\ \frac{\text{hr}\ °\text{C}}{\text{PCU}}$$

Since the capacitance term will be

$$C = Mc\ \frac{\text{PCU}}{°\text{C}}$$

the time constant will be

$$\tau = RC = \frac{Mc}{(h_c + h_r)A}\ \text{hrs}$$

Similarly, the resistance term associated with the use of the overall coefficient of heat transfer was determined to be:

$$R = \frac{1}{\text{UA}}\ \frac{\text{hr}\ °\text{C}}{\text{PCU}}$$

whence,

$$\tau = RC = \frac{Mc}{UA}\ \text{hrs}$$

Problems

1. The dc output current of an electronic controller is filtered with a 30-microfarad condenser and connected into a 600-ohm load thus:

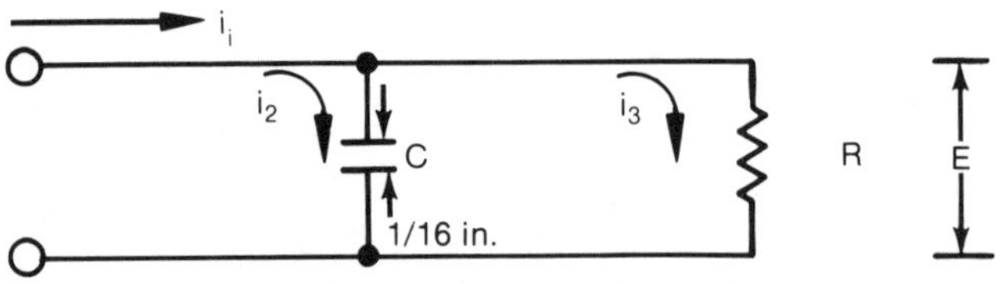

(a) Develop the transfer function for E/i_1.

(b) Evaluate the time constant.

2. In the heat conduction system described by Equation 4-14, calculate the time constant if the bar is 1 inch square and the distance b between T_1 and T_2 is 12 inches:
(a) For steel: weight = 0.282 lb/in.^{3}, thermal conductivity is 35 PCU ft/hr ft^2 °C, and the specific heat is 0.11 PCU/lb°C.

(b) For copper: 0.321 lb/in.^{3}, 220 PCU ft/hr ft^2 °C, 0.092 PCU/lb°C.

3. In the convective heat transfer system described by Equation 4-18, what is the time constant of a 1/2-inch diameter × 3 inch long bulb (area = 5.1 in.^{2}, volume = 0.59 in.^{3}) having the characteristics of mercury (density = 849 lbs/ft^3, specific heat = 0.0332 PCU/lbs°C):
(a) In water, with h = 150 PCU/hr ft^2 °C?

(b) In air, with h = 10?

4. In the radiant heat transfer system described by Equation 4-23, calculate the time constant for a cylinder of aluminum ($M = 100$ lbs, $c = 0.226$ PCU/lbs°K, $A = 10$ ft^2, $\varepsilon = 0.113$):
(a) If $T_1 = 600°$K, $T_2 = 800°$K.

(b) If $T_1 = 800°$K, $T_2 = 1000°$K.

4-4-3 Incompressible Fluid Systems

- *Streamline Flow*

Consider the liquid system shown in Figure 4-13 in which:

w = flow rate, lbs/sec
L = liquid level, ft
M = weight of liquid, lbs
A = tank sectional area,
ϵ = liquid density, lbs/ft^3
p = pressure, lbs/in.^{2}

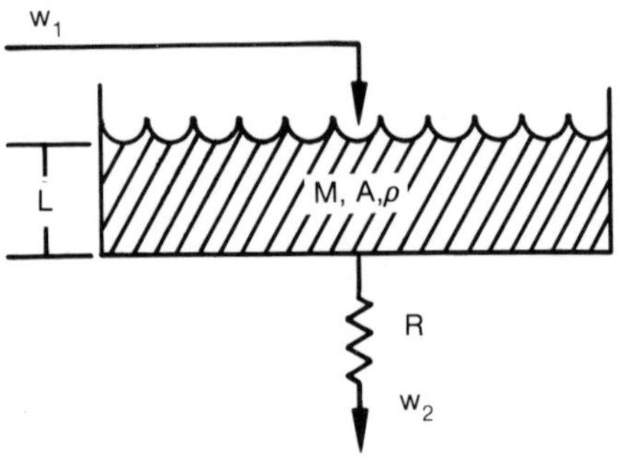

Figure 4-13. Liquid Level Process

Assume that resistance R is a capillary in which the flow is streamline and that the fluid discharges to atmosphere. From the unsteady-state equation,

$$w_1 - w_2 = C\frac{dL}{dt} \tag{4-24}$$

and, from the driving-force equation,

$$w_2 = \frac{p_1 - p_2}{R} \tag{4-25}$$

The driving-force must now be expressed in terms of the accumulation variable (L) before combining Equations 4-24 and 4-25. The pressure can be expressed in terms of the liquid head:

$$p = \frac{L\rho}{144} \qquad \frac{\text{ft}}{} \bigg| \frac{\text{lbs}}{\text{ft}^3} \bigg| \frac{\text{ft}^2}{\text{in.}^2} \qquad \frac{\text{lbs}}{\text{in.}^2} \tag{4-26}$$

Then, combining Equations 4-25 and 4-26 with $p_2 = 0$,

$$w_2 = \frac{L\rho}{144\ R} \tag{4-27}$$

Equations 4-24 and 4-27 are now combined and solved for the transfer function L/w_1:

$$w_1 - \frac{L\rho}{144\ R} = CsL$$

$$\frac{144\ R}{\rho}\, w_1 = \frac{144\ RC}{\rho}\, s + L$$

$$\frac{144\ R}{\rho}\, w_1 = \left(\frac{144\ RC}{\rho}\, s + 1\right) L$$

$$\frac{L}{w_1} = \frac{144\ R/\rho}{\dfrac{144\ RC}{\rho}\, s + 1}$$

In this case, then,

$$\tau = \frac{144\ RC}{\rho}$$

with the $144/\rho$ required to reconcile the units of R and C. By the final value theorem, ΔL will ultimately equal $144\ R\ \Delta\ w_1/\rho$.

From previous sections, it is known that the resistance term for streamline flow has the form

$$R = \left.\frac{p_1 - p_2}{w}\right|_{ss} \quad \frac{\text{sec}}{\text{in.}^2} \tag{4-28}$$

and the capacitance term is

$$C = \rho A \ \frac{\text{lbs}}{\text{ft}}$$

Obviously, the product RC will not give a time constant in units of time only; hence the need for the factor $144/\rho$. Checking units and remembering that 144 has units of in.^{2}/ft^2:

$$\tau = \frac{144\ RC}{\rho} = \frac{\text{in.}^2}{\text{ft}^2}\cdot\frac{\text{sec}}{\text{in.}^2}\cdot\frac{\text{lbs}}{\text{ft}}\cdot\frac{\text{ft}^3}{\text{lbs}} = \text{sec}$$

It is wise also to confirm that the numerator of the transfer function has the same dimensions as L/w_1 (whose units are ft sec/lbs)

$$\frac{144\ R}{\rho} = \frac{\text{in.}^2}{\text{ft}^2}\cdot\frac{\text{sec}}{\text{in.}^2}\cdot\frac{\text{ft}^3}{\text{lbs}} = \frac{\text{sec ft}}{\text{lbs}}$$

Thus, the final change in level resulting from a change in inflow can be defined as

$$\Delta L = \left(\frac{144\ R}{\rho}\right) \Delta w_1$$

Note again that the choice of definition for R and C, and hence τ, depends on the functioning of the specific system under consideration.

- *Turbulent Flow*

Resistance (R) can be redefined as pertaining to an orifice, through which the flow is turbulent. Previously, the linearized form of the turbulent flow equation was developed as

$$w = \frac{p_1 - p_2}{R}$$

in which

$$R = \left.\frac{2(p_1 - p_2)}{w}\right|_{ss} \quad \frac{\text{sec}}{\text{in.}^2} \tag{4-29}$$

Noting that $p_2 = 0$ and that p_1, expressed in terms of level, becomes $L\rho/144$:

$$w_2 = \frac{L\rho}{144\ R} \tag{4-30}$$

This is the same expression that applied to the streamline flow equation (i.e., Equations 4-30 and 4-27 are identical). Since the unsteady-state equation will not change, combining Equations 4-24 and 4-30 will give the same transfer function as Equations 4-24 and 4-27:

$$\frac{L}{w_1} = \frac{144\ R/\rho}{\dfrac{144\ RC}{\rho}\ s + 1}$$

The value of the time constant τ will be different, however, for streamline and turbulent flow, as evidenced by the methods of evaluating the resistance terms.

Comparison of Equations 4-28 and 4-29 shows that, for the same flow and pressure drop, the value of resistance R, hence time constant τ, will be twice as great for turbulent flow as for streamline flow.

Appendix C shows the application of the concepts of resistance and capacitance to a multi-tank system. Cross products of R and C appear and lead to very complex expressions; thus, the use of computer simulation for larger systems becomes mandatory. It can be observed however, that the analytical expressions of Appendix C provide direction as to process revisions to improve dynamics, where systems synthesis using a simulation is largely a trial-and-error procedure.

4-4-4 Compressible Fluid Systems

Consider the gas system shown in Figure 4-14. By unsteady-state material balance,

$$w_1 - w_2 = Csp_1 \tag{4-31}$$

and, by the driving-force equation,

$$w_2 = \frac{p_1 - p_2}{R} \tag{4-32}$$

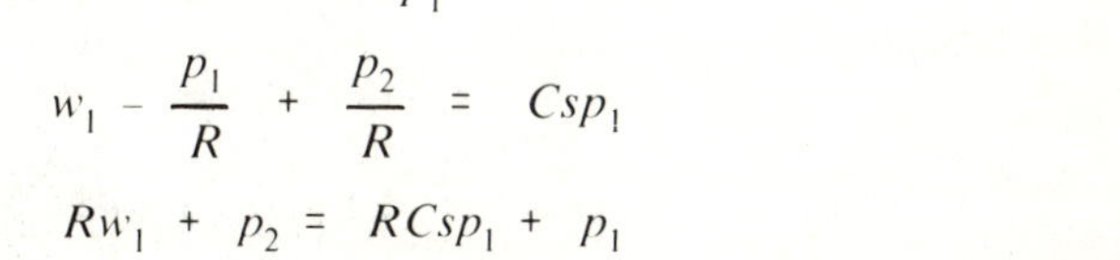

Figure 4-14. Gas Pressure Process

Combining Equations 4-31 and 4-32 and solving for the transfer function of p_1:

$$w_1 - \frac{p_1}{R} + \frac{p_2}{R} = Csp_1$$

$$Rw_1 + p_2 = RCsp_1 + p_1$$

$$Rw_1 + p_2 = (RCs + 1)p_1$$

$$p_1 = \frac{Rw_1 + p_2}{RCs + 1} \tag{4-33}$$

The block diagram form is shown in Figure 4-15.

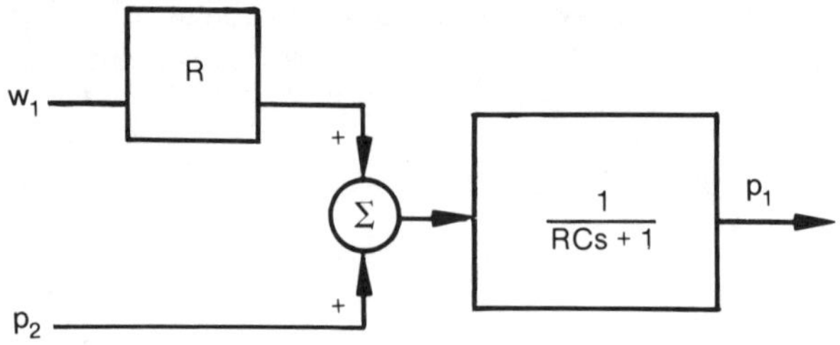

Figure 4-15. Gas Pressure Block Diagram

Note that the units of R must be such as to convert the units of flow into the units of pressure. Define:

w = flow rate, lbs/sec

p = pressure, lbs/in.^{2} absolute

R = resistance, sec/in.^{2}

C = Capacitance, in.^{2}

Then,

$$Rw = \frac{\text{sec}}{\text{in.}^2}\left|\frac{\text{lbs}}{\text{sec}}\right. = \frac{\text{lbs}}{\text{in.}^2} = p$$

In Section 3-6-2, it was established that the capacitance C for a compressible fluid (ideal gas) system is:

$$C = \frac{mV}{n\bar{R}T} \text{ in.}^2 \tag{4-34}$$

where m = molecular weight, lbs/mol

V = volume, ft^3

n = polytropic exponent dimensionless

$\bar{R}$ = gas constant, 19.3 lbs ft^3 /in.^{2} mol°K

T = temperature

Section 3 considered the three flow regimes for compressible fluids (streamline, turbulent, and critical) in terms of evaluating the resistance. They will be considered separately here.

- *Streamline Flow*

From Section 3-7-3,

$$R = \left.\frac{p_1 - p_2}{w}\right|_{ss} \quad \frac{\text{sec}}{\text{in.}^2} \tag{4-35}$$

where R was evaluated using an average density corresponding to the average of p_1 and p_2. Substituting Equations 4-34 and 4-35 into 4-33:

$$p_1 = \frac{\left.\dfrac{p_1 - p_2}{w_2}\right|_{ss} w_1 + p_2}{\left.\dfrac{p_1 - p_2}{w_2}\right|_{ss}\left(\dfrac{mV}{n\bar{R}T}\right)s + 1}$$

and

$$\tau = RC = \left.\frac{p_1 - p_2}{w_2}\right|_{ss}\left(\frac{mV}{n\bar{R}T}\right) \text{ sec}$$

- *Turbulent Flow*

In Section 3-7-3, the resistance term for turbulent flow was evaluated as

$$R = \left.\frac{2(p_1 - p_2)}{w}\right|_{ss} \quad \frac{\text{sec}}{\text{in.}^2} \tag{4-36}$$

In evaluating R, the density corresponding to upstream conditions is used if pressure drop is less than 10 percent of the upstream absolute pressure. For pressure drops of 10 to 50 percent of the upstream pressure, a modified form of the equation is used in which density does not appear directly (Section 3-7-3). Note that, as in the case of incompressible fluids, R for turbulent flow is twice that for streamline flow under the same conditions of pressure and flow rate.

Substituting Equations 4-34 and 4-36 into 4-33:

$$p_1 = \frac{\left.\dfrac{2(p_1 - p_2)}{w_2}\right|_{ss} w_1 + p_2}{\left.\dfrac{2(p_1 - p_2)}{w_2}\right|_{ss}\left(\dfrac{mV}{n\bar{R}T}\right)s + 1}$$

whence, (4-37)

$$\tau = \left.\frac{2(p_1 - p_2)}{w_2}\right|_{ss}\left(\frac{mV}{n\bar{R}T}\right)$$

- *Critical Flow*

The resistance term for critical flow was defined as:

$$R = \left.\frac{(T_1/m)^{1/2}}{8.32\ D^2}\right|_{ss} \quad \frac{\text{sec}}{\text{in.}^2} \tag{4-38}$$

where D = restriction diameter, ft.

Substituting Equations 4-34 and 4-36 into 4-33, and noting that p_2 is effectively zero (i.e., it does not influence flow rate):

$$p_1 = \frac{\left.\dfrac{(T_1/m)^{1/2}}{8.32\ D^2}\right|_{ss} w_1}{\left.\dfrac{(T_1/m)^{1/2}}{8.32\ D^2}\right|_{ss}\left(\dfrac{mV}{n\bar{R}T}\right)s + 1}$$

and

$$\tau = \left.\frac{(T_1/m)^{1/2}}{8.32\ D^2}\right|_{ss} \left(\frac{mV}{n\bar{R}T}\right)$$

- *Non-Ideal Gases and Vapors*

In the event that the Ideal Gas Law cannot be used for calculating capacitance, Section 3 states that the following form may be used:

$$C = V\frac{d\rho}{dp}$$

where the change in density (in lbs/ft³) and pressure (in lbs/in.²) are evaluated in the region of interest. This value of C can then be substituted into any of the foregoing equations for calculating time constants of compressible fluid systems. Thus, for example, the transfer function for turbulent flow conditions (Equation 4-37) would become

$$p_1 = \frac{\left.\frac{2(p_1 - p_2)}{w_2}\right|_{ss} w_1 + p_2}{\left.\frac{2(p_1 - p_2)}{w_2}\right|_{ss} \left(V\frac{d\rho}{dp}\right) s + 1}$$

and

$$\tau = \left.\frac{2(p_1 - p_2)}{w_2}\right|_{ss} \left(V\frac{d\rho}{dp}\right)$$

Problems

1. A tank having a cross-sectional area of 10 ft² discharges water ($\rho = 62.4$ lbs/ft³) through a restriction at 36,000 lbs/hr with a 5 lbs/in.² pressure drop. Calculate the time constant if:
(a) The restriction is a deionizer (streamline flow).

(b) The restriction is a heat exchanger (turbulent flow).

2. In the two cases above, what will be the ultimate change in level if the inflow is increased by 3600 lbs/hour?

3. Ethylene gas (mol. wt. = 28) is flow controlled at 1800 lbs/hr to a deoxidation unit having 10 ft³ free volume. Deox unit inlet pressure is about 300 psig; pressure drop across the bed is 5 lbs/in.² Assuming that the resistance of the bed can be lumped at the exit of the deox unit, develop the transfer function relating deox unit pressure to inflow and downstream pressure. If the unit operates at 127°C, evaluate the constants in the transfer function, assuming turbulent flow and adiabatic expansion (n = 1.3).

4. Superheated steam (mol. wt. = 18) is supplied at 360 psia and 275°C by a pressure controller to a manifold containing 20 jets in parallel, through which the stream exhausts to atmosphere after surface treating synthetic filaments. The 20 jets will pass a total of 2700 lbs/hour of steam under the above conditions. If the pressure control loop requires a process time constant of 2 seconds to hold pressure within tolerance on specified load changes, what volume must be provided between the control valve and the jets? Take dp/dp = 0.002 in.²/ft³ in the region of interest.

4-4-5 Composition and Mass Transfer — Composition Systems

Consider the composition system shown in Figure 4-16 in which:

w = flow rate, lbs/hour

M = holdup, lbs

X_A = weight fraction of component A

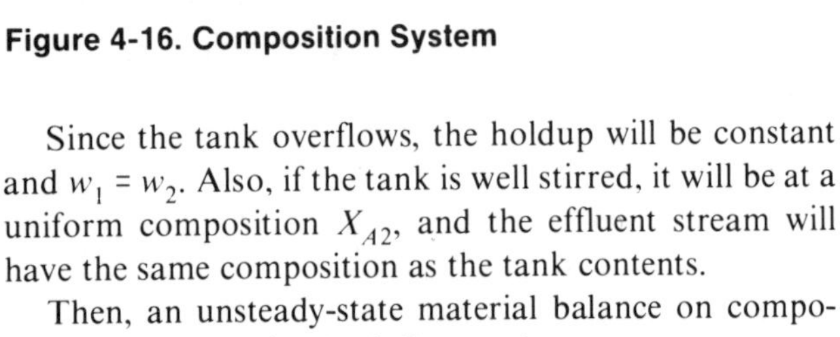

Figure 4-16. Composition System

Since the tank overflows, the holdup will be constant and $w_1 = w_2$. Also, if the tank is well stirred, it will be at a uniform composition X_{A2}, and the effluent stream will have the same composition as the tank contents.

Then, an unsteady-state material balance on component A yields (calling both flows w_1):

$$w_1 X_{A1} - w_1 X_{A2} = M\frac{dX_{A2}}{dt} = MsX_{A2}$$

Rearranging to obtain the transfer function X_{A2}/X_{A1},

$$X_{A1} = \frac{M}{w_1} sX_{A2} + X_{A2}$$

$$X_{A1} = \left(\frac{M}{w_1} s + 1\right) X_{A2}$$

$$\frac{X_{A2}}{X_{A1}} = \frac{1}{\frac{M}{w_1} s + 1}$$

and

$$\tau = \frac{M}{w_1} \text{ hours}$$

Note that the typical first-order transfer function was developed directly from the unsteady-state equation without the use of the driving-force equation. This situation resulted from the assumption that composition in the tank was uniform as a result of sufficient agitation. Without this assumption, some complex form of the driving-force equation would have been required to account for the 3-dimensional mass transfer by diffusion and convection resulting from concentration gradients. Solution of such driving-force equations is generally impractical or impossible. Accordingly, it is generally assumed that, at least for a first approximation, any tank that has agitation is well mixed.

Note also that the time constant is equal to the holdup divided by the throughput.

Suppose now that the influent stream contained a component B at concentration X_{B1}. A second unsteady-state material balance on this component could then be written:

$$w_1 X_{B1} - w_1 X_{B2} = MsX_{B2}$$

Since the equation duplicates the previous unsteady-state material balance except for the subscripts, the transfer functions can be expected to be the same:

$$\frac{X_{B2}}{X_{B1}} = \frac{1}{\frac{M}{w_1}s + 1}$$

Thus, the time constant will be the same for any and all components, regardless of their concentration in the influent stream.

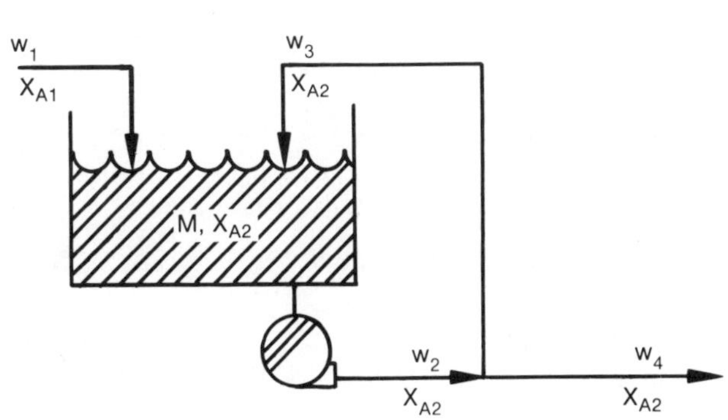

Figure 4-17. Recycle Composition Process

Consider now the system shown in Figure 4-17 in which holdup M is constant; i.e., $w_1 = w_4$. By material balance around the tank on component A, assuming a well mixed tank,

$$w_1 X_{A1} + w_3 X_{A2} - w_2 X_{A2} = M\frac{dX_{A2}}{dt} \quad (4\text{-}39)$$

but,

$$w_3 X_{A2} = w_2 X_{A2} - w_4 X_{A2} \quad (4\text{-}40)$$

Substituting Equation 4 40 into 4-39,

$$w_1 X_{A1} + w_2 X_{A2} - w_4 X_{A2} - w_2 X_{A2} = M\frac{dX_{A2}}{dt}$$

Cancelling the $w_2 X_{A2}$ terms, and letting $w_4 = w_1$, the transfer function X_{A2}/X_{A1} can be solved:

$$w_1 X_{A1} - w_1 X_{A2} = MsX_{A2}$$

$$X_{A1} = \left(\frac{M}{w_1}s + 1\right) X_{A2}$$

$$\frac{X_{A2}}{X_{A1}} = \frac{1}{\frac{M}{w_1}s + 1}$$

Note that the time constant is defined by the holdup and the net throughput (w_1 or w_4), and is independent of the recirculation stream w_3. The reason is evident: the pump and recirculaton stream represent an agitator and therefore do not influence system dynamics so long as agitation is sufficient for good mixing.

4-4-6 Mass Transfer Systems

- *Diffusion of Gases*

Consider the equimolar counterdiffusion of two gases A and B as shown in Figure 4-18,

where w_A = flow rate of component A, mols/hour
Y_A = mol fraction of component A
N = total number of mols, mols

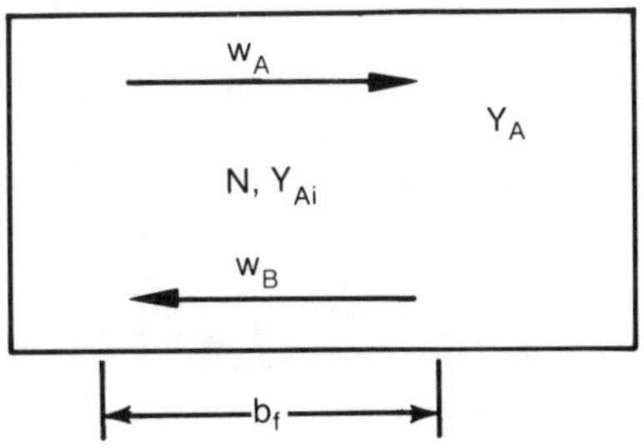

Figure 4-18. Counterdiffusion Process

Note that a mol of any substance is that quantity of the substance whose weight in pounds, grams, or any other mass unit is equal to its molecular weight.

Since the diffusion system is defined as an equimolar (i.e., the same molar flow rate in each direction), the number of mols (N) of gas in any region will remain constant. Only the mol fractions will change.

An unsteady-state material balance on component A in the interfacial region of thickness b_f yields

$$w_A = C\frac{dY_{Ai}}{dt} \quad (4\text{-}41)$$

in which it is evident that the capacitance C and the number of mols N are the same. The driving-force equation, developed earlier, had the form

$$w_A = \frac{Y_A - Y_{Ai}}{R} \quad (4\text{-}42)$$

and R was defined as

$$R = \frac{\bar{R}Tb_f}{DpA} \quad \frac{\text{hr}}{\text{mol}}$$

where $\bar{R}$ = gas content, 19.3 lbs/ft^3/in.^{2} mol °K
T = temperature, °K

b_f = film thickness, ft

D = diffusion coefficient, ft^2/hr

p = total pressure, lbs/in.^{2} absolute

A = diffusion area, ft^2

Combining Equations 4-41 and 4-42 to obtain the transfer function Y_{Ai}/Y_A,

$$\frac{Y_A - Y_{Ai}}{R} = CsY_{Ai}$$

$$Y_A = (RCS + 1)\ Y_{Ai}$$

$$\frac{Y_{Ai}}{Y_A} = \frac{1}{RCS + 1} \qquad (4\text{-}43)$$

The capacitance C is equal to the number of mols N; from the gas law, N can be expressed as

$$N = \frac{pV}{\overline{R}T} = C \quad \text{mol}$$

where V is the volume of the interfacial region in ft^3, whence:

$$\tau = RC = \left(\frac{\overline{R}Tb_f}{DpA}\right)\left(\frac{pV}{\overline{R}T}\right) \text{ hr}$$

$$\tau = \frac{b_f V}{DA} \text{ hr} \qquad (4\text{-}44)$$

Substituting Equation 4-44 into 4-43,

$$\frac{Y_{A1}}{Y_A} = \frac{1}{\frac{b_f V}{DA}s + 1}$$

For a vessel with a uniform cross section in the direction of diffusion,

$$\frac{V}{A} = b_f$$

Hence,

$$\tau = \frac{b2_f}{D} \text{ hr}$$

and

$$\frac{Y_{Ai}}{Y_A} = \frac{1}{\frac{b^2 f}{D}s + 1}$$

- *Mass Transfer Between Phases*

Consider a tray in a two-component distillation column as shown in Figure 4-19, in which

L = liquid flow rate, mols/hour

V = vapor flow rate, mols/hour

X = mol fraction of one component in liquid

Y = mol fraction of that component in vapor

M = holdup on tray (assumed constant), mols

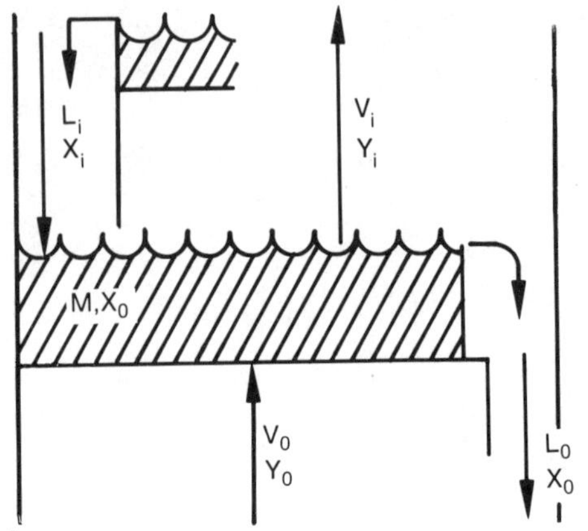

Raoult's Law

(Vapor pressure) (Liquid mol fraction)
= (Total pressure) (Vapor mol fraction)

Vapor Pressure

The pressure of vapor of that liquid at any temperature at which vapor and liquid can exist in equilibrium.

Figure 4-19. Distillation Column

The unsteady-state material balance was previously developed as

$$L_iX_i - L_oX_o + V_oY_o - V_iY_i = MsX_o$$

Assuming that there is an ideal solution (one that follows Raoult's Law), vapor-liquid equilibrium compositions can be related by the expression

$$pY_i = p_oX_o$$

or

$$Y_i = \frac{p_o}{p} X_o \qquad (4\text{-}45)$$

where p = total pressure

p_o = vapor pressure of the pure component of interest at the operating temperature

Any consistent set of units can be used to express p and p_o since they will always appear as a ratio. Equation 4-45 might be considered as a driving-force equation, since, being an equilibrium expression, it forces Y_i into correspondence with X_o.

$$L_iX_i - L_oX_o + V_oY_o - V_i\frac{p_o}{p}X_o = MsX_o$$

$$L_iX_i + V_oY_o = \left(Ms + L_o + V_i\frac{p_o}{p}\right)X_o$$

$$L_iX_i + V_oY_o = \left(L_o + V_i\frac{p_o}{p}\right)\left(\frac{M}{L_o + V_i\frac{p_o}{p}}s + 1\right)X_o$$

$$X_o = \frac{L_iX_i + V_oY_o}{\left(L_o + V_i\dfrac{p_o}{p}\right)\left(\dfrac{M}{L_o + V_i\dfrac{p_o}{p}}s + 1\right)} \qquad (4\text{-}46)$$

Note that the time constant,

$$\tau = \frac{M}{L_o + V_i\dfrac{p_o}{p}} \text{ hr}$$

is dependent on both liquid and vapor flow rates, and that the higher the vapor pressure p_o, the smaller will be the time constant. The block diagram form is shown in Figure 4-20.

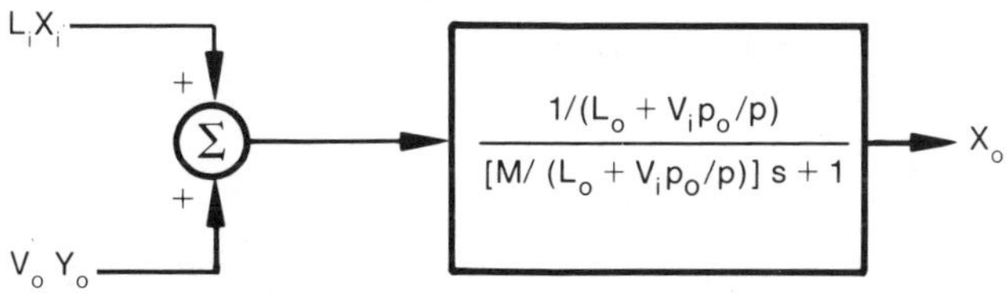

Figure 4-20. Distillation Column Block Diagram

Applying the final value theorem to Equation 4-46 or to the block diagram:

$$\Delta X_o = \left(\frac{1}{L_o + V_i\dfrac{p_o}{p}}\right) \Delta(L_iX_i + V_oY_o)$$

in which the first parenthetical expression is the steady-state gain term,

$$K = \frac{1}{L_o + V_i\dfrac{p_o}{p}} \text{ hr/mol}$$

To summarize the review of first-order systems, it might be said that first-order systems are self-regulating systems, the response of which is defined by a time constant. Directly or implicitly, they comprise the combination of the unsteady-state equation with the driving-force equation in such a fashion that the accumulation variable exerts a negative feedback influence on the driving-force equation, thereby ultimately bringing input and output rates into balance. Application of the final value theorem allows determination of the ultimate steady-state value of the accumulation variable that corresponds to a given step change in some one of the input values.

4-5 Dead Time

Dead time, also called transportation lag or distance-velocity lag, is the time required for energy or material to travel from one point to another without any change in its properties or characteristics. The last phrase is important, as it distinguishes dead time from other types of dynamic responses such as first-order lags, distributed-parameter systems, etc. If, for example, the composition of a fluid at the inlet of a pipe is suddenly changed, the composition "front" will travel through the pipe at the velocity of the fluid and will appear essentially intact at the pipe exit some time later. This is dead time. If, on the other hand, we were to vary the proportion of two reactants at the pipe inlet, the resultant composition would vary both with distance and time as the reaction progresses. This is a distributed-parameter system — one whose characteristics vary with time and distance.

Mathematically, if there is a time-varying function $\theta_1(t)$ at one point in a process, separated from a second point by a dead time a, then at the second point the response will be

$$\theta_2(t) = \theta_1(t-a)$$

that is, what is now happening at location 2 happened a units of time previously at location 1. In terms of Laplace notation, this becomes

$$\theta_2 = \theta_1 e^{-as}$$

For example, if a solar flare is seen right now ($t = 0$), the event actually occurred at 9.3×10^7 miles/ (1.86×10^5 miles/sec) = 500 seconds previously; i.e.,

$$\theta_2(t) = \theta_1(t - 500)$$

or

$$\theta_2 = \theta_1 e^{-500s}$$

One other example will serve to clarify the concept of dead time. Consider a composition system as shown in Figure 4-21, in which:

w = flow rate, lbs/min

X_A = weight fraction of A, dimensionless

M = holdup, lbs

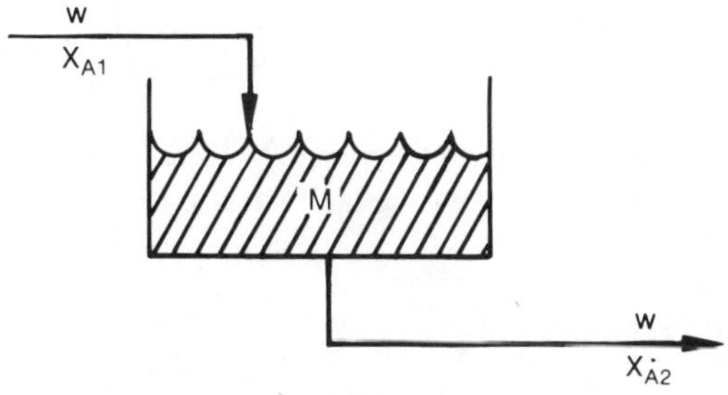

Figure 4-21. Dead Time Process

In the study of first-order systems, with the tank contents well mixed, it was established that the transfer function for this system was

$$\frac{X_{A2}}{X_{A1}} = \frac{1}{\frac{M}{w}s + 1}$$

and the time constant was

$$\tau = \frac{M}{w} \text{ min}$$

Suppose, however, that the tank contents are not mixed at all and that the influent stream is introduced carefully so as to avoid turbulence; then plug flow will occur in the tank. If, now, a change is made in X_{A1}, it will not appear in the effluent line as a change in X_{A2} until the entire contents of the tank have been replaced by the new composition. The displacement time is the dead time a, and is defined as

$$a = \frac{M}{w} \text{ min}$$

Then,

$$X_{A2} = X_{A1}e^{-\frac{M}{w}s}$$

Note that although the time constant and the dead time are both computed as holdup/throughput, the significance is entirely different. In the case of the first-order system, X_{A2} has attained 63.2 percent of its ultimate value after a time equal to M/w. In the case of dead time, however, X_{A2} does not change at all until a time equal to M/w.

Problems

1. A mix tank combines a recovered monomer stream of flow rate w_1 and composition X_{A1} with a monomer makeup stream (flow = w_2, composition X_{A2}) and discharges a stream of flow rate w_3 and composition X_{A3} to a reactor:
   (a) Develop the transfer function for X_{A3}.

   (b) For a change in recycle monomer composition of ΔX_{A1}, with $\Delta X_{A2} = 0$, what will be the ultimate value of ΔX_{A3}?

2. A railroad tank car, after having been unloaded by venting to a storage tank, contains a residual 1000 lbs of ethylene, due to the operating pressure in the storage tank. It is desired to recover the residual ethylene by purging the tank car with 500 lbs/hour of inert gas (nitrogen, mol.wt. = 28) and sending the effluent mixture to a reactor. If the reactor cannot tolerate less than about 5 percent ethylene in the gas leaving the tank car, how long can the inert gas purge be continued? Assume the contents of the tank car are well mixed. (See Section 3-7-5, Problem 4).

3. What is the time constant for the counter diffusion of air and carbon dioxide ($D = 0.64$ ft^2/hr) through an interfacial path of 0.8 ft?

4. Around a plate in a distillation column, the liquid rate is 500 mols/hour and the vapor rate is 750 mols/hour. Vapor pressure of the component of interest is 2 atmospheres, while the column operates at atmospheric pressure. Tray holdup is constant at 100 mols. Assuming an ideal solution,
(a) Evaluate the composition time constant.

(b) Evaluate the steady-state gain.

5. An infrared analyzer has a 50-foot run of 1/4 in. pipe (flow area = 0.104 in.^{2}) to the top of a distillation column operated at 5 atmospheres and 50° C. Sample flow rate is 0.5 cfm based on analyzer sample cell conditions of 1 atmosphere and 50° C. Neglecting heat losses and pressure drop, what is the dead time from column to analyzer if the pressure reducing valve is located
(a) At the analyzer?

(b) At the top of the column?

6. A suspension consisting of 30 percent oil and 70 percent water flows into a 60-ft^3 decanter at 21 gal/min. (A decanter is a vessel where the lighter and heavier portions of a liquid mixture are allowed to separate.) The decanter holdup is equally divided between the two phases. Acid in the water phase is neutralized by injection of 0.3 gal/min of caustic solution just upstream from the decanter, based on pH control. To prevent oil from fouling the pH cell electrodes, the cell is mounted in the water-phase effluent piping close to the decanter.

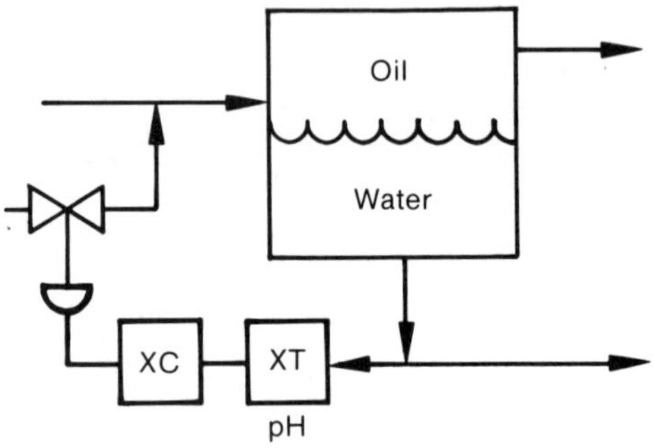

(a) What is the dead time in the pH control loop?

(b) Does location of the control valve in the caustic injection line influence the dead time? Why?

SECTION 5

CONTROL VALVES

5-1 INTRODUCTION

The final control element, or manipulated variable, in a control system can be a rheostat, a variable speed drive, a piston and cylinder positioning device, a hydraulic drive, a damper or one of a host of other devices. In the fluid processing industries, however, the final control element is almost always a control valve. The control valve is a variable resistor in the pipeline, with the resistance being set by the signal from a controller. The valve consists of two parts: an actuator that converts the controller signal into a position; and a variable restriction in the pipeline that is positioned by the actuator. Control valves are manufactured in many configurations, which are well described elsewhere. [26,27]

5-2 INHERENT VALVE FLOW CHARACTERISTIC

Each valve configuration has an inherent characteristic relationship between the fraction of valve lift and the relative flow through the valve. This relationship is called the valve flow characteristic. The choice of the flow characteristic, similar to sizing or noise abatement, is important to the valve selection since it plays an important part in the dynamic response of the control system.[28,29]

The equation for the flow through a contol valve was shown previously to be

$$w = Kx\,(p_1 - p_2)^{1/2} \tag{5-1}$$

The flow characteristic is a plot of the flow w through the valve at various stem positions x. Equation 5-1 shows flow w to be not only a function of x, but also of pressure drop $p_1 - p_2$. Therefore, pressure drop must be specified as a part of the valve characteristic. The valve flow characteristics shown in manufacturers' literature are known as inherent flow characteristics, since they were determined at constant pressure drop. After a valve is installed in a pipeline, pressure drop across the valve is seldom constant. The varying pressure drop changes the shape of the flow characteristic, which is then known as the installed valve flow characteristic.

The five basic inherent valve characteristics are shown in Figure 5-1. Only three of these characteristics (quick opening, linear, and equal percentage) are available in the

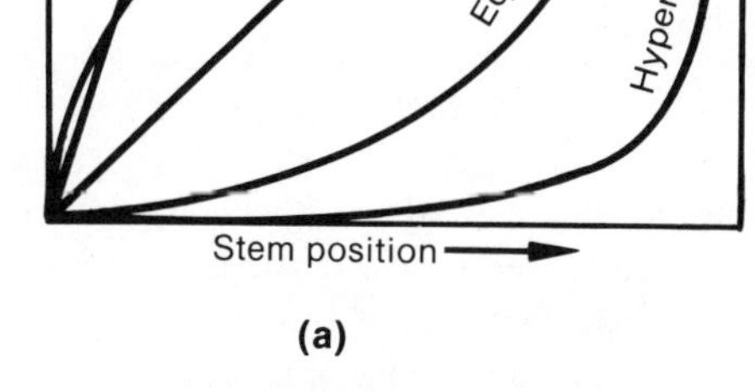

(a)

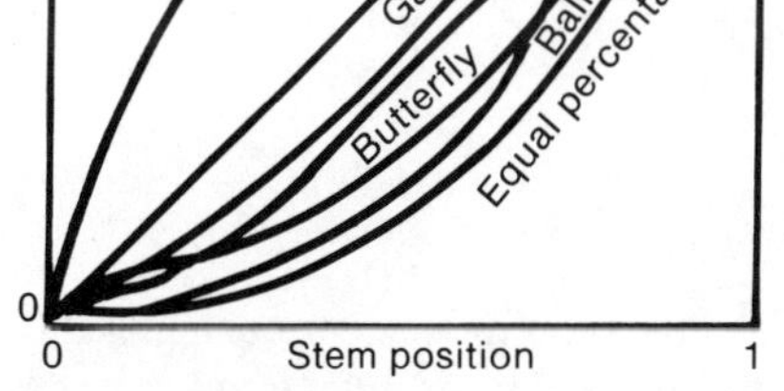

(b) Percent of Maximum Flow versus Stem Travel for Typical Valve Types

Figure 5-1. Basic Inherent Valve Characteristics[28]
(Courtesy ISA)

typical manufacturer's catalog. Other characteristics are available by the use of characterized valve positioners.

5-3 INSTALLED VALVE FLOW CHARACTERISTIC

While the inherent characteristic is valuable in describing the valve as supplied by the manufacturer, it is the installed characteristic that dictates the relationship between stem movement and flow when the valve is installed in the pipeline. Of course, installations exist where upstream and downstream pressures are constant, and, in this special case, the inherent characteristic is the installed characteristic.

Generally, the installed characteristic changes markedly with a factor that is the ratio of the pressure drop across the piping system. The variation is shown in Figure 5-2.

Figure 5-2 shows that both linear and equal percentage valve characteristics become quick-opening installed characteristics when only a small proportion of system pressure drop is across the valve. Cross-country liquid pipelines suffer from this situation, where valves are purchased the same size as the pipeline for energy savings. Then almost all of the pressure drop is across the pipeline, with a very small proportion across the control valve. Any inherent valve characteristic becomes an installed, quick-opening characteristic under these conditons. Controllability is poor with the quick-opening characteristic because of its very high gain at low flows and the very sharp change in gain at the nonlinear point.[30]

A difficult trade-off exists between a low pressure drop for energy savings and an adequate pressure drop for controllability.

5-4 VALVE GAIN

Equation 5-1 shows the relationship that describes flow through a control valve. Generally,

$$w = f\,(x, p_1 - p_2)$$

This can be expanded by the total derivative:

$$\Delta w = \frac{\delta w}{\delta x}\Delta x + \frac{\delta w}{\delta(p_1 - p_2)}\Delta(p_1 - p_2) \qquad (5\text{-}2)$$

By observation, for a linear installed valve flow characteristic,

$$\frac{\delta w}{\delta x} = \frac{w\big|_{max}}{1}$$

where x is valve stem movement normalized 0 to 1.

Consider an equal percentage valve characteristic as being representative of the nonlinear characteristics. The linearization represented by the partial derivative takes place over small deviations (Δs) or perturbations around an operating point or design point. The partial derivatives are tangents to the curve at the operating points, and, as such, are the gains required for the block diagram. The partial derivatives are taken as follows:

For an equal percentage valve characteristic,

$$w = w\big|_{max}\ (a)^{x-1}$$

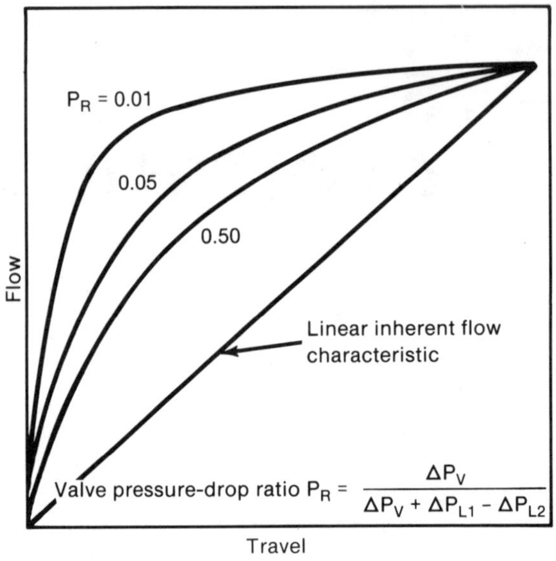

(a) Installed Flow Characteristic with Linear Inherent Flow Characteristic

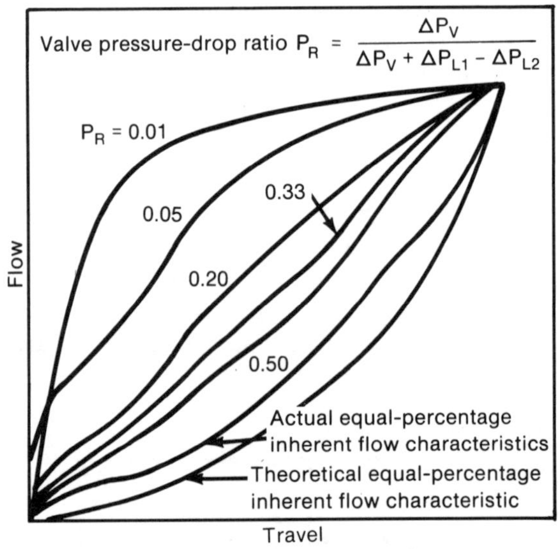

(b) Installed Flow Characteristic with Equal-Percentage Actual Inherent Flow Characteristic

Figure 5-2. Installed Flow Characteristics[28]
(Courtesy ISA)

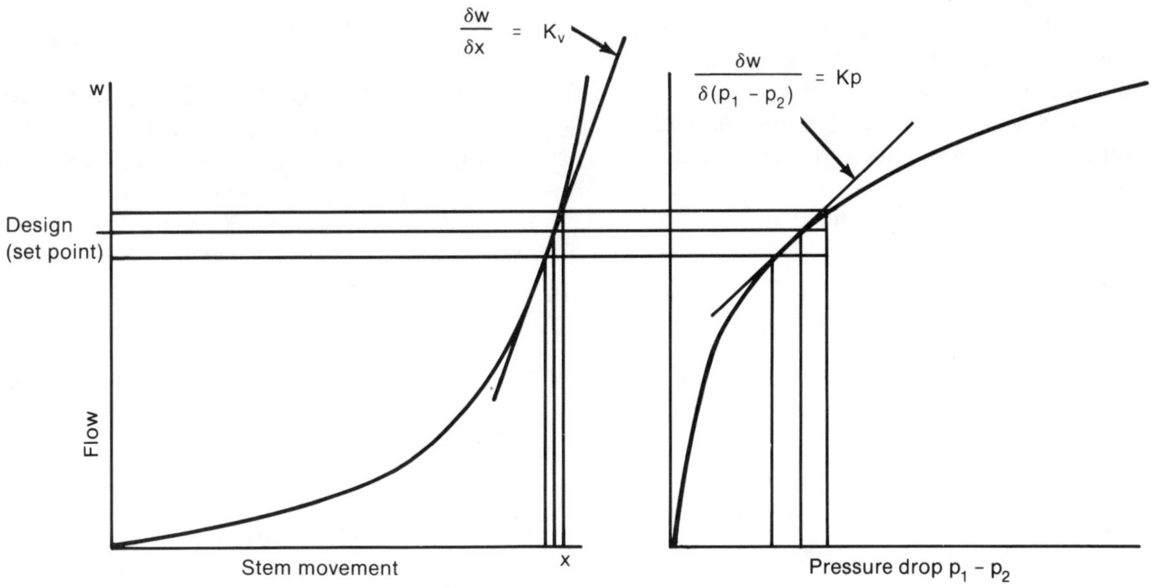

Figure 5-3. Nonlinear Equal Percentage Valve Characteristic and Square Law Pressure Drop Relationship

where a = valve rangeability. Then,

$$\frac{\delta w}{\delta x} = K_v = (\ln a)\, w \quad (5\text{-}3)$$

And from Section 3,

$$\frac{\delta w}{\delta(p_1 - p_2)} = K_p = \frac{w}{2(p_1 - p_2)}$$

Then, omitting the Δs for convenience, Equation 5-2 can be written

$$w = K_v x \times K_p (p_1 - p_2)$$

or

$$w = (\ln a)\,(w)\Big|_{ss} x + \frac{w}{2(p_1 - p_2)}\Big|_{ss} (p_1 - p_2)$$

It must be remembered that the steady-state (ss) values are to be evaluated at the design point, as shown in Figure 5-3, and cannot be cancelled with identical terms in the variables.

5-5 OPEN-LOOP GAIN

The open-loop gain is the product of all of the gains in the control loop. The controller gain is a variable gain used to adjust the value of the open-loop gain for stability. Section 6-6 shows the open-loop gain to defined as

$$K_c K_v K_p K_T = K_o \quad (5\text{-}4)$$

The maximum permissable open-loop gain is determined by the dynamics of the control system. The stability analysis in Section 10 shows that[31]

$$K_o \leq \frac{\tau_p}{\tau_m}$$

where K_o = Open-loop gain
K_c = Controller gain
K_V = Valve gain
K_p = Process gain
K_T = Transmitter gain
τ_p = Process time constant
τ_m = Measurement time constant

Stability over an operating range requires that the terms in Equation 5-4 remain as constant as possible. A review of Section 4 will show that some processes are represented by

$$K_p = \frac{K_1}{w} \qquad \tau_p = \frac{K_2}{w} \quad (5\text{-}5)$$

Substituting Equation 5-5 into Equation 5-4:

$$K_c K_V \frac{K_1}{w} K_T \leq \frac{K_2/w}{\tau_m}$$

or,

$$K_c K_V K_1 K_T \leq \frac{K_2}{\tau_m}$$

The right-hand side of the equation is constant, and constant stability across the operating range requires that the left-hand side be constant. It has been shown that the valve gain (K_V) is constant for a linear installed characteristic. The linear characteristic is then required to provide a constant open-loop gain.

Another type of process is represented by

$$K_p = \frac{K_1}{w} \qquad \tau_p = \tau_p$$

Then,

$$K_c K_V \frac{K_1}{w} K_T \leq \frac{\tau_p}{\tau_m} \tag{5-6}$$

Substituting Equation 5-3 into 5-6, where Equation 5-3 represents the equal percentage valve characteristic:

$$K_c (\ln a)\, w \frac{K_1}{w} K_T \leq \frac{\tau_p}{\tau_m}$$

$$K_c (\ln a)\, K_1 K_T \leq \frac{\tau_p}{\tau_m}$$

The valve gain is proportional to the flow through the valve for an equal percentage installed valve characteristic. The flow term cancels the flow term in the denominator of the process gain term, resulting in a constant open-loop gain. The equal percentage characteristic is required for this type of process. Finally, other processes are described by

$$K_p = K_p \qquad \tau_p = \tau_p$$

and

$$K_c K_V K_p K_T \leq \frac{\tau_p}{\tau_m}$$

Processes where neither gain nor time constant are functions of through-put flow (w) require a constant valve gain to provide a constant open-loop gain. The linear valve characteristic meets this requirement.

Problems

1. Consider the valve flow equation

$$w = Kx\,(p_1 - p_2)^{1/2}$$

Let $p_1 - p_2 = 25$ psi

$K = 10$

$a = 30$ (rangeability)

Calculate the valve gain if:

(a) The characteristic is linear and $X = 0.25$.

(b) The characteristic is linear and $X = 0.75$.

(c) The characteristic is equal percentage and $X = 0.25$.

(d) The characteristic is equal percentage and $X = 0.75$.

2. Consider the open-loop gain equation

$$K_c K_V K_p K_T = \frac{\tau_p}{\tau_m}$$

$$\text{Let } \tau_p = \tau_m$$

$$K_p = 5 \cdot \frac{°\text{C}}{\text{lb/sec}}$$

$$K_T = 0.02 \frac{\text{fraction}}{°\text{C}}$$

Let K_c be dimensionless and K_V have the dimension, lb/sec/fraction.
(a) What are the dimensions of the open-loop gain?

(b) Calculate the controller gains for the valve gains in Problem 1 (a),(b),(c),(d).

3. (a) What valve characteristic should be chosen if

$$\tau_p = \tau_m \quad \text{and} \quad K_p = \frac{10}{w}$$

(b) Using the data from Problem 2, what will be the value of controller gain K_c?

SECTION **6**

SIGNAL FLOW DIAGRAMS

6-1 INTRODUCTION

The signal flow diagram, or block diagram, is a device for representing all of the elements of the control system in common terminology. It has the advantages of stressing the action of the variables rather than the physical components in the system; providing mathematical relationships among variables; and showing the relationships among the system components. The block diagram for a given process may be drawn in many different ways and may be easily manipulated to provide insight into the problem at hand.

6-2 SIGNAL FLOW DIAGRAM ELEMENTS

A signal flow diagram is made up of blocks, representing operators or transfer functions, which operate on the intelligence carried to the block by a line to produce intelligence carried away from the block by another line. A generalized block is shown in Figure 6-1.

The signal flow diagram thus depicts the flow of intelligence through a system and the static and dynamic response of each component in the system in terms of the transfer function of the component. While there is no "right" or "wrong" diagram for a given process, the following rules are helpful in setting up the diagram:

(1) Boxes with a single input and a single output are used to represent suitable parts of the system.

(2) The transfer function KG, both static and dynamic parts, is placed in the box to represent the effect on an input of each part of the system.

(3) Lines with direction arrows are used to represent the flow of "information" from one box to the next. This information can be in any physical form representing the variables in the system.

(4) Information lines may be added in magnitude, provided the information is in the same units. A summing point is used for this purpose and consists of a circle with a Greek letter *sigma* (Σ) in it. The *sigma* is sometimes omitted for simplicity. Signs should be attached to each incoming line.

(5) Branches of information lines without summing points infer that the same value is transmitted to all branches.

(6) Only the addition or subtraction of variables can be handled; multiplication, division, or powers first require linearization over a small, limited range.

(7) Any variable can be chosen as an output. An input can be indicated at any point by the insertion of a proper summing function.

(8) Any number of simultaneous inputs and outputs can be handled by treating each pair separately and combining the results by superposition.

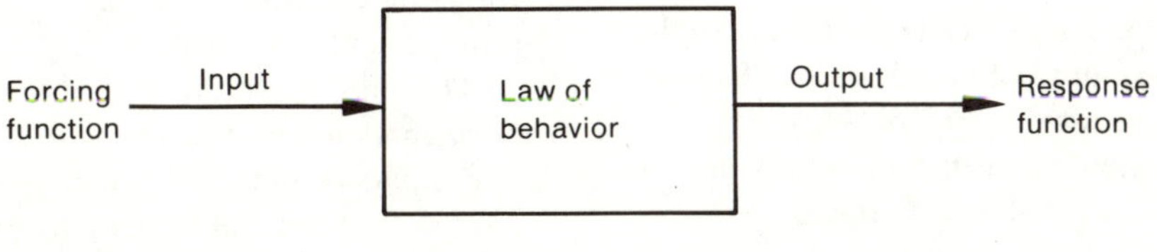

Figure 6-1. Generalized Block Diagram

The signal flow diagram symbols are shown in Figure 6-2.

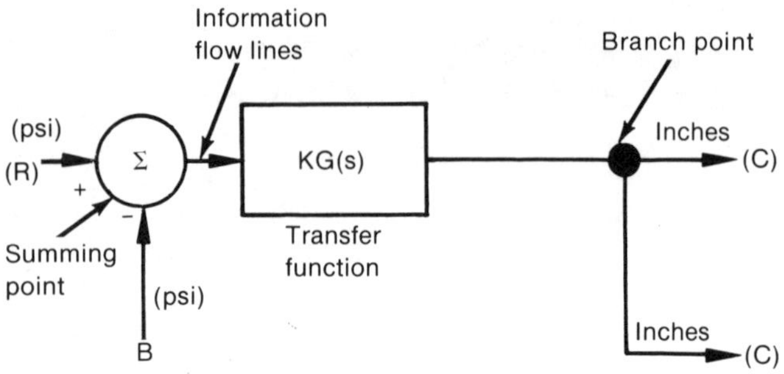

Figure 6-2. Signal Flow Diagram Symbols

Should the block shown in Figure 6-2 represent a temperature transmitter, it could be diagrammed as shown in Figure 6-3.

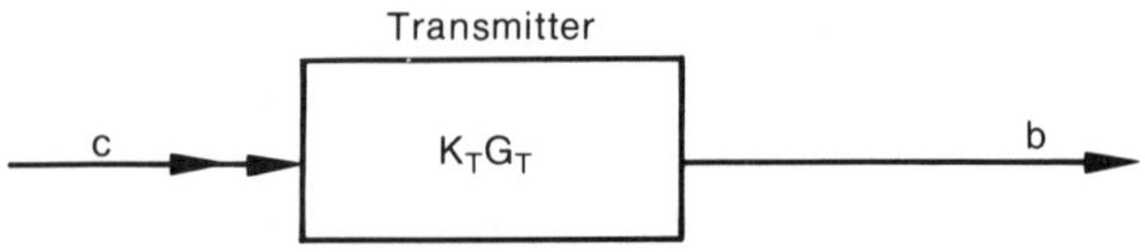

Figure 6-3. Generalized Block

The gain (K_T) of the transmitter in this example is the relationship between *c* and *b* under static conditions. The term is commonly denoted as steady-state gain.

Under steady-state conditions, the dynamic term (G_T) becomes unity and the relationship becomes

$$b = K_T c$$

The gain is merely the proportionality constant that relates the two variables. If the transmitter in the example were a temperature transmitter with a span of 0° to 240° F, and a pneumatic output of from 3 to 15 psi, the gain can be evaluated immediately:

$$K_T = \frac{b}{c}$$

$$= \frac{15\text{-}3}{240\text{-}0} \text{ psi}/°\text{F}$$

$$= 0.05 \text{ psi}/°\text{F}$$

Actually, the concept of a gain can be generally applied to systems with a single input and a single output. Consider the mechanical linkage shown in Figure 6-4.

The displacement gain between the input *x* and the output *y* can be written from similar triangles:

$$\frac{x}{l_1} = \frac{y}{l_2}$$

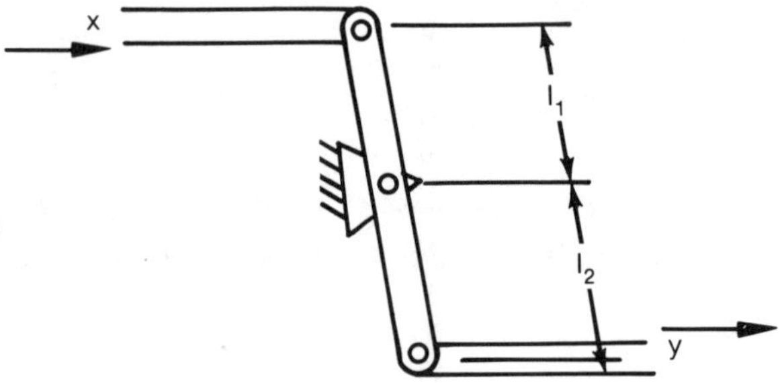

Figure 6-4. Linkage

where *x* and *y* are displacements. Thus, $K = l_2/l_1$, as shown in Figure 6-5.

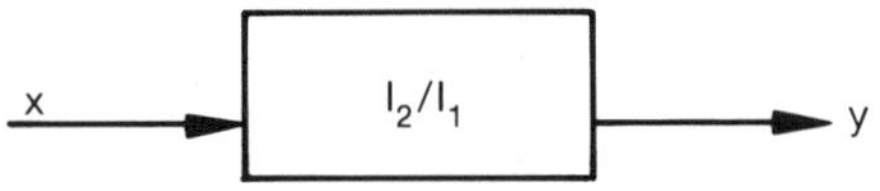

Figure 6-5. Displacement Block

The force gain between *x* and *y* can be determined by taking moments about the pivot point:

$$l_1 x = l_2 y$$

where *x* and *y* are now forces. Thus, as shown in Figure 6-6, $K = l_1/l_2$.

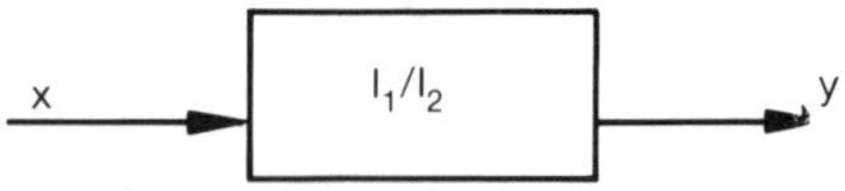

Figure 6-6. Force Block

It is now evident that the same physical system can have a different gain for each combination of defined variables.

One of the gain terms for this linkage will be greater than one, while the other will be less than one. A gain of less than one is called an attenuation.

6-3 CONTROLLER ERROR MEASUREMENT

The error measurement by a pneumatic controller is sometimes accomplished by this same type of linkage arrangement (see Figure 6-7).

The links transmitting *r* and *b* are constrained to limited horizontal movement. The output (*e*) is thus a function of two inputs (*r* and *b*). A single block on the diagram can have but a single input and a single output. This system will evidently require two blocks. Mathematically:

$$e = f(r, b)$$

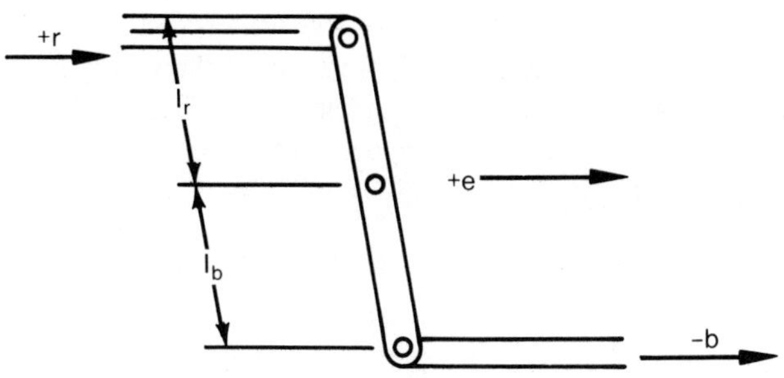

Figure 6-7. Controller Linkage

If the function (f) meets the mathematical restrictions that it be single valued and continuous, the total differential of e can be written:[5]

$$de = \left.\frac{\delta f}{\delta r}\right|_{b\,=\,\text{constant}} dr = \left.\frac{\delta f}{\delta b}\right|_{r\,=\,\text{constant}} db$$

or:

$$\Delta e = \left.\frac{\delta f}{\delta r}\right|_{b\,=\,\text{constant}} \Delta r + \left.\frac{\delta f}{\delta b}\right|_{r\,=\,\text{constant}} \Delta b \qquad (6\text{-}1)$$

This technique is commonly called linearization. Equation 6-1 now says that a change in the variable e is the algebraic sum of a change in r with b held constant and a change in b with r held constant. In this example, the partial derivatives can be simply evaluated by similar triangles. For b held constant:

$$\Delta e = \frac{l_b}{l_r + l_b} \Delta r$$

$$\frac{l_b}{l_r + l_b} = \left.\frac{\delta f}{\delta r}\right|_{b\,=\,\text{constant}} = K_1 \qquad (6\text{-}2)$$

For r held constant:

$$\Delta e = \frac{-l_r}{l_r + l_b} \Delta b$$

$$\frac{l_r}{l_r + l_b} = \left.\frac{\delta f}{\delta b}\right|_{r\,=\,\text{constant}} = K_2 \qquad (6\text{-}3)$$

The changes Δe, Δr, and Δb can be defined as deviations from the set point values e_s, r_s, and b_s. For the purpose of analysis, the set point values are defined to be zero:

$$\Delta e = e - e_s = e$$
$$\Delta r = r - r_s = r$$
$$\Delta b = b - b_s = b \qquad (6\text{-}4)$$

Substituting Equations 6-2, 6-3, and 6-4 into Equation 6-1:

$$e = K_1 r - K_2 b \qquad (6\text{-}5)$$

Equation 6-5 can be exhibited as the portion of the block diagram that describes the error-generating mechanism in the controller (see Figure 6-8).

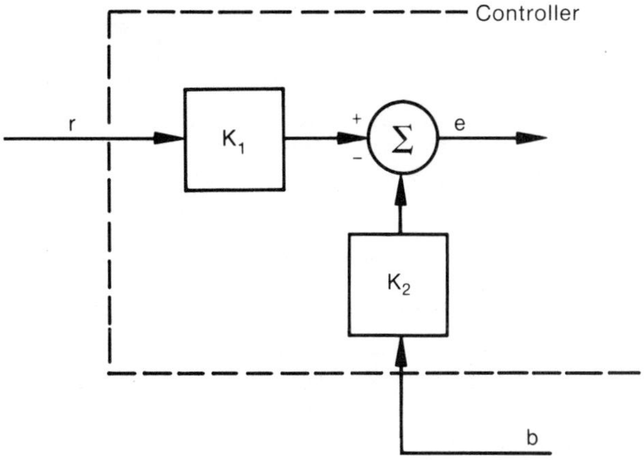

Figure 6-8. Controller Error Diagram

The partial derivatives of the total differential have thus become gains in the block diagram.

6-4 OPEN LOOP VERSUS CLOSED LOOP

Static gain considerations also emphasize the major advantage of the closed-loop control system over the open-loop system. Consider the open-loop system shown in Figure 6-9.

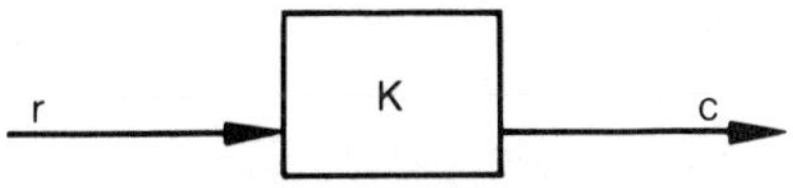

Figure 6-9. Open-Loop System

The open-loop system is a calibrated system. A common example is a wristwatch. The watch is set (calibrated) to the correct time. There is no way of determining its accuracy until it is again set. The accuracy is determined entirely by the precision of its parts. Any error due to a faulty part appears as an error in the output:

$$\frac{\Delta K}{K} = \frac{\Delta c}{c}$$

Another example of the calibrated system is a metering pump with a pneumatic variable speed drive. The drive is actuated by a remote manual loader. The scale on the manual loader can be calibrated to the flow through the pump. The accuracy with which the scale actually conforms to the flow is dependent entirely on the components of the system.

The dependence on the quality of the component parts can be reduced by closing a feedback loop around the system (see Figure 6-10).

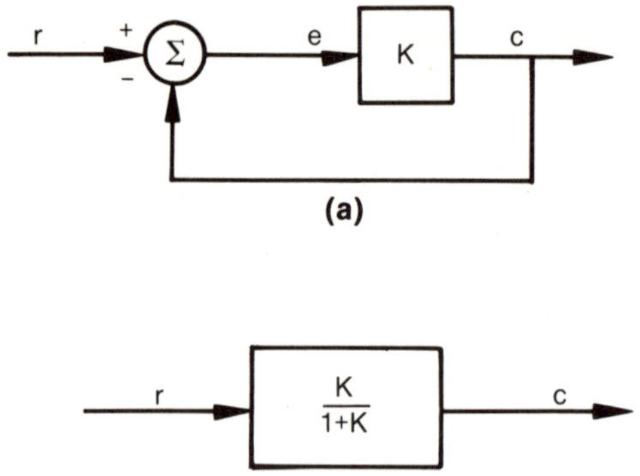

Figure 6-10. Closed-Loop System

Then:

$$c = \frac{K}{1 + K} r$$

$$c + \Delta c = \frac{K + \Delta K}{1 + K + \Delta K} r$$

Dividing through by c and substituting for r:

$$1 + \frac{\Delta c}{c} = \frac{(K + \Delta K)(1 + K)}{(1 + K + \Delta K)(K)}$$

$$\frac{\Delta c}{c} = \frac{(K + \Delta K)(1 + K) - (K)(1 + K + \Delta K)}{(1 + K + \Delta K)(K)}$$

$$\frac{\Delta c}{c} = \frac{1}{1 + K} \frac{\Delta K}{K}$$

since

$$1 + K + \Delta K \cong 1 + K$$

when

$$\Delta K << 1 + K$$

Thus the inaccuracies in the physical components are reflected in the controlled variable only after multiplication by a factor. If K can be made very large, the error in the controlled variable can be kept as small as is desired.

6-5 SIGNAL FLOW DIAGRAMS

Signal flow diagrams, or block diagrams, depict the flow of intelligence through a system and the dynamic response of each component of the system to this intelligence in terms of the transfer function of the component. The transfer function is often expressed initially in the generalized form shown in Figure 6-11, which signifies that

$$Y = KG(s)X$$

where K = steady-state gain and G = dynamic response.

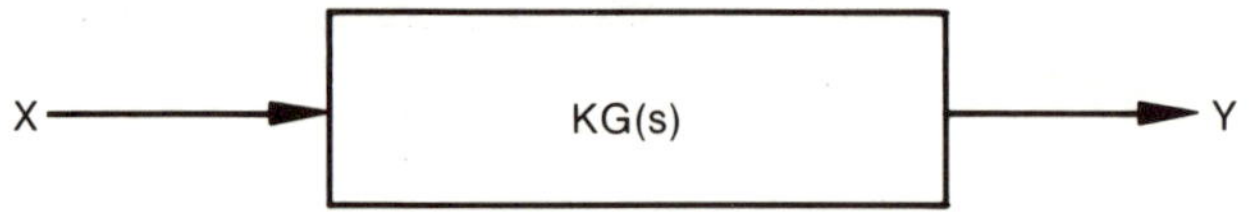

Figure 6-11. Typical Block

The parenthetical s, usually omitted, denotes that the transfer function is expressed in the Laplace notation.

As an example, consider a simple level control loop as shown in Figure 6-12,

where q = flow rate, ft^3/min
A = tank area, ft^2
L = level, ft
a = transmitter output, $lbs/in.^2$
b = controller output, $lbs/in.^2$

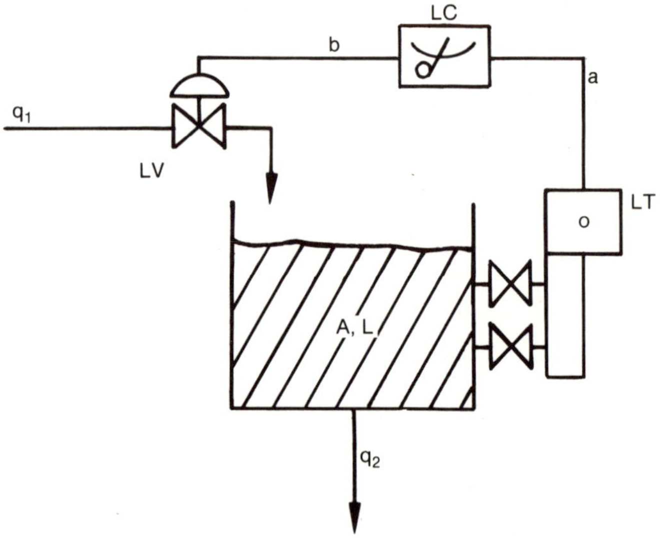

Figure 6-12. Liquid Level Control System

In constructing the signal flow diagram, the flow of intelligence (signals) and energy or material around the flow loop are considered. The KG notation for the transfer functions will be used initially for simplicity.

The difference between q_1 and q_2 will influence the level in a time-varying (dynamic) manner. This can be expressed as shown in Figure 6-13.

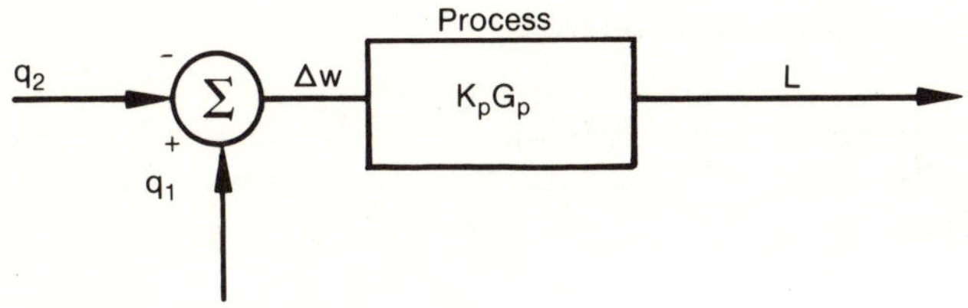

Figure 6-13. Summing Junction

The Σ signifies summation. The sign convention at the summation point indicates the negative feedback concept of the control loop; it means that the algebraic sum of q_1 and q_2 (or, more correctly, Δq_1 and Δq_2) must be zero if L is to be constant (or, if ΔL is to be zero). If there is a difference between q_1 and q_2, this difference will be acted upon by the steady-state process gain K_p (which converts flow into level) and the dynamic process response G_p. Since G_p is usually dimensionless, K_p must usually have units that convert flow in ft^3/min to level in feet. Note that the arrows show the direction in which information flows.

The diagram in Figure 6-14 shows that the level is sensed by the level transmitter, which has a steady-state gain of K_t, a dynamic response G_t, and an output signal a.

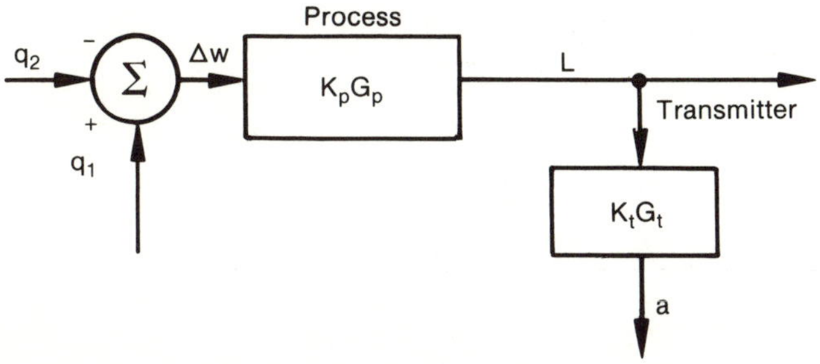

Figure 6-14. Process Block Diagram

The connection into the line marked L is not a summation; it indicates that information flows in two different directions. In other words, it states that $a = K_tG_tL$.

The controller compares the transmitted signal a with its set point L_{sp} and, assuming for simplicity that the controller has proportional control only, puts out a signal b equal to the error signal (e) multiplied by the controller gain K_c as shown in Figure 6-15.

Conventionally, the transmitter signal a is considered as being subtracted from the set point L_{sp}, hence the sign convention. The controller output activates the valve, which has the steady-state gain K_v and dynamic response G_v and results in a change in flow q_1 (see Figure 6-16).

If it is assumed that the disturbance of interest originates in q_2, L_{sp} and its summing point can be eliminated since concern is only with deviations and not with absolute values.

Note that the above signal flow diagram implies that the flow q_1 will vary in proportion to the controller output b. This situation would prevail only if pressure drop across the valve were constant and if the valve had linear trim. If it were necessary to accommodate nonlinear trim or to consider the effect of various pressure drops on q_1, linearized gain terms must be added to take these effects into account, as described in Section 5. The linearized valve flow equation is:

$$q_1 = \left.\frac{q_1}{2(p_1 - p_2)}\right|_{ss} (p_1 - p_2) + \left.\frac{q_1}{x}\right|_{ss} x$$

where p_1 and p_2 are pressures upstream and downstream from the control valve, and x is the fraction of valve capacity corresponding to the controller output at stated steady-state conditions. Substituting gain terms for the steady-state (gain) terms above to simplify the expression:

$$q_1 = K_w (p_1 - p_2) + K_x x$$

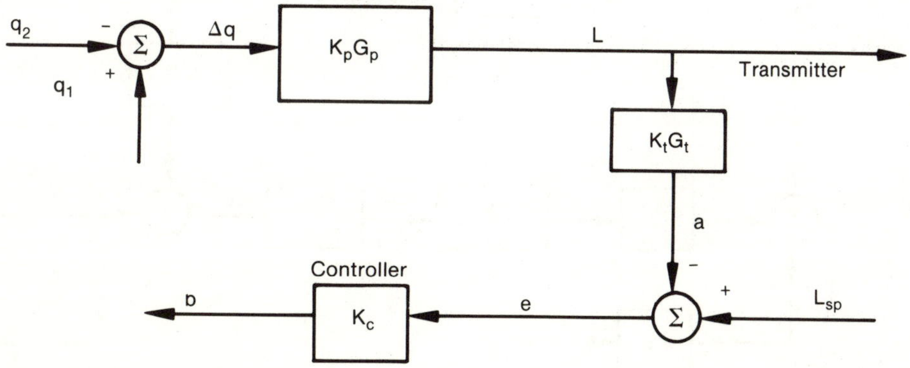

Figure 6-15. Process and Controller Diagram

The signal flow diagram can then be modified to that shown in Figure 6-17.

In this form of the block diagram, K_v will have a different value and a different set of dimensions than it did in the previous form. Previously, the dimensions were (ft³/min)/(psi controller output); now it becomes (fraction of valve capacity)/(psi controller output), and the K_x block provides units of (ft³/min)/(fraction of valve capacity). K_x is evaluated as the slope of the valve lift capacity curve in the region of interest.

6-6 OPEN-LOOP GAIN

In a closed-loop block diagram, the product of all of the steady-state gain terms must be dimensionless except where there is a pure integration term. In the example, there is such a term. The unsteady-state material balance is

$$q_1 - q_2 = AsL$$

or, in transfer function form,

$$\frac{L}{q_1 - q_2} = \frac{1}{As} = K_p G_p$$

Then,

$$K_p = \frac{1}{A} \quad \frac{1}{\text{ft}^2} \quad \text{or} \quad \frac{\text{ft}}{\text{ft}^3}$$

and

$$G_p = \frac{1}{s} \ \text{min}$$

Thus, the dynamic response G_p is not dimensionless, having units of time; it must therefore be incorporated into the steady-state gain units to keep the product dimensionless.

$$K_t K_c K_v K_x K_p \, (G_p)$$

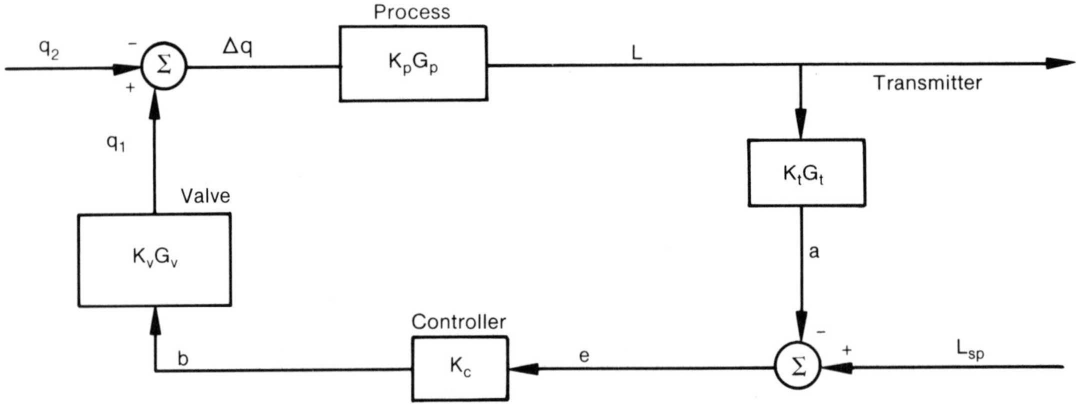

Figure 6-16. Control System Signal Flow Diagram

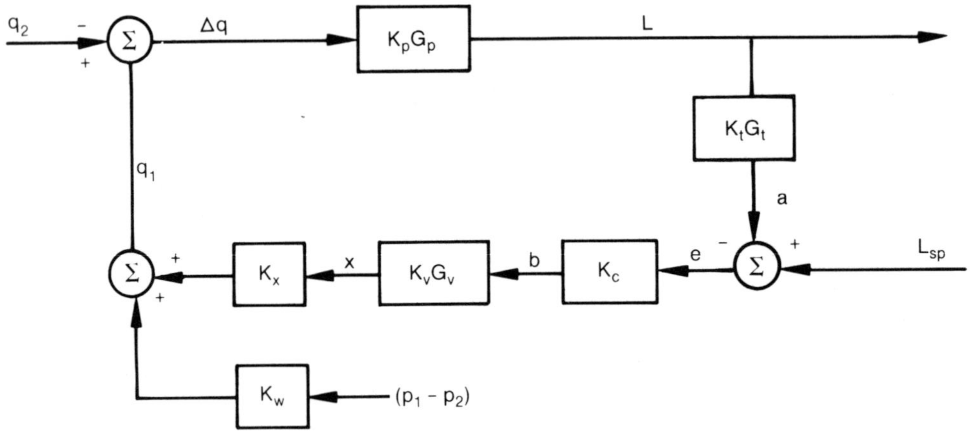

Figure 6-17. Liquid Level System Signal Flow Diagram

Substituting the dimensions of each gain term,

$$\underset{K_t}{\frac{\text{lbs/in.}^2}{\text{ft}}} \times \underset{K_c}{\frac{\text{lbs/in.}^2}{\text{lbs/in.}^2}} \times \underset{K_v}{\frac{\text{frac. of cap.}}{\text{lbs/in.}^2}} \times \underset{K_x}{\frac{\text{ft}^3/\text{min}}{\text{frac. of cap.}}} \times \underset{K_p}{\frac{\text{ft}}{\text{ft}^3}} \times \underset{G_p}{\text{min}}$$

it can be seen that they all cancel. Thus, where there is a pure integration term, it is necessary to include the dynamic gain term to cancel out the units of time.

If the G_p term had been a first-order lag such as

$$\frac{1}{\tau s + 1} \tag{16}$$

the τ in minutes times s in min^{-1} would have given a dimensionless product for G_p, and it would not have influenced the product of the steady-state gains.

Problems

1. Consider the vessel:

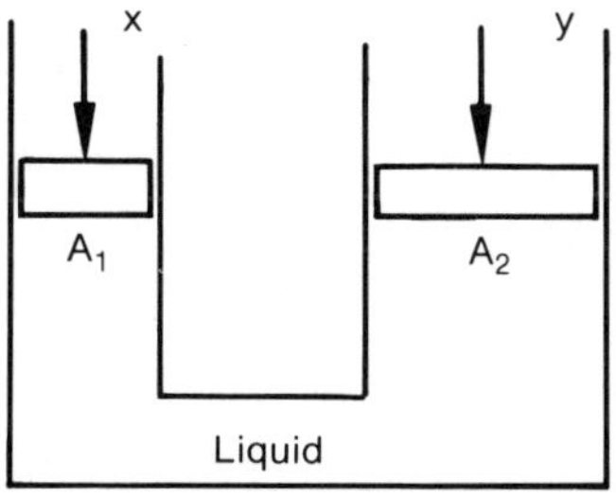

This system is represented by:

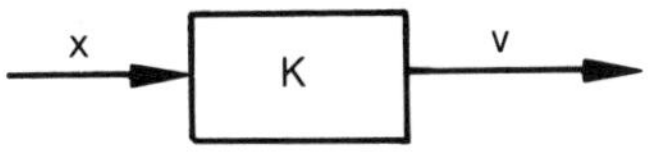

(a) What is the force gain of the system?

(b) The displacement gain?

2. A pH measuring system has a range of 7 to 10 pH units and is used in 3 to 15 psi service. What is the gain of this unit?

3. Suppose, in the level control system given as an example in this section, a level transmitter, controller, and electropneumatic converter had been used. Knowing that the controller input signal is 0 to 0.5 volts ac and the controller output is 1 to 5 mA dc:

   (a) Draw the signal flow diagram incorporating the electropneumatic converter.

(b) Assign dimensional units to all the gain terms.

4. A jet mixer is used to heat a cold stream (w_1) at temperature T_1 by adding it to a hot stream (w_2) at temperature T_2 to give a mixture temperature T_3, using a temperature transmitter, a controller ($K_c G_c$), and a control valve. Assuming constant pressure drop across the control valve and that the disturbance is a change in T_1:

(a) Construct the signal flow diagram.

(b) If the temperature element is located far enough downstream from the jet mixer that dead time is significant, show where the dead time is incorporated into the block diagram.

6-7 OPEN- AND CLOSED-LOOP CONCEPTS

6-7-1 The Open Loop: Allowable Loop Gain

Consider the simpler form of the block diagram developed for the level control system described previously in Figure 6-16. Now, open the loop between $K_v G_v$ and the summation point (considering L_{sp} to be constant), as shown in Figure 6-18. Working around the open loop from Δq to q_1, there are four steady-state gain terms in series relating q_1 to Δq thus:

$$(\Delta)q_1 = q \quad (K_p K_t K_c K_v)$$

Without mathematical proof, it can be stated that in any control loop there is a maximum allowable loop gain that cannot be exceeded without causing the loop to become unstable. The maximum allowable controller gain depends on two factors: the combined dynamic response of the elements that comprise the control loop, and the amount of overshoot that is deemed tolerable when the system responds to an upset.

The control system performance contours shown in Figure 6-19 show large deviations or overshoots at both very low and very high gains, with an area of acceptable

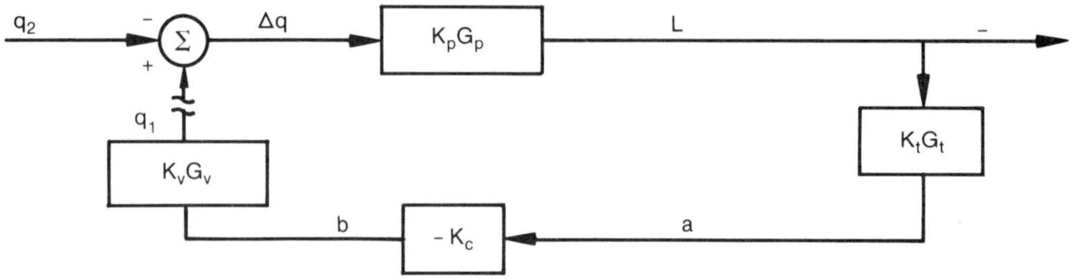

Figure 6-18. Open-Loop Signal Flow Diagram

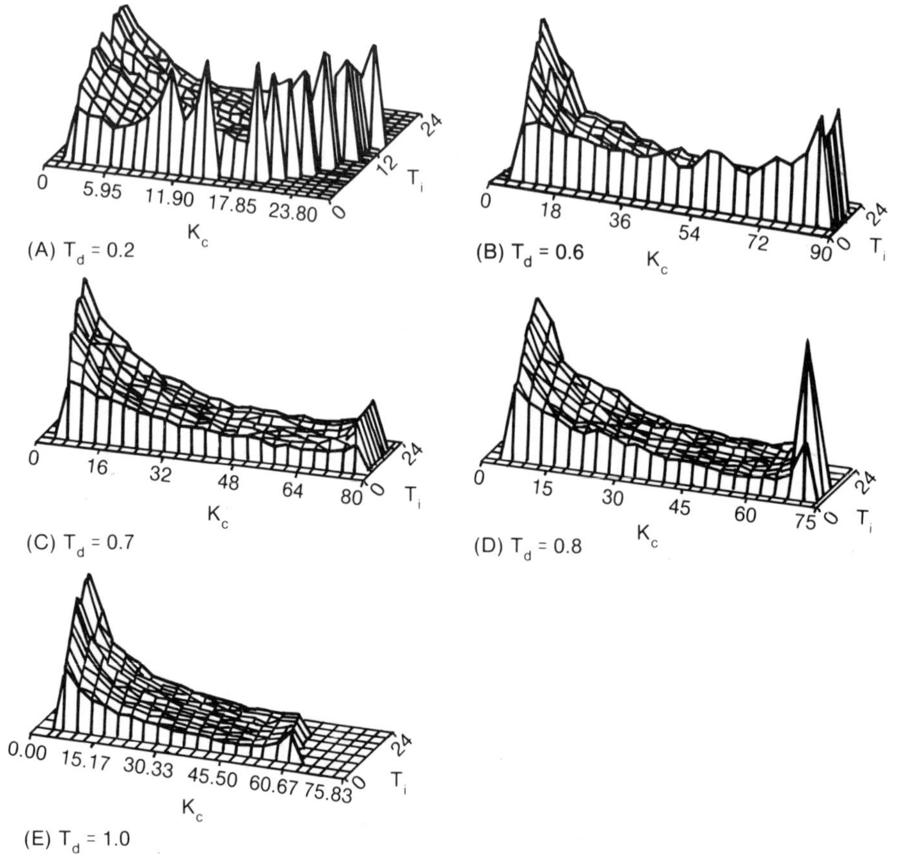

Figure 6-19. Control System Performance Based on Minimum IAE Deviations[111]

control system performance in between. Very low gain approaches the uncontrolled situation shown in Figure 2-10, and the deviation approaches the system response to the disturbance without control. A primary objective of controller tuning is to increase system gain by increasing controller gain in order to minimize deviations as shown in the flat portion of each contour. As gain is increased too far, higher overshoots occur because of approaching instability. It is this overshoot that limits the maximum gain setting.

From Equation 6-6, the open-loop gain is the product of K_p, K_t, K_c, and K_v. Obviously, K_p is a function of the process characteristics; these characteristics may or may not come under the control of the control system engineer. If they depend on the design of a piece of equipment, they could be changed if necessary to improve control performance. If, however, they depend on the physical properties of the process material (molecular weight, specific heat, diffusivity), they may have to be accepted as they are and the resulting K_p must be lived with. The value of K_t will depend on the span of the transmitter, and the value of K_v will depend on valve capacity. These are usually dependent on the specified range of operating conditions and thus are subject only to limited modification. The controller gain (K_c) is, then, the only term in the open-loop gain that is readily available for adjustment. The controller gain can be adjusted so that the open-loop gain does not exceed the maximum allowable loop gain.

6-7-2 The Closed Loop

Consider a generalized control loop. A consolidated expression can be developed that relates the output response to the input disturbance, as shown in Figure 6-20.

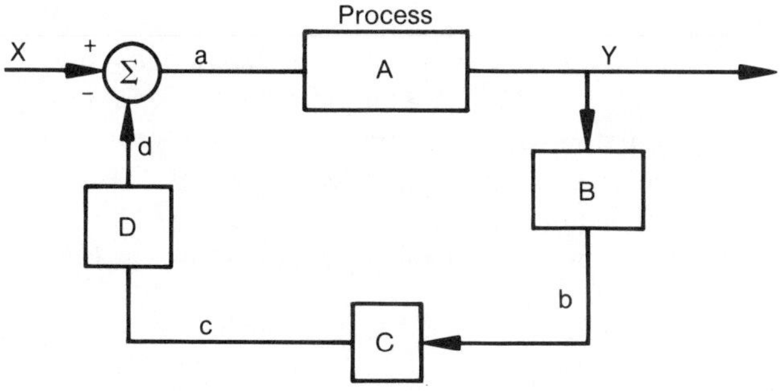

Figure 6-20. Generalized Closed-Loop Diagram

To develop a single function relating Y to X and taking the relationship stepwise, starting from Y, it can be seen that:

$$b = BY$$

$$c = Cb = CBY$$

$$d = Dc = DCBY$$

$$a = X - d = X - DCBY$$

$$Y = Aa = A(X - DCBY)$$

$$Y = AX - ABCDY$$

$$Y(1 + ABCD) = AX$$

$$\frac{Y}{X} = \frac{A}{1 + ABCD}$$

See Figure 6-21.

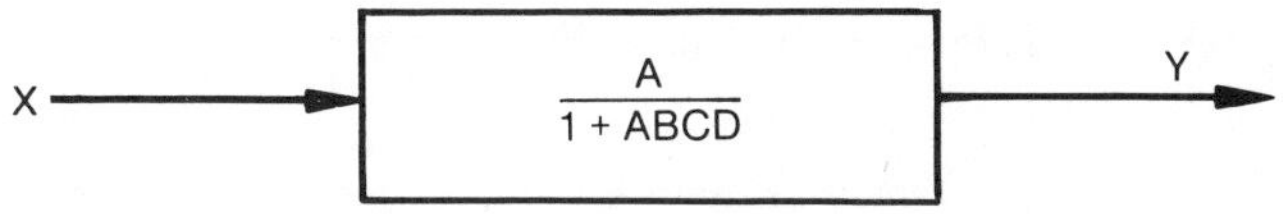

Figure 6-21. Combined Signal Flow Diagram

6-8 RULE FOR COMBINING BLOCK DIAGRAM

A simpler method of arriving at the final transfer function is that "the closed-loop transfer function is the feedforward divided by 1 plus the loop." The "feedforward" refers to the forward relationship between X and Y, namely A. The "loop" refers to the product of the components in the open loop (which can be opened at any point), namely $ABCD$, or $DABC$, etc. The signal flow diagram must be reduced to a single loop for this method to apply. For example, if inner loops result from internal feedback and if there is a block outside the loop, the method shown in Figure 6-22 would be used.

The M and N blocks can be consolidated by the method mentioned above;

$$\frac{\text{feedforward}}{1 + \text{loop}} = \frac{M}{1 + MN}$$

with the diagram as shown in Figure 6-23.

Now the "feedforward" relating X to Y is

$$\frac{LAM}{1 + MN}$$

and "1 + the loop" is

$$1 \quad = \frac{AMBCD}{1 + MN}$$

"Feedforward over 1 + the loop" becomes

$$\frac{Y}{X} = \frac{\dfrac{LAM}{1 + MN}}{1 + \dfrac{ABCDM}{1 + MN}} = \frac{LAM}{1 + MN + ABCDM}$$

which is diagrammed in Figure 6-24.

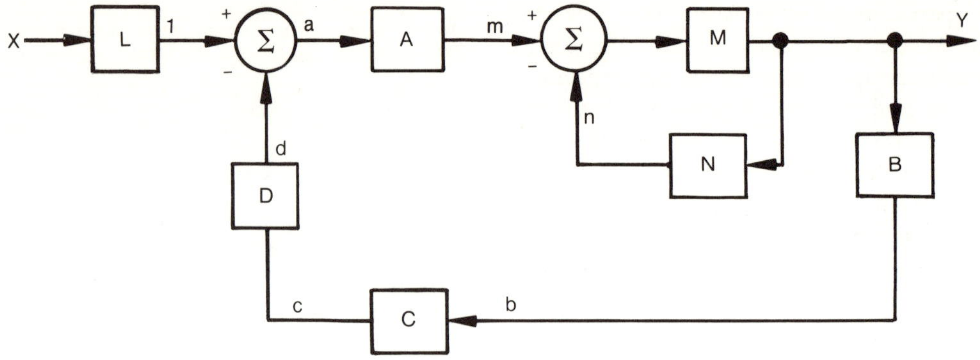

Figure 6-22. Inner Loop Signal Flow Diagram

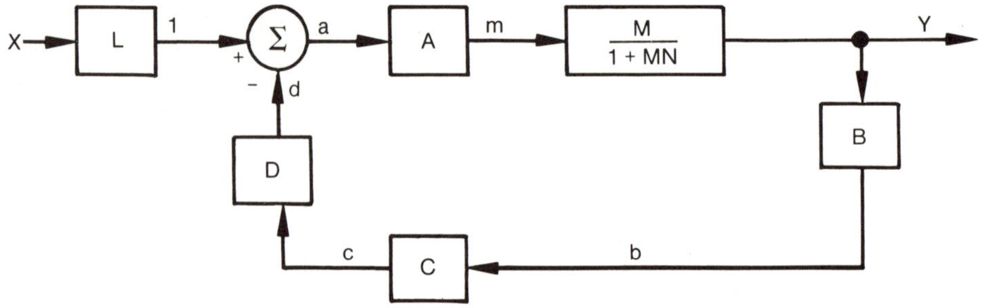

Figure 6-23. Consolidated Signal Flow Diagram

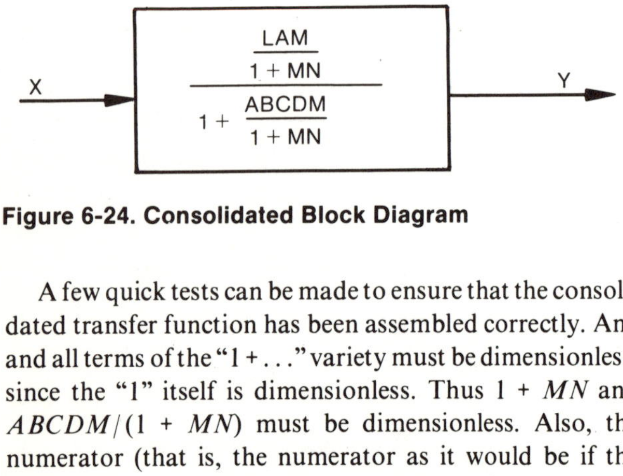

Figure 6-24. Consolidated Block Diagram

A few quick tests can be made to ensure that the consolidated transfer function has been assembled correctly. Any and all terms of the "1 + . . ." variety must be dimensionless, since the "1" itself is dimensionless. Thus $1 + MN$ and $ABCDM/(1 + MN)$ must be dimensionless. Also, the numerator (that is, the numerator as it would be if the transfer function were reduced to a simple fraction) must have the proper dimensions to convert from the units of X to the units of Y. The check for dimensional consistency must include the dynamic gains as well as the steady-state gains to ensure that transfer functions representing pure integration can contribute their "time units" to the dimensional check.

Applying the method of collapsing the block diagram to the level control problem results in

$$\frac{L}{q_2} = \frac{K_p G_p}{1 + K_p G_p K_t G_t K_c K_v G_v}$$

In the previous section, it was shown that $K_p = 1/A = 1/\text{ft}^2$ and $G_p = 1/s = \text{min}$. As a check of dimensions, then, with q_2 expressed in ft^3/min,

$$L = q_2 K_p G_p = \frac{\text{ft}^3}{\text{min}} \times \frac{1}{\text{ft}^2} \times \text{min} = \text{ft}$$

Section 6-5 confirmed that the denominator was dimensionless, so there is a good chance that the consolidation was done correctly.

A few additional comments on block diagram manipulation may be in order. The manipulation of blocks in a diagram is no more than a graphical form of algebra, so manipulations must satisfy the rules of algebra. A few examples will demonstrate the rules.

Presently, the straight-through path shows in Figure 6-25 that

$$Y = AX$$

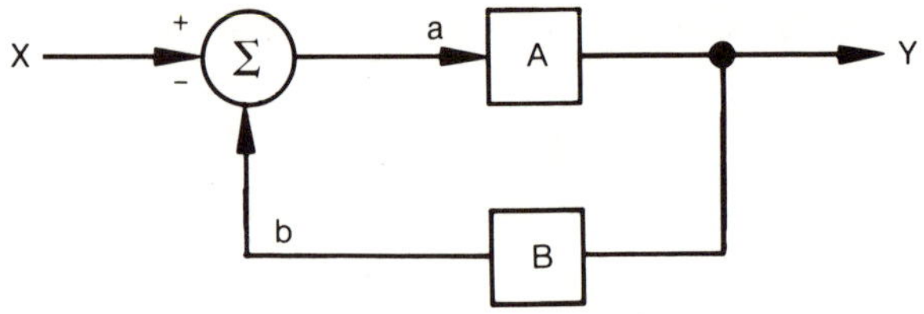

Figure 6-25. Typical Signal Flow Diagram

How can the diagram be manipulated so that there is unity feedback (i.e., no block in the feedback path)? At present, it can be seen that

$$a = X - b = X - BY$$

and this relationship must be maintained. Thus, if B is removed from the feedback path, B must be put in the feedforward path between the summing point and A. This, however, would then make $Y = BAX \neq AX$. Therefore, a $1/B$ block would be required between X and the summing point, so that

$$Y = \frac{1}{B} BAX = AX$$

The unity feedback form of the diagram would then be as shown in Figure 6-26.

Consolidating this diagram into a single transfer function consisting of "feedforward over 1 plus the loop," gives

$$\frac{Y}{X} = \frac{\frac{1}{B}BA}{1 + AB} = \frac{A}{1 + AB}$$

Inspection of the original block diagram with B in the feedback path shows that the consolidated transfer function would also be $A/(1 + AB)$. Thus, the two diagrams represent the same system.

A few other examples are shown in Figure 6-27 that will demonstrate the algebraic rules for block diagram manipulation.

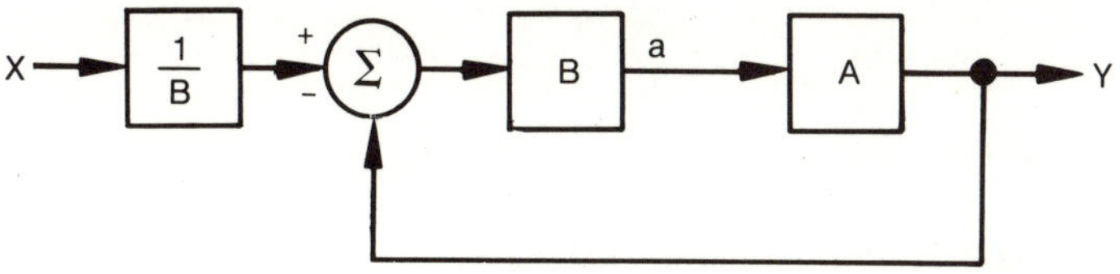

Figure 6-26. Unity Feedback Diagram

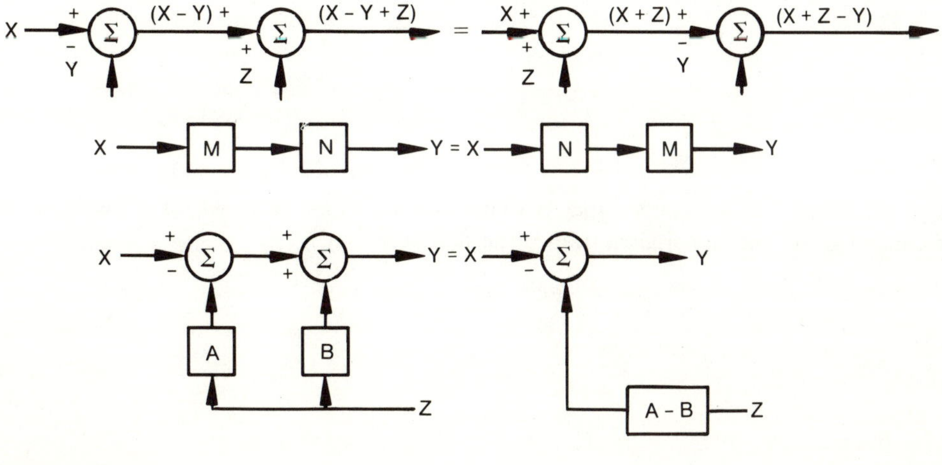

Figure 6-27. Block Manipulation

Problems

1. A polymerization reactor is fed with a constant flow of reactant X and a viscosity-controlled flow of reactant Y. The transmitter has a range of 0 to 6000 poise, the valve (linear trim) has a capacity of 10 lbs/hour of reactant Y per psi change in controller output. Laboratory tests show that product viscosity will change 1000 poise per lbs/hour change in the flow rate of reactant Y. If the allowable loop gain is 2.0,

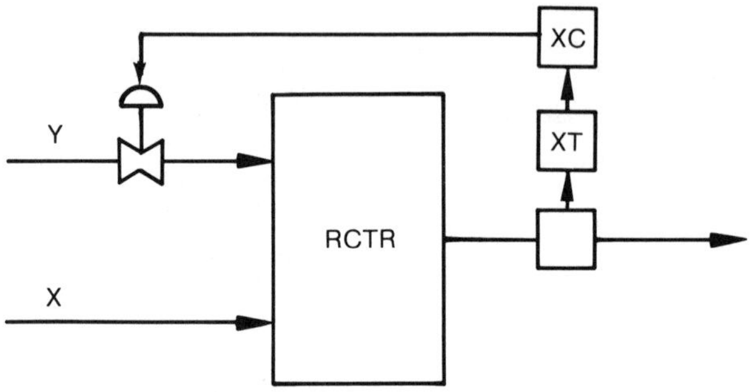

(a) What is the maximum controller gain K_c?

(b) What is the corresponding proportional band? (PB = $100/K_c$)

2. In the problem above, what would be the effect on the maximum controller gain if

(a) A suppressed-range transmitter (3000 to 6000 p) had been used?

(b) A control valve with half the above capacity (with parallel bypass valve set for a fixed flow to make up for the smaller control valve) and the original transmitter had been used?

3. A positive displacement recycle gas compressor discharges into a receiver, on which the process exerts a variable flow demand. The compressor operates at constant speed and a constant suction pressure. The receiver pressure is controlled by a PT-PC-PV loop that bypasses gas from the receiver to the suction of the compressor.

   (a) Develop the signal flow diagram relating receiver pressure to process demand flow.

   (b) Consolidate the block diagram to a single closed-loop transfer function.

4. In problem 3 above, assume that flow is expressed in lbs/min and pressure in atmospheres.

   (a) What are the units of the process steady-state and dynamic gain terms K_p and G_p?

   (b) Assuming pneumatic instruments, assign units to the various steady-state gain terms around the loop and check that their product is dimensionless.

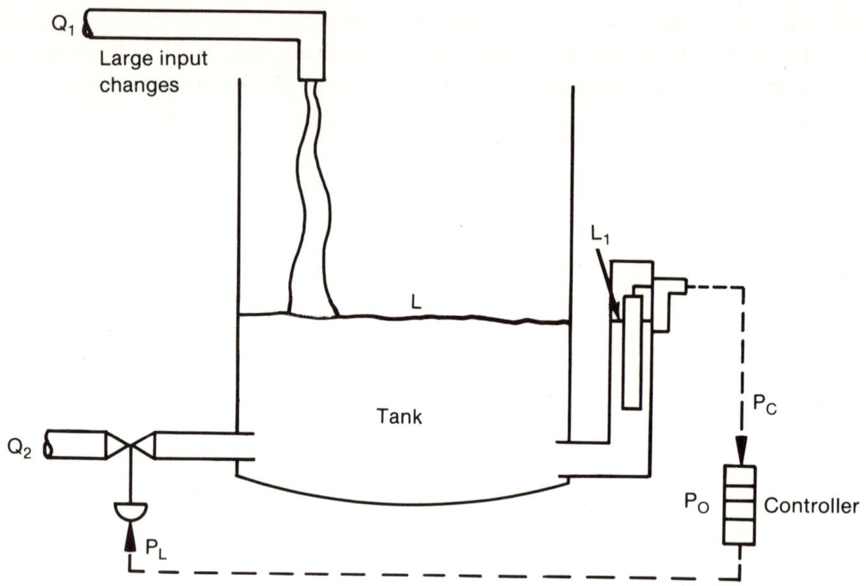

a. Tank is equipped with: side arm, displacement-type level transmitter, controller, and valve in effluent pipe.

b. Input is subject to large variations in flow. (±30%).

c. Level must be kept within ±3 inches.

5. Draw the signal flow diagram for the control system pictured above. Is this system self-regulating or nonself-regulating?

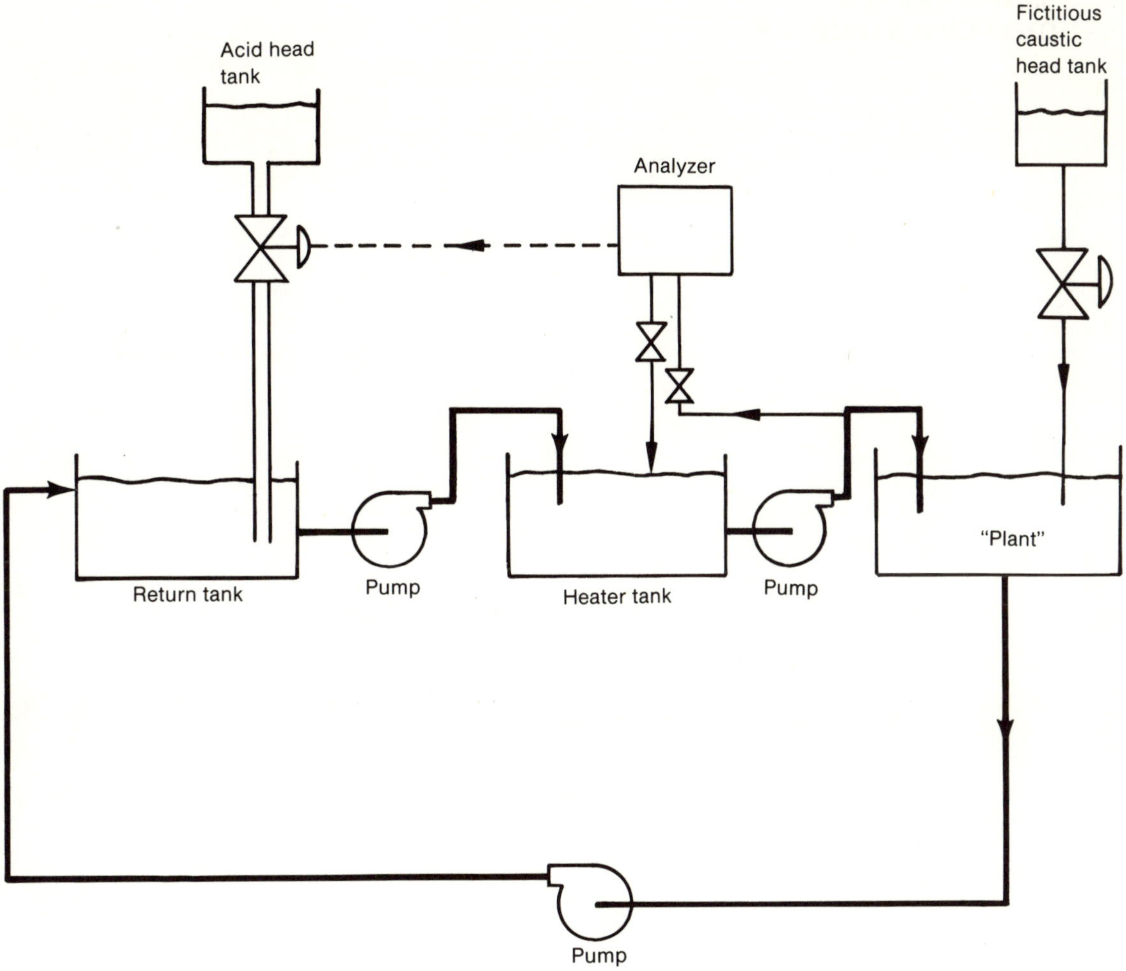

a. An acid bath is piped through a plant, and in the process it is partially neutralized. Neutralization is shown by basic material from a fictitious caustic head tank. The salt and water produced are removed in the "plant."

b. The acidity is restored by the addition of acid to the return tank.

c. Allowing whatever mixing can occur in the *unagitated* return tank, the pumps, and the heater tank, an analyzer then checks acidity and controls the acid valve.

6. Draw the signal flow diagram for the control system illustrated above.

STUDENT SUMMARY NOTES and QUESTIONS FOR INSTRUCTOR

SECTION 7
SINUSOIDAL ANALYSIS

7-1 INTRODUCTION

The use of sinusoidal methods of analysis and testing of dynamic systems has gained widespread popularity because of the simplicity in treating complex control systems. Appendix C shows transient analysis to be limited to second-order systems because of algebraic complexity. Analysis of larger systems in the time domain requires the use of computer simulation. Sinusoidal methods make possible the quantitative analysis of complex systems by simple, graphical techniques.

7-2 SINUSOIDAL TESTING

The dynamic analysis techniques discussed in previous sections represent methods by which the static and dynamic characteristics of both processes and instruments can be determined by the basic laws of mechanics, physics, and chemistry. These same characteristics can be determined experimentally. One experimental method is the employment of sinusoidal signals, the use of which is broadly called frequency response.

Figure 7-1 shows the way in which a sine wave is generated mathematically. The projection of a point on a unit circle will define a distance from the x-axis, y, which is related to the angle θ by $y = A \sin \theta$, where A = circle radius or amplitude.

As θ increases in the counterclockwise direction, the magnitude of y varies from +1 to −1 going through zero and describes the curve to the right. Since the sine functions are based on the relationships of a circle, they have been evaluated and tabulated, and they are included in most technical calculator capabilities.

Suppose that $\theta = \omega t$, where, ω is a constant and t is time. There is then a continuous periodic time function. Three characteristics are used to describe this function:

(1) *Amplitude.* The maximum value of y occurring at the peak, or $-y$ at the valley is the amplitude in whatever units the sine wave is representing — inches, degrees centigrade, gpm, etc.

(2) *Frequency or Period.* There are four numbers for any given sine wave that can be used to describe this characteristic. From any one, all the others can be derived. Two definitions are necessary in understanding them.

(a) A cycle is defined as the shortest continuous part of the function that is periodically repeated.

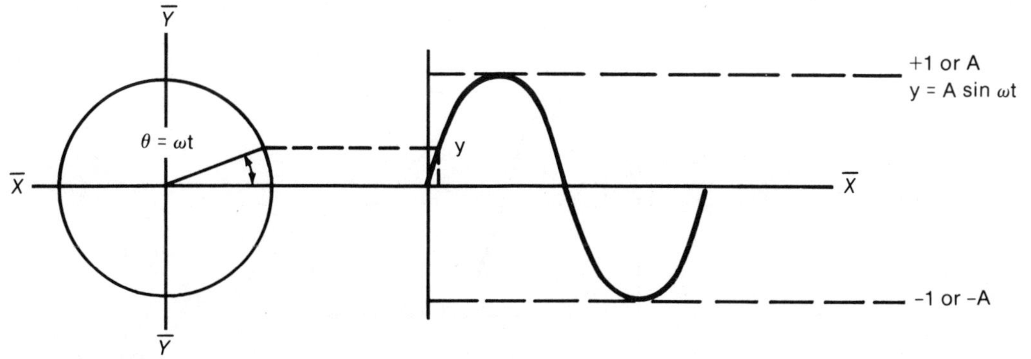

Figure 7-1. Generation of Sinusoid

(b) A radian is defined as the angle subtended by an arc equal in length to the radius. There are 2π (6.28) radians in a circle. Radians are dimensionless. Figure 7-2 shows a table relating the various forms.

(Note that ω is the angular rate at which the radius defining the generating point rotates about the circle, while τ is the time to sweep 1.0 radian. The frequency f is the number of complete circle sweeps per unit time, and T is the time required to sweep one circle. The reason for using ω and τ at all is that they have only the dimension of time and can be used in natural functions without converting units.)

(3) *Phase.* In Figure 7-1 the angle θ was zero when time t was zero, i.e., the wave started at zero time with an amplitude y of zero. This was entirely arbitrary. The amplitude y could have had any value within $\pm A$ at time equal to zero. In Figure 7-3, $y = y_0$ at time zero. Note that the angle θ_0 is equivalent to the amplitude y_0, also shown by the mathematical expression (i.e., at $t = 0$, $y = A \sin \theta_0 = y_0$).

The angle θ_0 is called the phase angle of this sine wave. It may be expressed in degrees, in radians, or as a fractional part of the period T.

In relating one sine wave to another of the same frequency, interest usually lies in the difference in the phase angles of the two waves. This can be pictured as the angle between the two rotating radii, which are generating the waves. Since the angular rates of the two radii are the same, the phase difference will be constant. Usually, zero time is arbitrarily chosen to make one phase angle zero, so that the difference becomes the angle of the second.

$$y_1 = A \sin \omega t \qquad y_2 = B(\sin \omega t + \theta_0)$$

$$\text{Phase difference} = \theta$$

Going back for a moment to a restriction on the type of linear systems under discussion, it was stated that only linear systems (that is, systems with constant physical parameters or characteristics) could be handled by frequency response methods directly. Linear systems have two properties that are of major importance:

(1) If a linear system or component is excited with a sine wave of a given frequency, its output can contain only that frequency.

(2) In linear systems, the property of superposition holds. This means that the response of a linear system to a series of inputs is exactly the same as the sum of the responses to each input applied individually.

The basic technique consists of changing input to a physical system, say, a differential pressure transmitter, in sinusoidal fashion and measuring the effect on the system. The resulting variation in system output is recorded and is a measure of the system response. The technique applies

Symbol	Name	Unit (X)*	F(f)	F(T)	F(ω)	F(τ)
f	Frequency	$\frac{\text{cycles}}{\text{minutes}}$	f	$\frac{1}{T}$	$\frac{\omega}{2\pi}$	$\frac{1}{2\pi\tau}$
T	Period	$\frac{\text{minutes}}{\text{cycle}}$	$\frac{1}{f}$	T	$\frac{2\pi}{\omega}$	$2\pi\tau$
ω	Angular Frequency	$\frac{\text{radians}}{\text{minutes}}$	$2\pi f$	$\frac{2\pi}{T}$	ω	$\frac{1}{\tau}$
τ	Angular Period	$\frac{\text{minutes}}{\text{radians}}$	$\frac{1}{2\pi f}$	$2\pi T$	$\frac{1}{\omega}$	τ

*X — Any units of time may be used

Figure 7-2. Frequency and Period

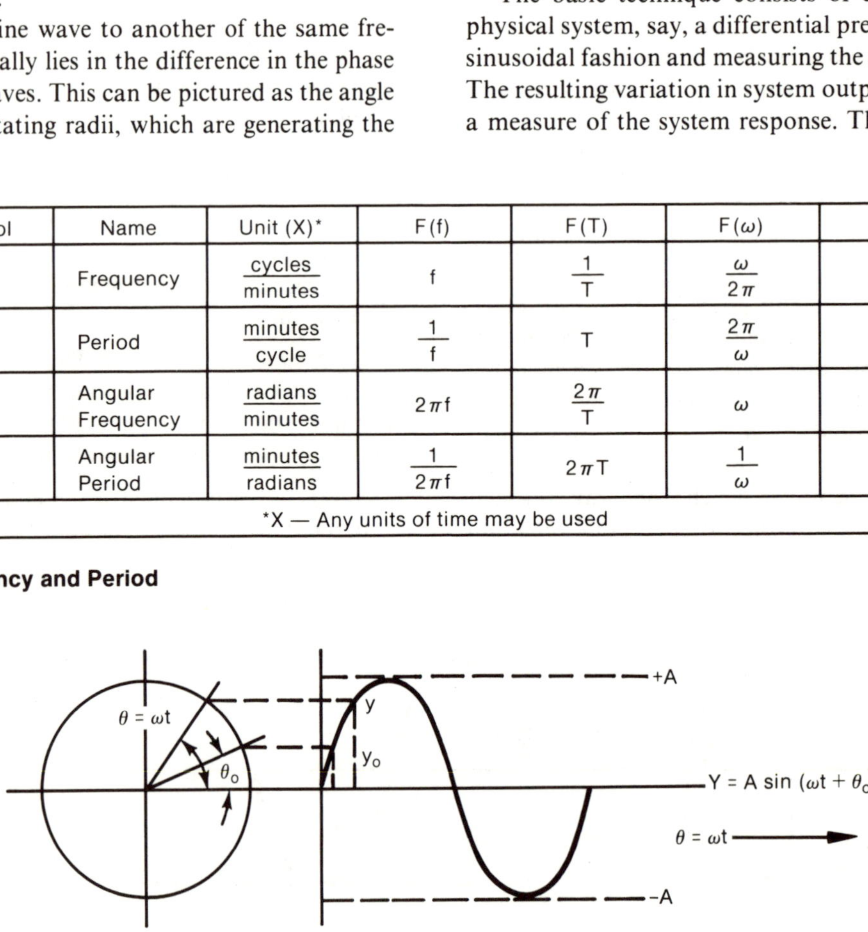

Figure 7-3. Phase Angle

only to linear systems, and thus the technique is applied as a small deviation or perturbation over a small, linear range of a nonlinear system. For this reason, input sine wave amplitude will be some ±5 to ±10% of the range of the equipment being tested.

A low frequency sine wave is applied to the input of the system shown in Figure 7-4. The sine wave shown has continued for sufficient time for the initial transient to have disappeared.

The signal is applied by a sine wave generator (Figure 7-5), and both inlet and outlet signals are recorded.

Since the system is linear, the output frequency will be identical to the input frequency at all times. At low frequencies, the output sine wave will exactly match the input sine wave. As the frequency of the input signal is increased, the output signal will begin to lag the input in phase. Further increase in frequency will cause greater phase lag and will also cause the output amplitude to decrease. The ratio of output signal amplitude to input signal amplitude is called magnitude ratio or, if output is less than input, attenuation.

An increase in frequency will cause a significant phase lag and attenuation, as shown in Figure 7-4. Further increases in frequency will finally reach a point where output amplitude is zero, or where the system being tested cannot respond to the input signal at all.

Testing has shown that magnitude ratio and phase lag are not constant but vary with the frequency of the input signal. This is known as the frequency response of this particular component. Both magnitude ratio and phase angle are functions of frequency and are quite reproducible and independent of input amplitude with the linear span.

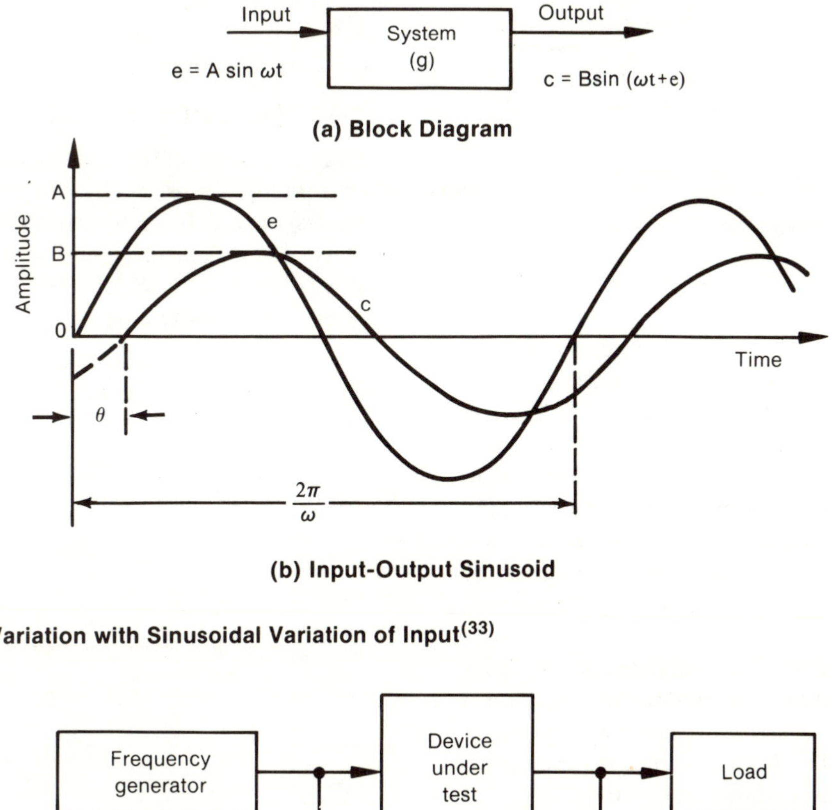

(a) Block Diagram

(b) Input-Output Sinusoid

Figure 7-4. Output Variation with Sinusoidal Variation of Input[33]

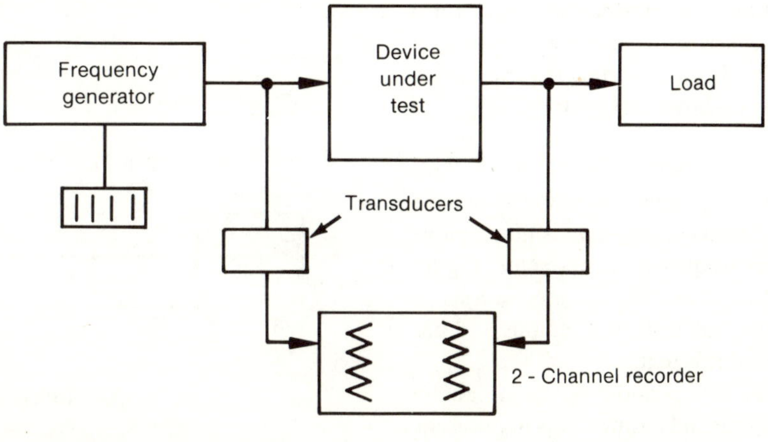

Figure 7-5. Frequency Response Test Arrangement

They can be plotted in a graphical representation of the dynamic behavior of the component being tested, as shown in Figure 7-6.[32,33,34]

Figure 7-6 is called a Bode (pronounced Bōdē) plot after its originator, and it permits evaluation of system time constant at the break point of the magnitude ratio curve, as shown.

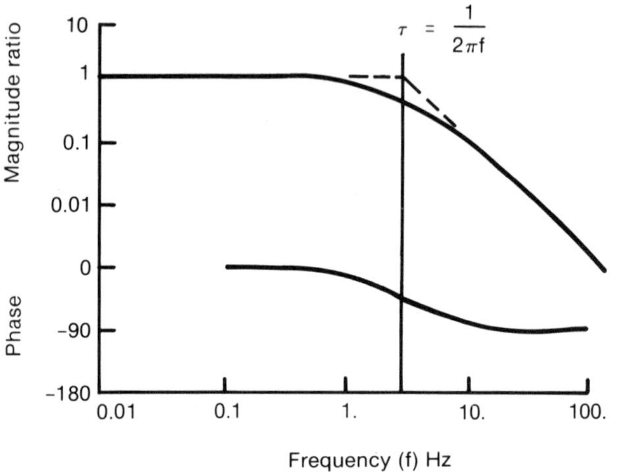

Figure 7-6. Frequency Response

7-3 SINUSOIDS

Sinusoidal analysis, or frequency domain analysis, represents a useful tool for performing productive, simple calculations on complex control systems. As such, it is a powerful technique. Intuitively, it is troublesome. First, seldom is a process in the real world disturbed by a pure sine wave, and one has to question the results based on the premise of a sine wave disturbance. Second, we live in a world of time, not frequency, and the relationship of numbers calculated in the frequency domain to the real world is fuzzy.

The first question can be discussed in terms of Fourier analysis, which says that any complex periodic wave form can be represented by a series of sines and cosines of different frequencies. The Fourier series, representing a few common waveforms, is shown in Figure 7-7.

Note that discontinuities in the basic wave are represented by high frequency components in the series. This explains why the sharp parts of inputs to physical systems are often smoothed out because the system is not capable of transmitting the higher frequency components. Figure 7-7 shows that any input wave can be expressed as a series of sinusoids, but the actual calculation is tedious and uneconomical for purposes of this text.

Figure 7-7 shows that any periodic function can be represented by a series of sine and cosine terms of different frequencies, but that physical processes cannot transmit the high frequencies demanded by sharp discontinuities. Thus the physical process can be adequately represented by only a few of the terms necessary for rigorous representation. Carrying this reasoning to its end point, the process can be described well enough by the first term in the series, a sine wave, for the purposes of analysis.

The question of the relationship of parameters calculated in the frequency domain to the real world time domain is somewhat more difficult. The mathematical development that follows is based on the relationship of the domains:

Time	Complex	Frequency
$\frac{d}{dt}$	s	$j\omega$

Even acknowledging this formal relationship, the relationship of frequency and phase to the real world seems a bit remote. At worst, frequency response can be looked upon as a powerful mathematical technique from which practical results can be obtained quite simply.

7-4 FREQUENCY RESPONSE

The significance of the location of dynamic elements in a control loop is most readiy understood in terms of frequeny response. A Bode diagram is a plot of the frequency response of an element or group of elements versus the frequency of the sinusoidal input signal. The frequency response consists of two parts:

(1) The magnitude ratio, or dynamic gain; i.e., the ratio of the amplitude of the output sine wave to the amplitude of the input sine wave.

(2) The phase angle, i.e., the number of degrees by which the output signal lags or leads the input signal.

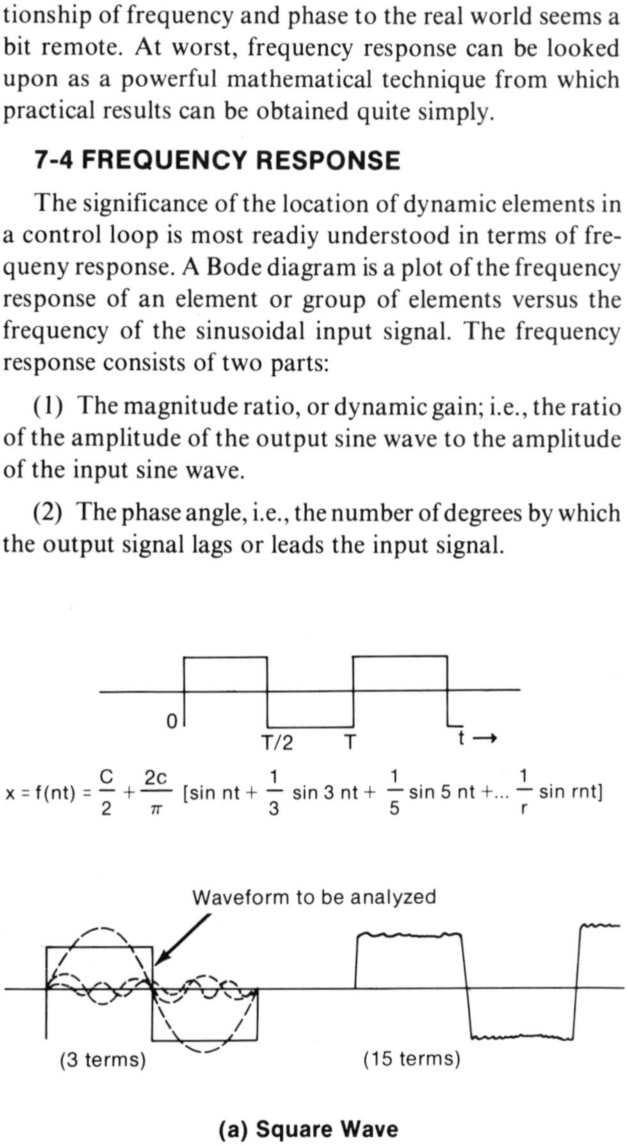

Figure 7-7. Fourier Series Representation

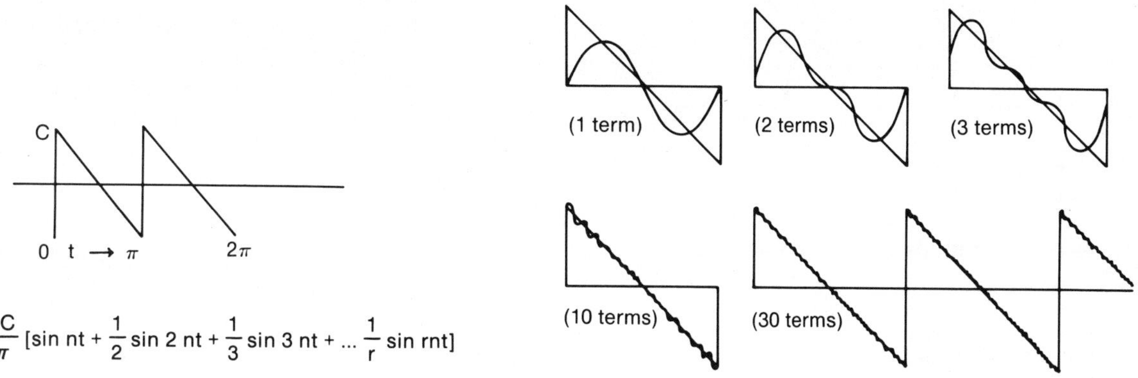

$$x = f(nt) = \frac{C}{2} + \frac{2C}{\pi}\left[\sin nt + \frac{1}{2}\sin 2\,nt + \frac{1}{3}\sin 3\,nt + \ldots \frac{1}{r}\sin rnt\right]$$

(b) Sawtooth Wave

Figure 7-7. Fourier Series Representation (continued)

The input frequency and the magnitude ratio are plotted on logarithmic scales; the phase angle is plotted on a linear scale.

The basis for the significance of frequency response analysis is that any input disturbance can be described by some combination of sinusoidal signals of various amplitudes and frequencies. Input signals with sharp corners (such as square waves, pulses, step changes) contain high frequency components; other inputs (such as slow random changes) may contain only low frequency components. With knowledge of how an element of a control loop responds over a wide frequency spectrum, its performance can be evaluated in terms of the types of disturbances it is likely to encounter in an actual installation.

In a hi-fi set, for example, fidelity to 20,000 to 25,000 cycles/second might be desirable. Human beings cannot hear sounds in this range, but they can associate sounds that have wave fronts equivalent to these frequencies with various musical instruments. Control instruments can be evaluated on the same basis.

Bode diagrams for a component or a system may be developed from frequency response test data or from evaluation of the appropriate transfer function using complex number theory. Bode diagrams for multicomponent systems can be developed by combining graphically the magnitude ratios and phase angles of the individual components.

- *The First-Order Lag*

As an example of the method of obtaining the frequency response from the transfer function, consider the first-order lag:

$$KG(s) = \frac{K}{\tau s + 1}$$

where K is the steady-state gain and the dynamic gain $G(s) = 1/(\tau s + 1)$. It is necessary to evaluate $G(s)$ in terms of its magnitude ratio M and its phase angle θ. First, substitute

$$s = j\omega$$

in which $j = (-1)^{1/2}$ = imaginary number and ω = frequency in radians/unit time. (Note that $\omega = 2\pi f$ where f is the frequency in cycles/unit time; i.e., there are 2π radians in a circle.) Then,

$$G(j\omega) = \frac{1}{j\omega\tau + 1}$$

The definition of dynamic gain in terms of magnitude (or, specifically, magnitude ratio) and phase angle implies the use of polar coordinates in which a point is located in terms of its radial distance from the origin and the angle the radius makes with the x-axis. In this case in the complex plane, the horizontal axis is the "real" axis (r) and the vertical axis is the "imaginary" axis (i). The property of the imaginary number (j) is such that when it is multiplied by a real term (A), the product jA is a complex number having a zero real part and a positive imaginary part of magnitude A; i.e., it would be plotted along the imaginary axis $(0 + jA)$.

The magnitude ratio M and the phase angle θ can then be defined in the complex plane as follows (see Figure 7-8):

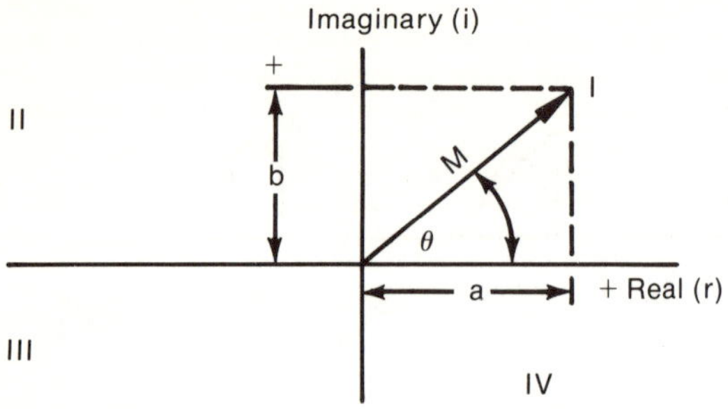

Figure 7-8. Magnitude Ratio and Phase Angle

From geometrical considerations, M and θ can be evaluated as

$$M = (a^2 + b^2)^{1/2} \tag{7-1}$$

and

$$\theta = \tan^{-1} \frac{b}{a} \tag{7-2}$$

where a = value of the real part of the solution

b = value of the imaginary part of the solution

$G(j\omega)$ must therefore be separated into its real and imaginary components (a and b above). The imaginary part will be multiplied by j; the real part will contain no j. The first step is to multiply $G(j\omega)$ by the conjugate of the denominator and, recognizing that $j^2 = -1$, separate the result into real and imaginary terms:

$$G(j\omega) = \frac{1}{j\omega\tau + 1} \quad \frac{j\omega\tau - 1}{j\omega\tau - 1}$$

$$= \frac{j\omega\tau}{j^2\omega^2\tau^2 - 1}$$

but $j^2 = -1$, so

$$G(j\omega) = \frac{j\omega\tau - 1}{-\omega^2\tau^2 - 1}$$

The j has now been eliminated from the denominator; the expression can then separate into two fractions, one of which will not contain j (i.e., the real part) and one of which will be multiplied by j (i.e., the imaginary part).

$$G(j\omega) = \underset{\text{(real)}}{\frac{-1}{-\omega^2\tau^2 - 1}} + \underset{\text{(imaginary)}}{\frac{j\omega\tau}{-\omega^2\tau^2 - 1}}$$

$$G(j\omega) = \frac{1}{\omega^2\tau^2 + 1} - j\,\frac{\omega\tau}{\omega^2\tau^2 + 1} \tag{7-3}$$

The first term is now the real part of the solution (i.e., a in Equations 7-1 and 7-2) and the second term is the imaginary part (b in Equations 7-1 and 7-2). The j coefficient of the imaginary term can be dropped since it serves only to identify the imaginary term (the one plotted against the imaginary axis) and does not assume a value in the calculation of M and θ.

M and θ can now be evaluated in terms of the real and imaginary terms in Equation 7-3.

$$M = (a^2 + b^2)^{1/2}$$

$$M = \left[\left(\frac{1}{\omega^2\tau^2 + 1}\right)^2 + \left(\frac{-\omega\tau}{\omega^2\tau^2 + 1}\right)^2\right]^{1/2}$$

$$= \left[\frac{1}{(\omega^2\tau^2 + 1)^2} + \frac{\omega^2\tau^2}{(\omega^2\tau^2 + 1)^2}\right]^{1/2}$$

$$= \left[\frac{1 + \omega^2\tau^2}{(\omega^2\tau^2 + 1)^2}\right]^{1/2}$$

$$M = \left[\frac{1}{\omega^2\tau^2 + 1}\right]^{1/2}$$

and

$$\theta = \tan^{-1}\left(\frac{b}{a}\right)$$

$$\theta = \tan^{-1} \frac{\left(\dfrac{-\omega\tau}{\omega^2\tau^2 + 1}\right)}{\left(\dfrac{1}{\omega^2\tau^2 + 1}\right)}$$

$$\theta = \tan^{-1} \frac{-\omega\tau}{1} \tag{7-4}$$

In evaluating θ, it is important to keep track of the signs of the numerator and denominator separately, since they determine the quadrant of the complex plane in which the vector M is located. That is the reason for showing the "1" in the denominator of Equation 7-4. In this case, since b is negative and a is positive, the vector will fall in the fourth quadrant. If tan θ were merely defined as equal to $-\omega\tau$, it could have resulted from $+\omega\tau/-1$, which would place the vector in the second quadrant ($+b$, $-a$). It is customary to express phase angles in the first quadrant in terms of the lead angle (0 to +90°) and in the fourth quadrant as lag angles (0 to −90°).

A plot of the frequency response of a system described by a first-order time constant is shown as Figure 7-9. The dimensionless frequency $\omega\tau$ is used in the Bode plot. The break point frequency is at the point $\omega\tau = 1.0$.

Thus the technique of evaluating the time constant from the plotted break point of experimental data, demon-

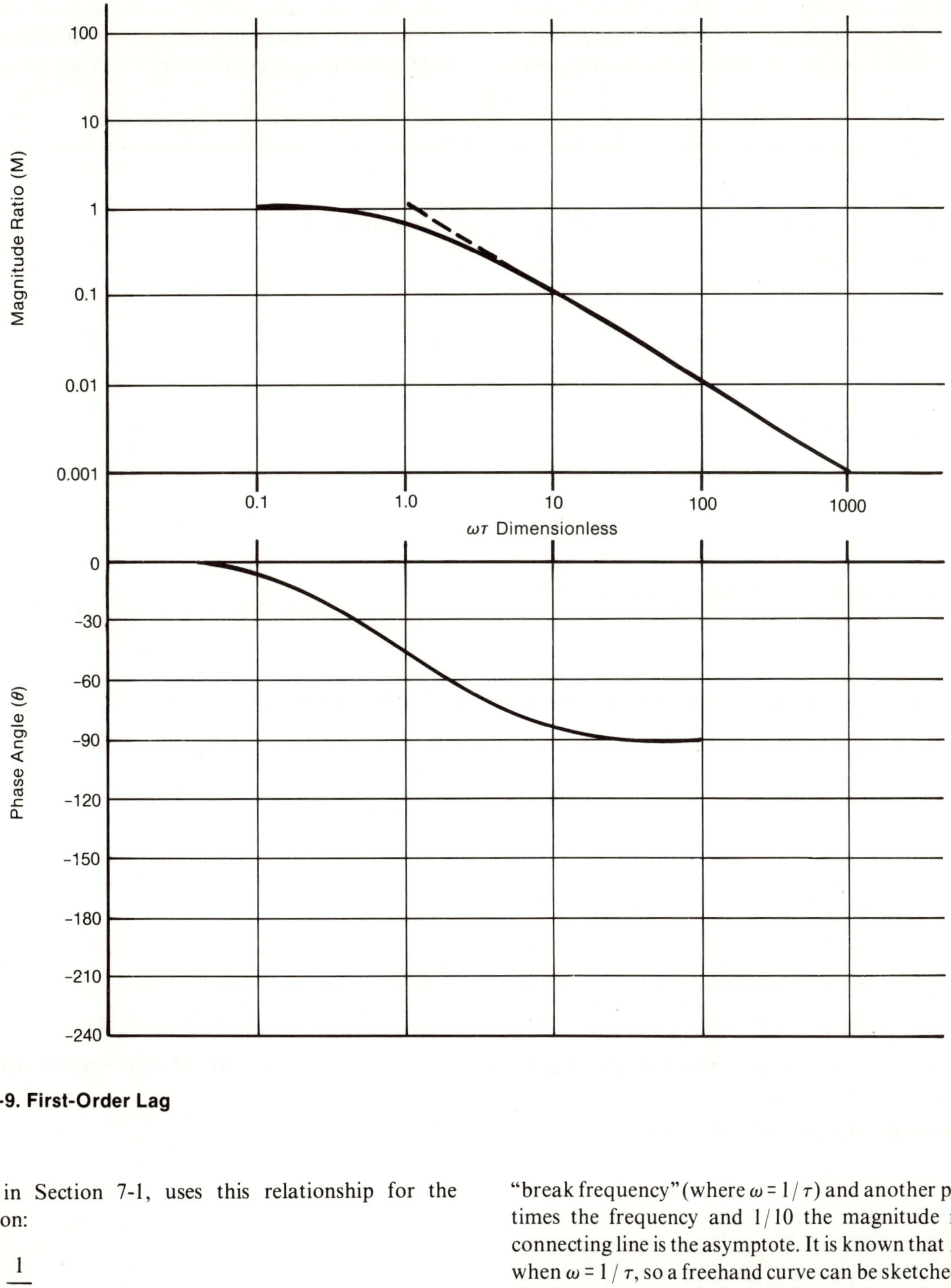

Figure 7-9. First-Order Lag

strated in Section 7-1, uses this relationship for the evaluation:

$$\tau = \frac{1}{\omega}$$

where dimensional homogeneity requires that ω be in radians/unit time.

Note that the magnitude ratio curve can be constructed from the asymptote rather easily. If a point is marked at the "break frequency" (where $\omega = 1/\tau$) and another point at 10 times the frequency and 1/10 the magnitude ratio, the connecting line is the asymptote. It is known that $M = 0.707$ when $\omega = 1/\tau$, so a freehand curve can be sketched starting with $M = 1.0$ at $\omega\tau = 0.1$, passing through $M = 0.707$ at $\omega\tau = 1.0$, and closing into the asymptote at $\omega\tau = 10$.

What does the frequency response curve signify? First, at frequencies (in radians/unit time) that are low, relative to the time constant, the input signal comes through with little or no attenuation (i.e., the dynamic gain is essentially unity), and the phase lag is small. When $\omega = 1/\tau$, about 30 percent of the input signal (i.e., $M = 0.707$) has been lost and a 45° phase lag has been acquired. As the input frequency is increased further, the first-order lag continues to reduce its output until it becomes essentially negligible, and the phase lag approaches 90° as shown in Figure 7-7.

Problems

1. If a first-order lag has a time constant of 0.159 min, what is the break frequency:

   (a) In radians/minute?

   (b) In cycles/minute?

2. The transfer function for nonlimited reset action is

$$G(s) = 1 + \frac{1}{\tau s}$$

   which, when evaluated in the complex plane, gives real part = 1 and imaginary part = $-1/\omega\tau$.

   (a) Write the expressions for M and θ.

   (b) Evaluate M and θ at $\omega = 0$, $1/\tau$, and ∞.

3. The transfer function for nonlimited derivative action is $G(s) = \tau s + 1$, which, when evaluated in the complex plane, gives real part = 1 and imaginary part = $+\omega\tau$.

   (a) Write the expressions for M and θ.

   (b) Evaluate M and θ at $\omega = 0$, $1/\tau$, and ∞.

4. From the above evaluations of reset and derivative action, what would you say of their responses to
   (a) Low frequency (slow) input variations?

   (b) High frequency (fast) input variations?

7-5 CONTROLLER ACTIONS

7-5-1 Derivative Action (Lead-Lag Network)

The previous section developed the Bode diagram for a first-order lag. This system (sometimes called a lag network) was characterized by a magnitude ratio that was unity at low frequencies (relative to the time constant), fell off rapidly above the break frequency, and had a phase angle that started off at 0° and reached −90° at high frequencies.

The problems developed the Bode diagram for nonlimited reset and derivative action. These were defined as nonlimited because the magnitude ratio could, in principle, reach infinity at one end or the other of the frequency spectrum.

The transfer function, $G(s) = \tau s + 1$, used to describe nonlimited derivation action is sometimes called a lead network because the phase angle goes from 0 to +90° (i.e., the output phase angle leads the input). In real controllers, nonlimited derivative action is not achieved; the 90° leading phase angle may be approached, but infinite gain at high frequencies is not attained. Instead, derivative action is characterized by a lead-lag network in which the high frequency gain is limited by the physical characteristics of the controller.

The derivative action in a real controller usually has the form

$$G(s) = \frac{\tau_d s + 1}{\frac{\tau_d}{\alpha} s + 1}$$

where τ_d = derivative time constant

α = constant (typically 6 to 55), which may or may not depend on the controller gain setting (proportional band)

The numerator of the transfer function is the lead portion and will have a break frequency of $\omega = 1/\tau_d$. The denominator is the lag portion of the network and will have a break frequency of $\omega = 1/(\tau_d/\alpha) = \alpha/\tau_d$.

A combination of the two portions of the transfer function will show how the overall transfer function for the lead-lag network looks.

If two transfer functions have gains of M_1 and M_2, the combined effect of the gains (M_c) will be

$$M_c = (M_1)(M_2)$$

The magnitude ratios are plotted as log functions on the Bode diagram; therefore,

$$\log M_c = \log M_1 + \log M_2$$

Thus, graphical addition can be used to combine transfer functions on a Bode diagram, provided that the treatment of the algebraic signs is consistent. The phase angles, which are plotted against a linear scale, can also be added directly (algebraically).

Taking $\tau_d = 10$ and $\alpha = 10$, numerator and denominator of the lead-lag network can be plotted separately and then combined as shown in Figure 7-10.

Note that derivative action (lead-lag network) features a magnitude ratio that increases with frequency as a result of the lead network until the lag portion of the network comes into effect. The magnitude ratio then becomes constant. Similarly, the phase angle initially starts to lead the input, then returns to zero under the influence of the lag. It can be shown by application of the initial and final value theorems that at low frequencies the lead-lag network has a

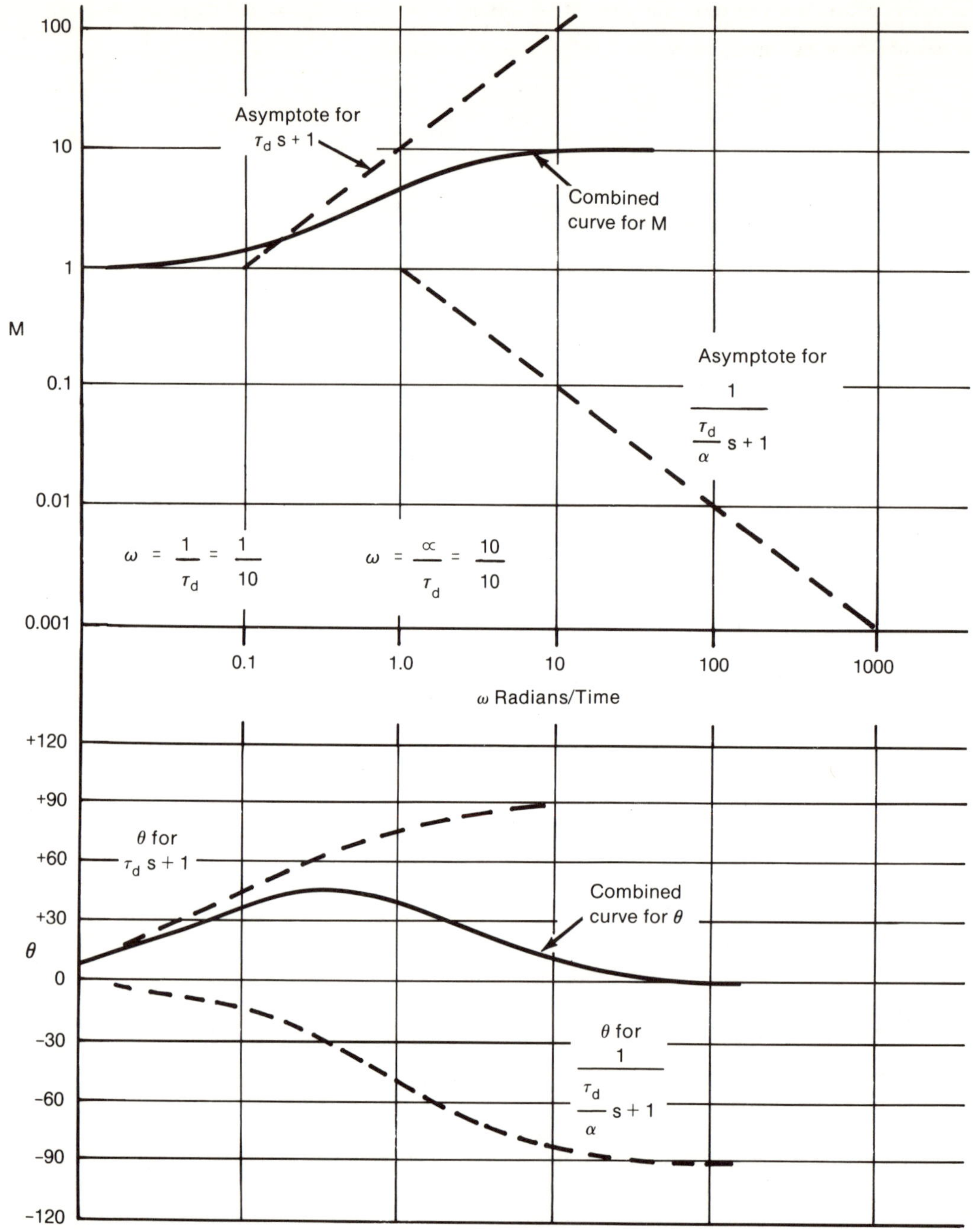

Figure 7-10. Lead-Lag Network

gain of unity, while at high frequencies the gain will be equal to α.

7-5-2 Reset Action (Lag-Lead Network)

Reset action in a real controller usually has a transfer function of the form

$$G(s) = K_i \left(\frac{\tau_i s + 1}{K_i \tau_i s + 1} \right)$$

where τ_i = integral time constant

K_i = reset gain (typically 100 to 500), which may or may not depend on the controller gain setting (proportional band)

By application of the initial and final value theorems, it can be shown that the transfer function above (a lag-lead network) will have a gain equal to K_i at low frequencies and a gain of unity at high frequencies.

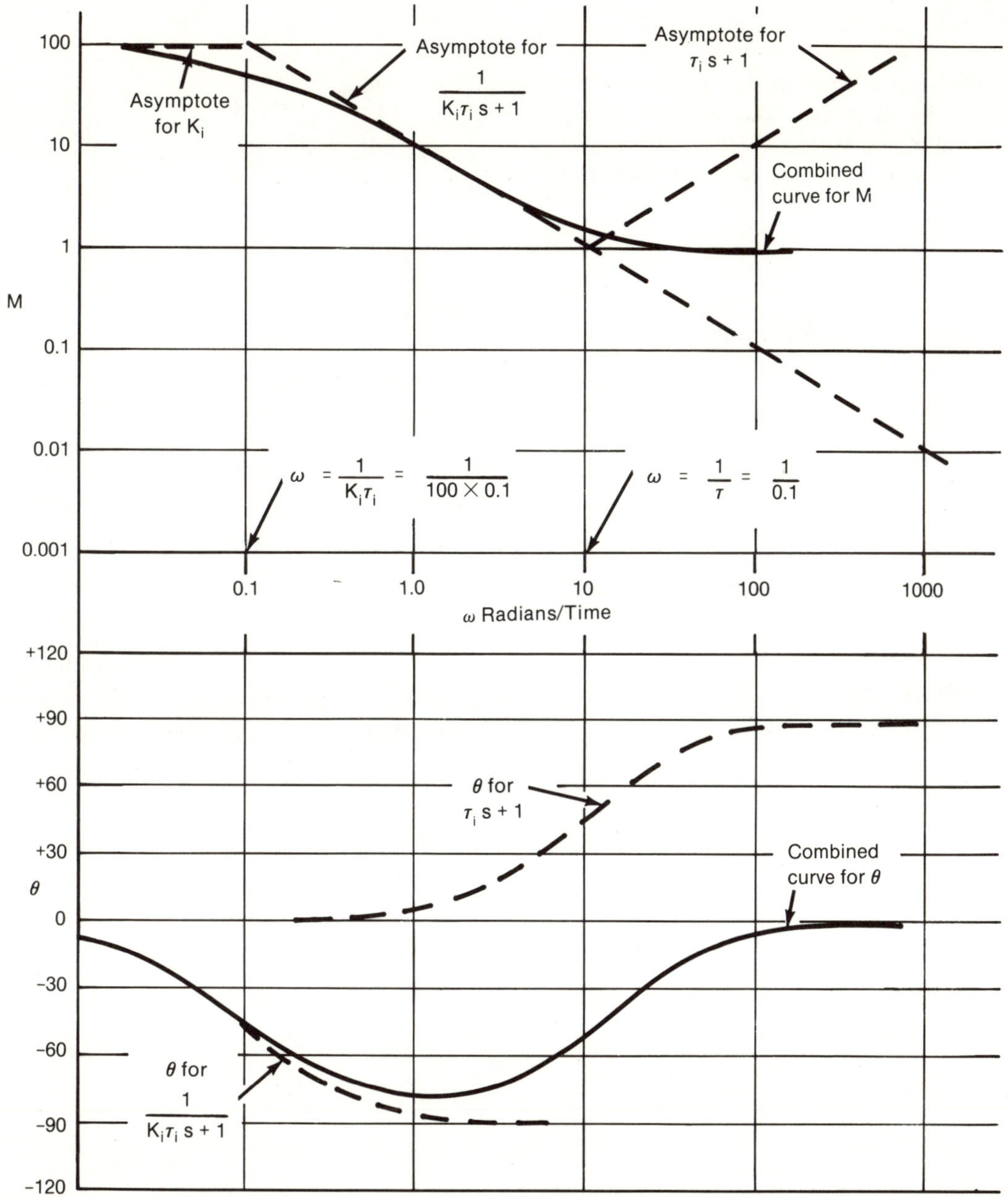

Figure 7-11. Lag-Lead Network

Taking $\tau_i = 0.1$ and $K_i = 100$, the Bode diagram for the lag-lead transfer function can be developed according to Figure 7-11.

Here, note that reset action (lag-lead network) features a large magnitude ratio at low frequencies, which diminishes at intermediate frequencies and is prevented from going below unity only by the increasing high frequency gain of the lead network. Phase angle lags initially, then returns toward zero under the influence of the lead network.

The distinction between lead-lag and lag-lead networks is evident in terms of the response of the phase angle to increasing frequencies on the Bode diagram. We can also distinguish the difference by inspection of the transfer function. Remember that the break frequency occurs at $\omega = 1/\tau$, where τ is the effective time constant (i.e., τ_d/α or $K_i\tau_i$ in these cases). Then the first (lowest) break frequency will occur at the largest effective time constant. If the largest effective time constant is in the numerator (the lead term), the magnitude ratio and phase curves will first trend upwards (increasing magnitude ratio and leading phase). If the largest effective time constant is in the denominator, the magnitude ratio and phase curves will first trend downwards (decreasing magnitude ratio and lagging phase).

Thus, in the lead-lag network, τ_d is greater than τ_d/α; hence, the lead occurs first because τ_d is in the numerator. In the lag-lead network, $K_i\,\tau_i$ is greater than τ_i; hence, the lag occurs first.

7-5-3 Dead Time

The transfer function was developed in Section 4 as

$$G(s) = e^{-as}$$

where a = dead time in units of time. Letting $s = j\omega$,

$$G(j\omega) = e^{-ja\omega}$$

or, using Euler's law,

$$G(j\omega) = \cos a\omega - j \sin a\omega$$

Since the complex function is already separated into real and imaginary parts, the magnitude ratio and phase angle can be evaluated readily:

$$M = [(\cos a\omega)^2 + (-\sin a\omega)^2]^{1/2}$$

$$= [\cos^2 a\omega + \sin^2 a\omega]^{1/2}$$

$$= [1]^{1/2}$$

$$M = 1$$

$$\theta = \tan^{-1}\left(\frac{-\sin a\omega}{\cos a\omega}\right)$$

$$= \tan^{-1}(-\tan a\omega)$$

$$\theta = -a\omega$$

M is always equal to 1; i.e., dead time in no way attenuates the amplitude of the disturbance. This confirms the definition of dead time in Section 4 when the description was qualified by the statement that no change could occur in the properties or characteristics of the material or the energy being transported.

Note also that phase angle (lag) increases linearly with frequency. There is no 90° limit such as the first-order lag exhibited. Evaluating θ at four convenient points:

at	$\omega = 0$	$\theta = -a(0)$	= 0 radians	= 0°
	$\omega = 1/a$	$\theta = -1/1$	= –1 radian	= 57.3°
	$\omega = \pi/a$	$\theta = -\pi$	= 3.14 radians	= –180°
	$\omega = \infty$	$\theta = -a(\infty)$	= –∞ radians	= –∞°

The Bode diagram for dead time is shown in Figure 7-12.

Problems

1. The Moore direct derivative unit has the transfer function

$$G(s) = \frac{\tau_d s + 1}{\frac{\tau_d}{6} s + 1}$$

What are the break frequencies in radians/minute if:

(a) $\tau_d = 0.5$ min?

(b) $\tau_d = 5.0$ min?

2. (a) Sketch freehand the Bode diagrams for the two values of τ_d above (superimpose on same coordinates).

(b) What is the effect of increasing derivative time τ_d?

3. The Moore proportional-plus-reset controller has the transfer function

$$G(s) = 200 \left(\frac{\tau_i s + 1}{\dfrac{200\,\tau_i}{K_c} s + 1} \right)$$

What are the break frequencies in radians/minute if:

(a) $K_c = 2.0$, $\tau_i = 0.1$ min?

(b) $K_c = 2.0$, $\tau_i = 0.5$ min?

(c) $K_c = 10$, $\tau_i = 0.5$ min?

(Note that at high frequencies, gain = K_c)

4. Sketch freehand the Bode diagrams for the three sets of values of K_c and τ_i above.

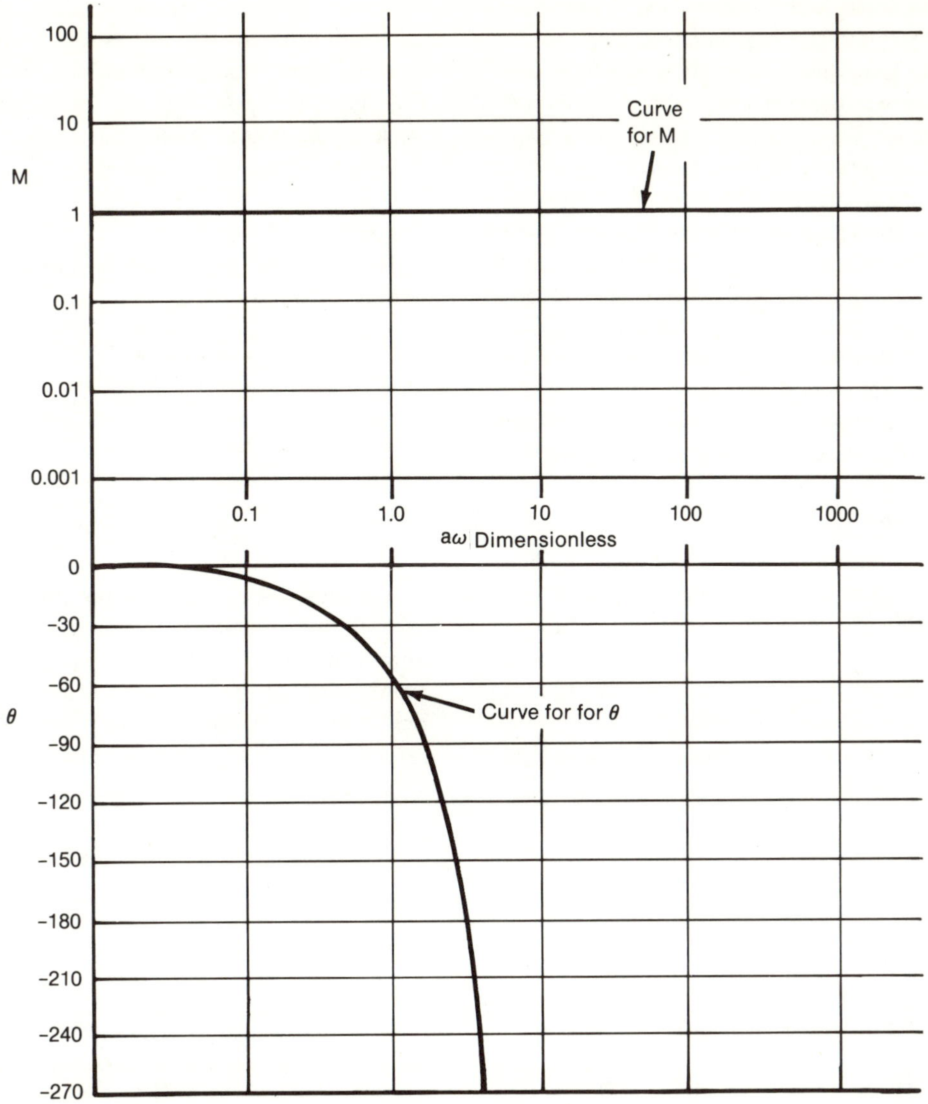

Figure 7-12. Dead Time

7-6 CONTROL LOOP DYNAMICS

7-6-1 General Concepts

A control loop will cycle continuously if the total phase lag around the loop is 360° and the loop gain is unity. Under these conditions, a disturbance entering the loop at any point will travel around the loop and reappear at the point of origin at its original amplitude some time later (the time depending upon the natural period or natural frequency of the closed loop). It will thus appear to the control loop as a new disturbance, and will continue its cycle around the loop. If the loop gain were less than unity, the disturbance would be attenuated during each cycle and would eventually decay to zero. If the loop gain were greater than unity, the disturbance would be amplified during each cycle and would continue to increase until some factor imposed a limit on the amplitude (i.e., control valve size, overflow line on a tank, rupture disc, etc.).

Thus, the limit of stability can be defined as the point at which phase lag reaches 360° when loop gain is unity. The summing junction section of a proportional controller contributes 180° phase lag in itself; i.e., the controller output is exactly out of phase with the input (error) signal. The controller must oppose the error signal in order to exert corrective action. Therefore, it is customary to omit the 180° phase lag contributed by the proportional controller and to state that the limit of stability has been reached when the phase lag contributed by all other components in the loop (including reset and derivative action in the controller) reaches 180° and the loop gain is unity.

Previously the term "maximum allowable loop gain" was used; this gain was deemed to be dependent upon the dynamics of the components in the loop and on the allowable overshoot during recovery from an upset. Now it can be seen that the maximum allowable loop gain would have to be such that the loop gain would be less than unity when phase lag reaches 180° (excluding the phase lag contributed by the controller proportional action). The loop gain could, however, be greater than unity when phase lag is less than 180°. In fact, the loop gain would have to be greater than unity at some time in order to cause overshoot.

It is time to consider the significance of the location and nature of various dynamic elements in a control loop.

7-6-2 Dead Time

Most instrumentation people have found by bitter experience that dead time is the worst offender as far as control loop stability and performance are concerned. Why?

Consider a reactor system in which two low viscosity streams react to form a high viscosity polymer. Assume that viscosity is a measure of polymer quality, and that a control loop is provided as shown in Figure 7-13.

Since the reactor is unmixed, plug flow occurs, resulting in a dead time (a_1) that corresponds to the transit time of the fluid from point 1 to point 2. There is also a dead time (a_2) that corresponds to the transit time of the sample stream from point 2 to point 3.

Neglecting for the moment all dynamics except dead time, the signal flow diagram can be drawn as shown in Figure 7-14.

By inspection, the closed-loop transfer function can be written as follows:

$$\frac{\mu_3}{w_A} = \frac{K_p e^{-a_1 s}}{1 + K_p K_t K_c K_v e^{-a_1 s} e^{-a_2 s}}$$

$$\frac{\mu_3}{w_A} = \frac{K_p e^{-a_1 s}}{1 + K_p K_t K_c K_v e^{-(a_1 + a_2)s}} \quad (7\text{-}5)$$

The closed-loop response (i.e., the denominator) contains the combined dead times regardless of where they appear in the loop.

Assume that $a_1 = 1$ minute and $a_2 = 0.5$ minute. From the previous section, the phase lag will be 57.3° when ω (in radians) $= 1/a$, and 180° when $\omega = \pi/a$. Then, for the values above:

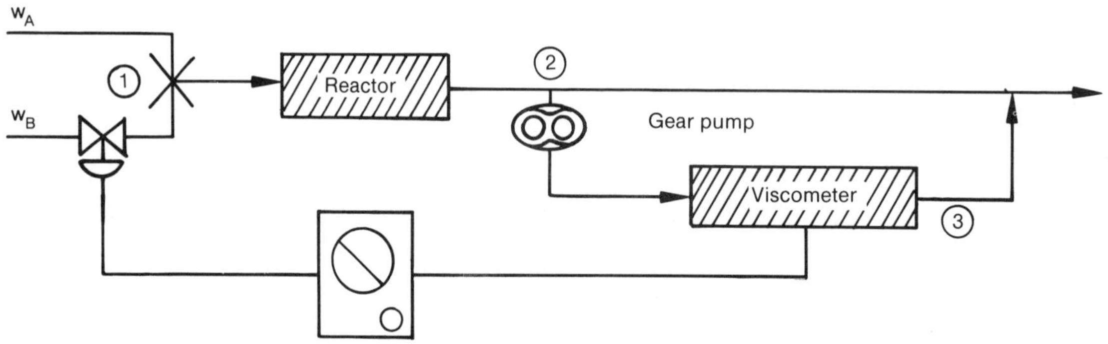

Figure 7-13. Reactor Control

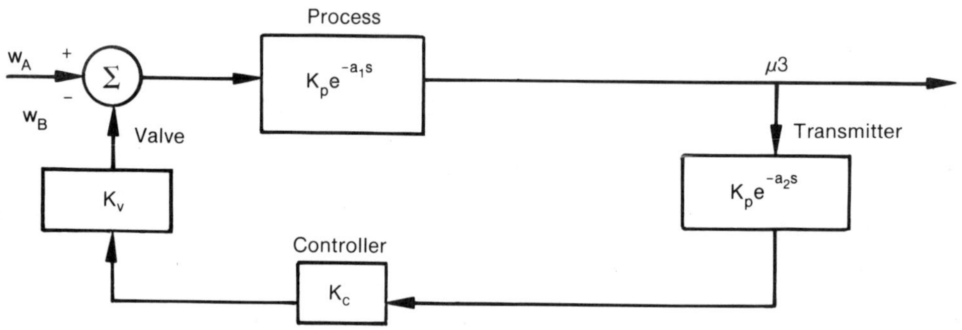

Figure 7-14. Reactor Signal Flow Diagram

$\omega_1 = 0$ $\qquad\theta = 0°$

$\omega_1 = 1/1 = 1$ $\qquad\theta = 157.3°$

$\omega_1 = \pi/1 = 3.14$ $\qquad\theta = -180°$

$\omega_2 = 0$ $\qquad\theta = 0°$

$\omega_2 = 1/0.5 = 2$ $\qquad\theta = -57.3°$

$\omega_2 = \pi/0.5 = 6.28$ $\qquad\theta = -180°$

It is also known that the magnitude ratio is always unity for dead time. The combined Bode diagram for the two dead times can now be sketched as shown in Figure 7-15. It is possible, of course, to combine the dead times initially and plot the phase curve for $(a_1 + a_2) = 1.5$ minutes immediately.

Inspection of Figure 7-15 discloses why dead time has such an adverse effect on control performance. There is no dynamic attenuation of the disturbance by the dead time, hence the loop gain is the product of the open-loop gains (as shown in Equation 7-5). Further, the slope of the phase lag curve is steep, and the 180° point is reached at frequencies that are low relative to the dead time. Thus, the period of the disturbance must be long (i.e., the frequency must be low) with respect to the dead time if control is to be effective. Short-term (high frequency) disturbances will come and go before the controller is aware of them. The control response to a short-term disturbance may thus lead readily to sustained oscillation unless controller gain is set relatively low.

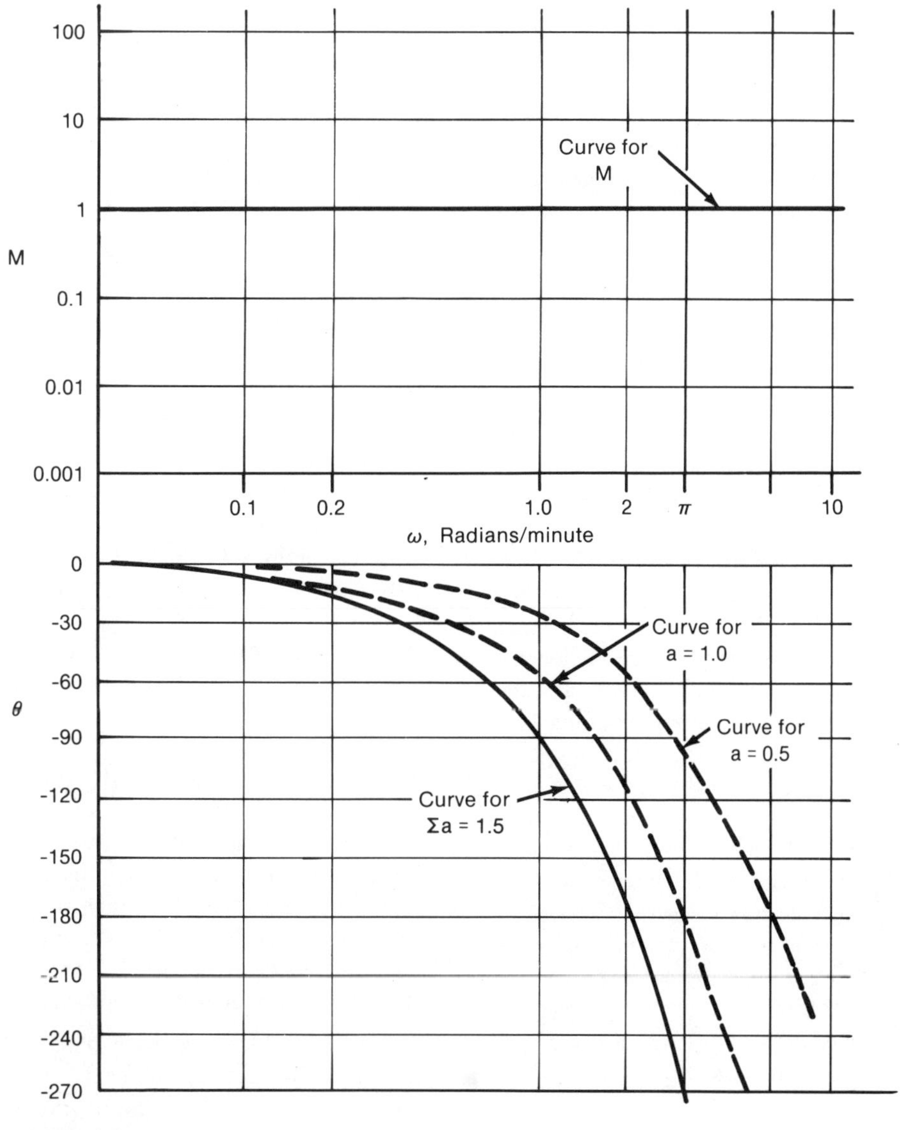

Figure 7-15. Combined Dead Times

Note again that the distribution of dead times around the loop has no influence on control performance. Any combination of dead times that totaled 1.5 minutes would have resulted in the same Bode diagram in Figure 7-15. A different distribution of dead times would merely have influenced the time when a disturbance appeared at various points in the loop; the total transmit time around the loop would have been the same.

The distribution of dead times may, however, be important in considering methods of improving control performance. If, for example, the reactor had been designed for 1.4 minutes holdup (dead time), to allow time for the reaction to go to completion, then an additional 0.1 minute dead time in the viscosity measurement (giving a total of 1.5 min) would not significantly influence controllability. If, on the other hand, the reactor dead time were 0.1 minute and the viscometer dead time were 1.4 minutes (still totalling 1.5 min), it would be a good idea to reduce the volume of the sampling system as a means of improving control performance.

Recall from previous sections that reset action provides high gain at low frequencies, but that it contributes up to nearly 90° phase lag at these same low frequencies. It would be expected, then, that the integral time constant (i.e., minutes per repeat) would have to be rather large in systems dominated by dead time in order to avoid contributing too much phase lag in the region where the dead time has already started contributing phase lag. Experience confirms that little reset action is tolerable in control loops dominated by dead time.

Derivative action, on the other hand, contributes high gain at high frequencies and phase lead in the same frequency range. Its value, however, is limited in systems dominated by dead time. Because of the steepness of the dead time phase lag curve, the phase lead contributed by the rate action has relatively little influence on the frequency at which the −180° phase angle is attained. Meanwhile, the dynamic gain contributed by derivative action raises the magnitude ratio above unity in the same frequency region where the phase lag is approaching 180°. Thus, little or no derivative action is tolerable in systems dominated by dead time.

The worst control situation arises where systems dominated by dead time also are characterized by a high steady-state process gain (K_p). In processes of this type, the allowable loop gain frequently results in proportional band settings in the range of 1000 to 2000 percent (controller gains of 0.1 to 0.05). Since, as has already been seen, the system will tolerate very little reset action, control as a whole is very ineffective. Only long-term drifts can be handled; short-term transients pass through the process unaffected.

• *Controller Parameters*

From the foregoing, the transfer function of a 3-mode controller can be written:

$$e \longrightarrow \boxed{200\left(\frac{\tau_i s+1}{\frac{200\,\tau_i}{K_c}s+1}\right)\left(\frac{\tau_d s+1}{\frac{\tau_d}{\alpha}s+1}\right)} \longrightarrow c$$

A shorthand means of describing this controller is by designating it a PID controller, where P is proportional action, I is integral or reset action, and D is derivative or rate action.

The zero frequency gain of the reset action and the high frequency gain limitation of the derivative action are of very little interest when selecting controller parameters (K_c, τ_i, τ_d) for stability, since stability is limited to the frequency range where the combined valve-process-transmitter Bode plot phase angle crosses 180°. Thus the foregoing controller transfer function can be simplified to

$$e \longrightarrow \boxed{K_c\left(1+\frac{1}{\tau_i s}\right)\qquad(\tau_d s+1)} \longrightarrow c$$

Remember that the Bode plot is a convenient graphical device for combining the transfer functions of the elements in the control loop. The control loop will become unstable if the magnitude ratio is greater than 1 where the phase angle exceeds 180°. It's as simple as that. There are no stated areas of stability or instability on the Bode plot.

The reason for having a controller is to adjust its functions to compensate as best as possible for the dynamic response of the other elements in the control loop. Thus the rate and reset functions are adjusted to move the 180° crossing point to as high a frequency as possible, which will ensure the fastest response of which the loop is capable. Then the proportional band is adjusted to a gain that will cause the magnitude ratio to be less than 1 at the 180° point.

Mechanically, the dynamic controller functions (reset and rate) are set by moving their individual phase curves horizontally on the Bode plot, after which the proportional band is set by moving the magnitude ratio curve vertically.

Given a Bode plot on which the combined magnitude ratio and phase relationships for all the elements in the control system except the controller have been plotted, then the controller parameters may be set as follows:

(1) *Reset.* Move the reset phase relationship horizontally until the reset contributes 10° of phase lag at the point where the combined curve contributes 170° of phase lag.

The reset parameter can be read from the magnitude ratio break point.

(2) *Rate.* Move the rate phase relationship horizontally until the rate action contributes 60° of phase lead at the 180° phase lag frequency. Read the rate parameter from the break point on the magnitude ratio curve.

(3) Proportional band. Move the magnitude ratio curve vertically so that the ratio is less than 1 at the frequency where the combined phase curve (including reset and rate) crosses 180°. The controller gain is thus calculated

$$K_c K_v K_p K_T \leq \frac{1}{\text{magnitude ratio}}$$

Since the gains and the magnitude ratio are known, the controller gain (K_c) can be calculated. (Note that the reset function must be combined with the rest of the loop on the Bode plot before rate is selected, and rate must be combined before proportional band is selected.)

Problems

1. Sketch freehand the Bode diagram for a combination of reset action (Section 7-5) and the dead time curve in this section that would provide stable control.

2. Estimate the value of τ_i, the integral time constant, from your Bode diagram of problem 1 above.

3. Sketch freehand the Bode diagram for a combination of derivative action and the dead time curve in this section that would give stable control action (if stable control is possible).

4. Estimate the value of τ_d, the derivative time constant from your Bode diagram of problem 3.

7-7 DYNAMIC ELEMENTS IN THE LOOP

7-7-1 Dead Time and First-Order Lag

In the previous section, the effect of dead time in a control loop was shown to be a rapid accumulation of phase lag without a corresponding decrease in magnitude ratio. The point of control instability ($M = 1, \theta = -180°$) was thus reached at disturbance frequencies that were relatively low in relationship to the dead time.

A first-order lag, on the other hand, is characterized by a maximum phase lag of 90° and a continually decreased magnitude ratio (above the break frequency). It would seem, then, that the addition of a large first-order lag somewhere in the loop could be utilized to reduce the magnitude ratio (dynamic gain) well below unity before the phase lag reached 180°, thus allowing a substantial increase in controller gain.

Consider the reactor system described in the previous section, where the process dead time (a_1) was 1.0 minute and the measurement dead time (a_2) was 0.5 minute. The combined ($a_1 + a_2$) magnitude ratio and phase angle curves are shown on Figure 7-16, with the corresponding Bode plots of three different first-order lags (τ = 1, 10, and 100 minutes).

Inspection of the Bode diagram shows that the addition of a 1-minute first-order lag to the 1.5-minute dead time

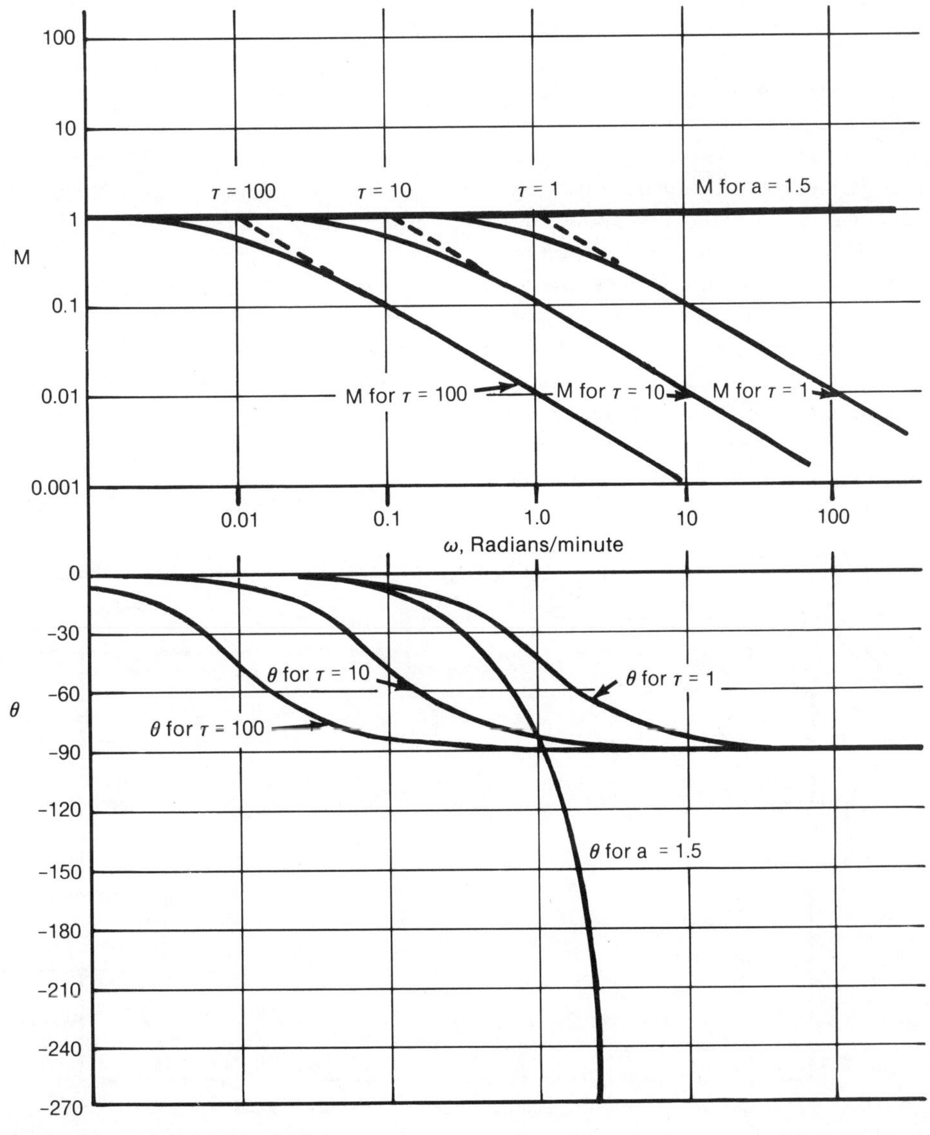

Figure 7-16. Dead Time and First-Order Lags

would result in a slight improvement in control stability. At a frequency of about 1.60 radians/minute, the magnitude ratio would have dropped to about 0.5, while the phase angle would have reached about −180° as a result of the −60° phase angle contributed by the first-order lag. The controller gain could thus be increased by a factor of 1/0.5 = 2 (compared to its value in a pure dead time system) without exceeding unity gain at −180° phase angle.

Consider now the combination of a 10-minute lag and the 1.5-minute dead time. The combined phase lag will reach 180° at a frequency of about 1.25 radians/minute, at which point the magnitude ratio has decreased to about 0.10. Controller gain could then be increased by a factor of 1/0.1 = 10.

With a 100-minute lag and a 1.5-minute dead time, phase lag would reach 180° at about 1.1 radians/minute, at which point the magnitude ratio would be about 0.01. Controller gain could then be increased by a factor of 1/0.01 = 100.

Thus, the degree of improvement in stability that can be affected by the addition of a first-order lag to an otherwise pure dead time system depends on the ratio of lag to dead time. If the lag is large compared to the dead time, controller gain can be increased. When the lag is about 7 times the dead time (i.e., 10/1.5), the influence of dead time is minor. At about 10/1, the dead time can generally be neglected as a serious influence on controllability, and at 70/1 its effect is hardly noticeable.

Note that it has not yet been decided where in the loop to locate the first-order lag. From the standpoint of stability alone, the location is immaterial, since the open-loop Bode diagram would be the same regardless of the locations of the lag and dead time(s) in the loop. From the standpoint of control performance, however, the location is most important.

Consider the block diagram of the reactor system, and assume that the disturbance is a change in flow of component w_A. Five possible locations will be considered for the first-order lag $G = 1/(\tau s + 1)$ (see Figure 7-17).

Inspection of the signal flow diagram shows that control performance (i.e., the ability of the control system to minimize changes in μ_3 resulting from changes in w_A) will not be enhanced by inserting the first-order lag in boxes 1, 2, or 3. A lag in any of these locations will merely attenuate the signal to the viscometer, the controller, or the valve. Stability would be improved by the lag, as was evidenced by the Bode diagram, but short-term changes in w_A would result in corresponding changes in μ_3. The valve controlling w_B would hardly know about these changes, however, since the change in μ_3 would be attenuated by the lag at locations 1, 2, or 3. If the lag were at location 1 or 2, the controller chart record would look good because the recorder would see only an attenuated version of what was happening to μ_3.

Insertion of the lag at location 4 or 5 would clearly be the best move. In either case, the open-loop transfer function relating μ_3 and w_A would be

$$\mu_3 = \frac{w_A K_p e^{-a_1 s}}{\tau s + 1}$$

Installation of a well-mixed tank at location 4 would average out short-term variations in the flow rate w_A. Recall from the study of composition systems that the unsteady-state material balance for such a system was

$$w_A X_{A1} + w_B X_{B1} - (w_A + w_B) X_{A2} = M \frac{dX_{A2}}{dt} \quad (7\text{-}6)$$

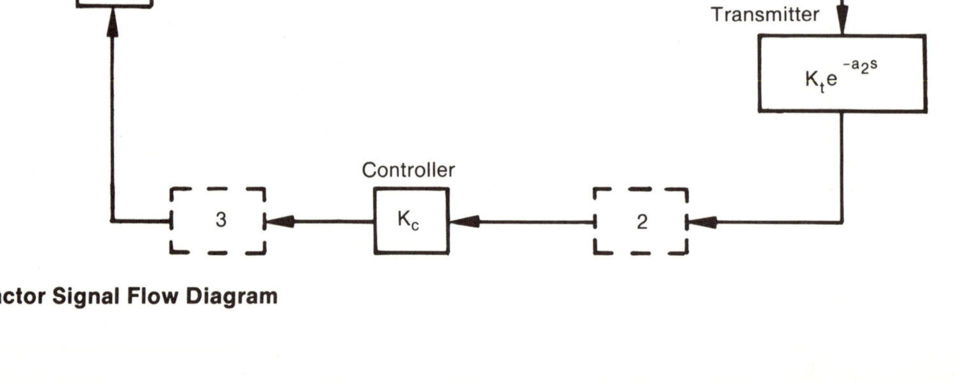

Figure 7-17. Reactor Signal Flow Diagram

where M = holdup in tank

X_{A1}, X_{B1} = concentrations of A and B in feed streams

X_{A2} = concentration of component A in the tank

Equation 2 can be rearranged to obtain the transfer function for X_{A2}, as

$$\frac{w_1 X_{A1} + w_B X_{B1}}{w_A + w_B} = \left(\frac{M}{w_A + w_B} s + 1\right) X_{A2}$$

$$X_{A2} = \frac{\dfrac{w_1 X_{A1} + w_B X_{B1}}{w_A + w_B}}{\dfrac{M}{w_A + w_B} s + 1} = \frac{\dfrac{w_A X_{A1} + w_B X_{B1}}{w_A + w_B}}{\tau s + 1}$$

The tank holdup M can thus be selected to yield a time constant that will limit the maximum rate of change of X_{A2}, the reactor feed composition, to a value that is slow compared to the dead time. If τ is about 10 [i.e., if $M = 10 (w_1 + w_2)$], the Bode diagram shows that the effect of dead time will be relatively minor.

By the same analysis, installation of a well-mixed tank at location 5 would allow variations in w_A to vary reactor effluent viscosity, but would then blend these short-term variations into a product whose viscosity μ_3 would change only slowly.

The preference for one or the other of locations 4 and 5 would depend on process considerations. If streams w_A and w_B reacted on contact, for example, then a tank at location 4 would in reality become the reactor; and the dead time a_1 would be eliminated. On the other hand, if viscosity μ_3 is very high, it might be impractical to mix effectively the contents of a tank at location 5. If both restrictions obtain, it would probably be best to eliminate the first-order lag entirely and accept the wide proportional band and slow reset action imposed by a pure dead time system. Except for noise suppression, it is seldom beneficial to overall control performance to insert lags in the feedback part of the closed loop (i.e., between the measured variable and the manipulated variable).

Problems

1. Sketch freehand the Bode diagram combining a 1.5-minute dead time and a 1-minute lag.

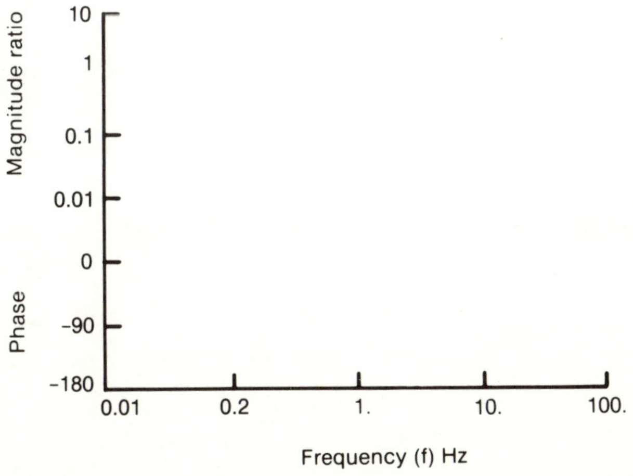

2. Repeat problem 1 for a 1.5-minute dead time and a 10-minute lag.

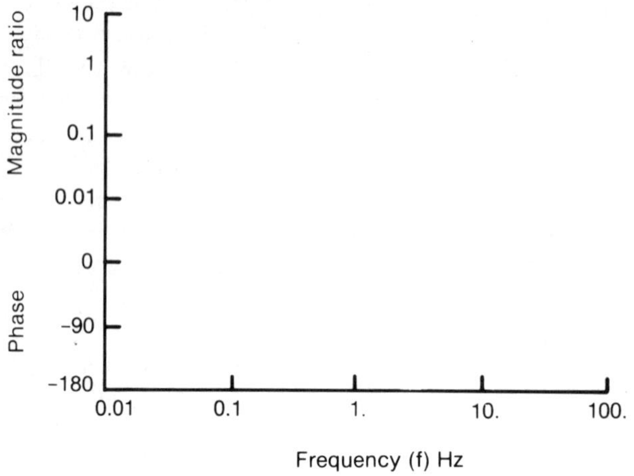

3. Repeat problem 1 for a 1.5-minute dead time and a 100-minute lag.

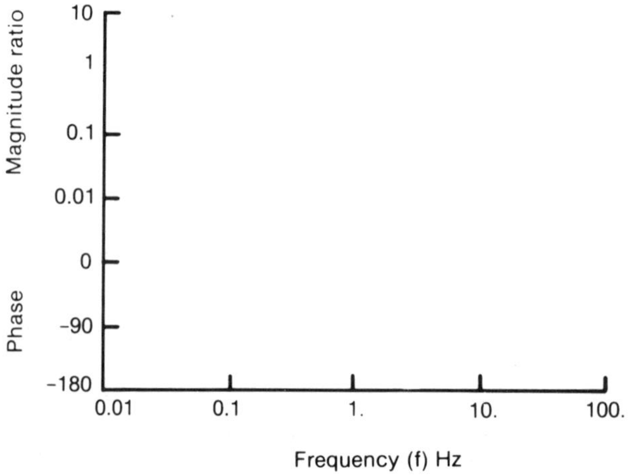

4. If the dead time had been 15 minutes in the problems above:

(a) What values of time constants would have given the same magnitude ratio and phase lag relationships as the three curves developed above?

(b) At what frequency (approximately) would the phase lag reach 180°?

7-7-2 Combinations of First-Order Lags

Consider a control loop in which only the process and measurement lags are significant, as shown in the signal flow diagram notation in Figure 7-18.

Assume that one of the time constants is 0.1 minute and the other is 1.0 minute. The corresponding break frequencies will occur at ω = 10 and 1 radian/minute, so the open-loop Bode diagram can be sketched as shown in Figure 7-19.

Inspection of the Bode diagram shows that the magnitude ratio will be well below 0.01 by the time the phase angle reaches −180°. It would, therefore, be expected that controller gain could be at a relatively high value before instability occurred. In theory, the controller gain could be set at infinity without instability, since the maximum total phase lag contributed by two first-order lags cannot exceed 180°. Hence, the system could do no worse than oscillate at a constant amplitude; it could not diverge.

In practice, the control valve, for example, will have some sort of dynamic response, however small the time constant. At some higher frequency, then, the valve will start to contribute phase lag, and instability will occur at high controller gain. There is thus imposed on almost any real system a maximum controller gain, beyond which instability will occur as a result of minor parasitic lags.

It has not yet been determined whether the process or the measurement time constant is the larger; and, in fact, it makes no difference as far as stability is concerned. It does, however, make a difference as far as control performance is concerned.

From the block diagram, it can be seen that the open-loop transfer function relating Y to X is

$$Y = \frac{XK_p}{\tau_p s + 1}$$

From the Bode diagram, note that if τ_p is the 1-minute time constant, the higher frequency signals will be well attenuated by the process itself. For example, a fast disturbance equivalent to a frequency of 10 radians/minute will be reduced to about 10 percent of its input amplitude by the process itself, leaving relatively little for the control loop to do. The transmitter, with a 0.1-minute time constant, would reduce the signal to about 70 percent. The controller would thus see 70 percent of the change in Y (at 10 radians/minute) and could act effectively.

Suppose now that the situation were reversed; i.e., that τ_p = 0.1 min and τ_t = 1.0 min. The process would now attenuate a disturbance equivalent to ω = 10 radians/minute to 70 percent of its input amplitude. In other words,

$$\Delta Y = 0.70\ K_p\ \Delta X$$

The transmitter, meanwhile, would attenuate a disturbance of this frequency to about 10 percent of its input value. Thus, the controller would be unaware of the real magnitude of the variations in Y, and its corrective action would only be 10 percent as effective as if it could see the real disturbance.

Remember that the controller gain, which was established only by stability considerations, would be the same in either case. It becomes obvious, then, that it is highly desirable to have the process time constant much larger than the measurement lag (and the valve lag, since the same analysis would apply if the valve lag were the dominant instrument lag). It is customary in hand calculations of control performance to neglect any lags that are less than about 10 percent of the second largest lag. While these small time constants do have an influence on control performance, their omission does not materially affect the results of the analysis.

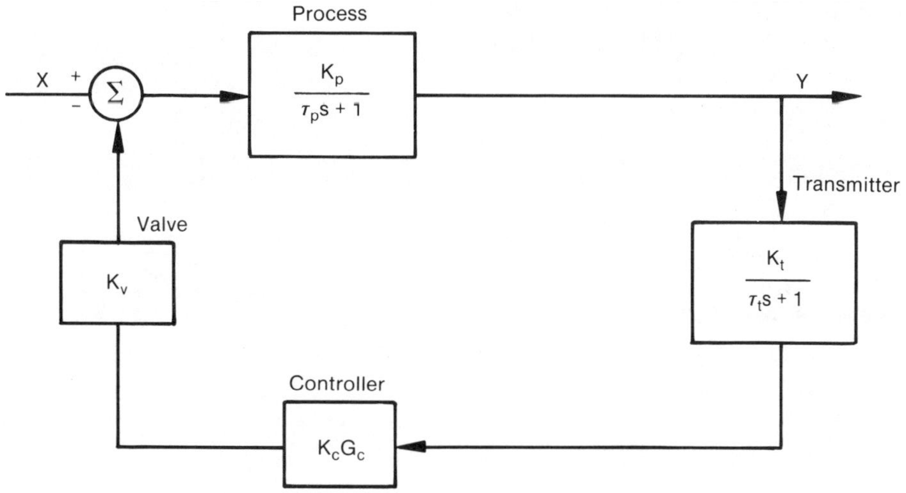

Figure 7-18. First-Order Lag Signal Flow Diagram

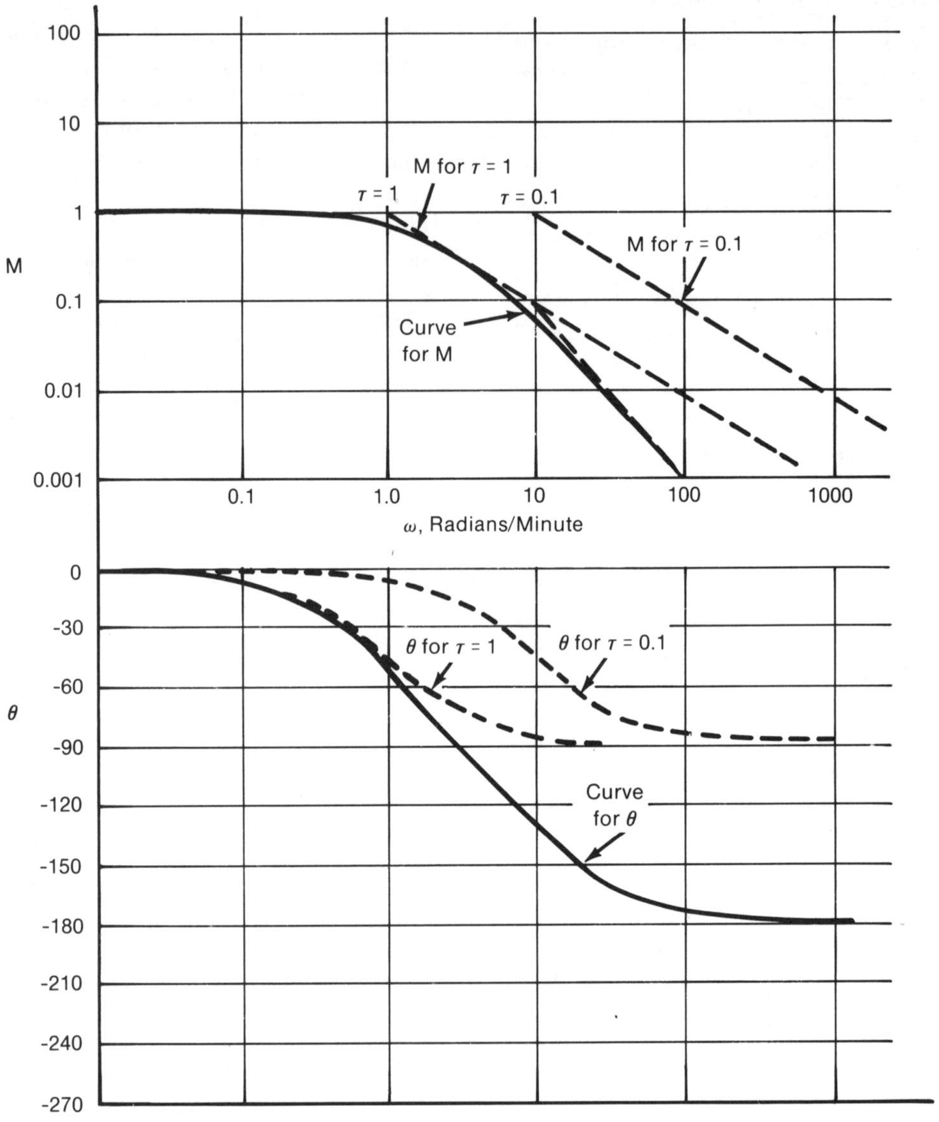

Figure 7-19. Two First-Order Lags

7-7-3 Derivative Action

Assume a situation where the measurement time constant is 1.0 minute and the process time constant is 0.1 minute. Returning to the signal flow diagram, the closed-loop transfer function can be written by inspection:

$$\frac{Y}{X} = \frac{\dfrac{K_p}{\tau_p s + 1}}{1 + \dfrac{K_p K_t K_c K_v G_c}{(\tau_p s + 1)\ (\tau_t s + 1)}} \tag{7-7}$$

Assume that the controller dynamic response is

$$G_c = \tau_d s + 1 \qquad (7\text{-}8)$$

Recall that this expression represents pure derivative action — a lead term. In reality, a real controller would have rate action equivalent to a lead-lag network, but the lag term has been omitted for simplicity in this illustration. Substituting Equation 7-8 into 7-7,

$$\frac{Y}{X} = \frac{\dfrac{K_p}{\tau_p s + 1}}{1 + \dfrac{K_p K_t K_c K_v (\tau_d s + 1)}{(\tau_p s + 1)\,(\tau_t s + 1)}} \qquad (7\text{-}9)$$

If the controller is now adjusted so that

$$\tau_d = \tau_t$$

it can be seen that the derivative action lead term will cancel the transmitter lag term, leaving

$$\frac{Y}{X} = \frac{\dfrac{K_p}{\tau_p s + 1}}{1 + \dfrac{K_p K_t K_c K_v}{\tau_p s + 1}}$$

Rearranging,

$$\frac{Y}{X} = \frac{K_p}{\tau_p\, s + 1 + K_p K_t K_c K_v}$$

$$\frac{Y}{X} = \frac{\dfrac{K_p}{1 + K_p K_t K_c K_v}}{\dfrac{\tau_p}{1 + K_p K_t K_c K_v}\, s + 1} \qquad (7\text{-}10)$$

Thus, by the use of derivative action, the (dominant) measurement lag has been compensated for. In terms of the Bode diagram, a lead term with $\tau_d = 1$ has been inserted, compensating for the attenuation and phase lag introduced by the relatively long measurement lag. The controller gain K_c could now be increased until the "effective" time constant of the process was reduced to the point where parasitic time constants (such as, perhaps, the valve) became significant, at which point the limits of stability would again be approached. Note again that the relative values of the time constants, rather than the absolute values, govern the limit of stability.

One might wonder whether derivative action could be used to compensate for the process lag, if it were the dominant lag. The answer is no, for several reasons. First, inspection of Equation 7-9 shows that cancelling $\tau_p s + 1$ with $\tau_d s + 1$ in the denominator of the transfer function would still leave a $\tau_p s + 1$ term in the numerator; on rearrangement, the process lag would reappear in the denominator. Second, τ_p depends on the characteristics of the process; they cannot be changed by a controller setting. Third, the best control performance occurs when the process lag is clearly dominant (except in the special case where the disturbance enters the loop after the process); hence, there is no advantage in reducing the process time constant.

Note that the numerator of Equation 7-10 represents the steady-state closed-loop response of Y to changes in X. As K_c is increased, the numerator decreases, thus reducing the change in Y for a given change in X.

Tivy[114] has applied derivative action in his experience as shown in Table 7-1.

7-7-4 Summary

The significance of the location of dynamic elements in a control loop can be summarized as follows:

(1) It is the relative rather than the absolute values of dynamic terms that govern stability and control performance.

(2) Stability depends on the relative values of the various dynamic terms; control performance depends on their location in the control loop.

(3) Best control performance is obtained when the process lag is clearly dominant (except where the disturbance enters after the process).

(4) Dead times totaling less than 10 percent of the largest time constant have only minor effect on control performance.

(5) Instrument lags of less than 10 percent of the process lag have relatively little influence on control performance.

(6) Lags of less than 10 percent of the second largest lag can generally be neglected.

(7) Derivative action can be used to compensate for instrument lags when they become significant relative to the process lag.

Table 7-1
Application of Derivative Action in Control of Various Processes[114]

Application	Measurement	Predominant Difficulty	Valve Controls	Notes	Use Derivative
Fractionating Column	Temp.	Transfer & Transportation Lags	Reflux or Reboiler Oil	Smaller deviations from control point improve product quality	Yes
Furnace Outlet	Temp.	Transfer & Transportation Lags	Fuel Supply	Often ability to hold exact control point permits a higher operating temperature and better results	Yes
Heat Exchanger	Temp.	Transfer & Transportation Lags	Heating Medium	Symmetrical type - Choice depends on economics	Yes or No
Heat Exchanger	Temp.	Transfer & Transportation Lags	Heating Medium	Mixed flow type — A nonsymmetrical response possible	No
Reactor in Fluid Catalyst Unit	Temp.	None	Hot Catalyst Flow	High demand capacity points to ease of control — Also rapid changes in rate of flow of catalyst are undesirable	No
Absorber Tower	Press.	None	Overhead Vapors	High demand capacity points to easy control	No
Debutanizer Tower	Press.	Transfer & Transportation Lags	Reboiler Heating Oil	Close control increases yield quality — This is a almost as difficult as temperature control	Yes
Most Pump Suctions	Press.	None	Pump Speed	Often controlled satisfactorily by a simple regulator	No
Pipe Line Pump Suction	Press.	Inertia of Fluid in Pipe Line	Pump Speed	This is quite a difficult control problem and should be treated as such	Yes
Gas Engine Driven Pump	Speed	None	Fuel to Engine	With properly engineered valves and regulation this is an easy application	No
Steam Driven Reciprocating Pump	Speed	Inertia of Flywheel can be High	Steam to Engine	Slow Speed Engines require heavy flywheels. Control consequently becomes difficult	Yes

Problems

1. Below is the Bode diagram of Figure 7-19. Sketch freehand and label three curves showing the new Bode diagram when each of the following dynamic terms is added to the two-lag system (do not add the dynamic terms successively to each other; add each to the two-lag system):
   (a) A first-order valve lag with $\tau_v = 0.01$ minute.
   (b) A dead time with $a = 0.1$ min.
   (c) A pure lead term with $\tau_d = 0.1$ min.
2. From the new Bode diagram, estimate the magnitude ratio when $\theta = -180°$.

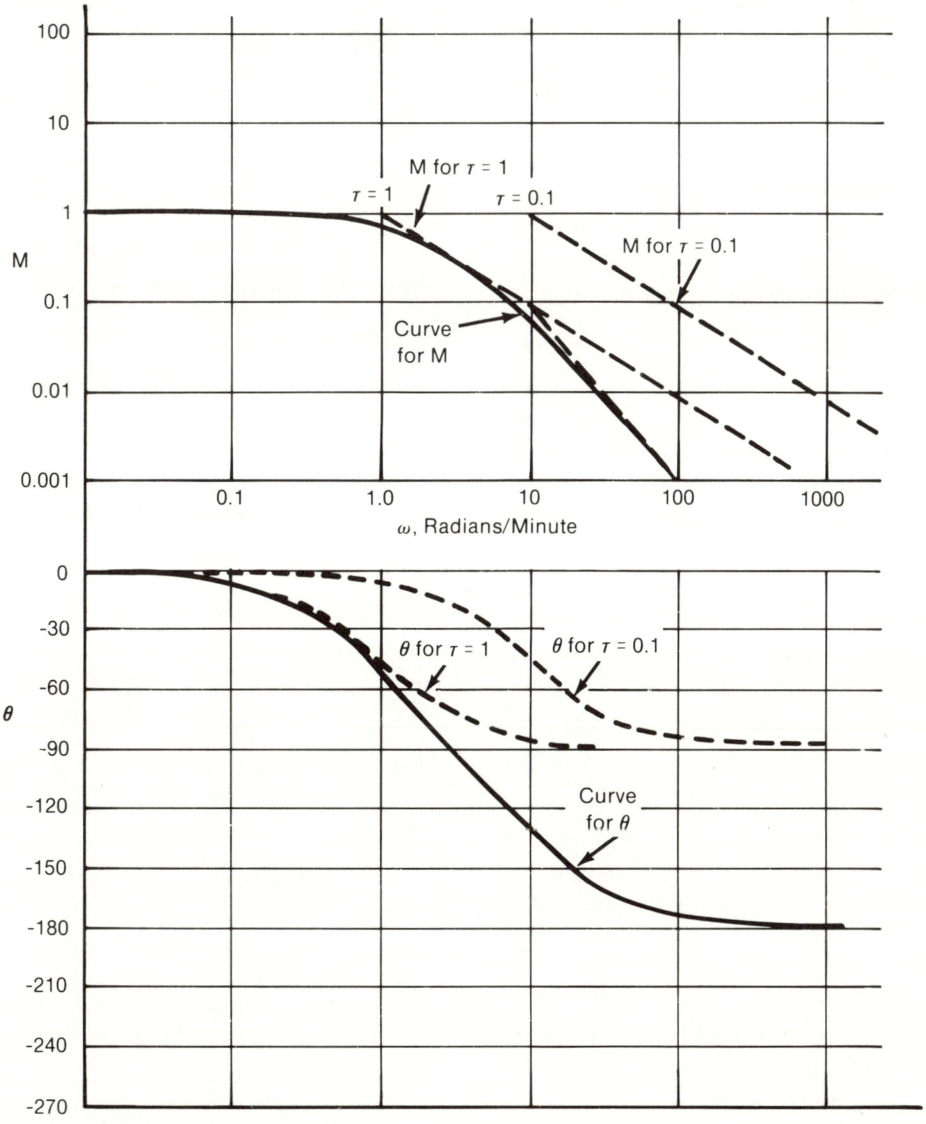

Figure 7-19. Two First-Order Lags

7-8 CONTROLLER TUNING

Preliminary controller settings can be predicted from knowledge of system gains and dynamics. Final settings for optimum control performance must, however, be made eventually on the operating process because there are always differences between estimated and actual system parameters. A few methods of predicting controller settings and of making the final adjustments on the operating plant will be considered.

7-8-1 Preliminary Estimation of Controller Settings

There are several methods of predicting approximate controller settings from preliminary knowledge of system gains and dynamics. The reliability of the prediction varies more or less directly with the complexity of the computations involved. A few of the more common methods will be described here.

- *Frequency Response Technique*

In this method, the Black-Nichols chart (which is a map on the complex plane of the relationship of open- and closed-loop magnitude ratios and phase angles) is used to convert the open-loop Bode diagram to the closed-loop Bode diagram. The technique allows selection of a controller gain and integral (reset) time constant that will produce the desired degree of overshoot on recovery from an upset. Derivative action is generally selected to compensate for the dominant instrument lag. This method, however, requires knowledge of frequency response beyond the scope of this text.

- *Routh's Stability Criterion*

In this technique, the denominator of the closed-loop transfer function (which represents the overall system response) is multiplied out and arranged to form a descending series of the powers of s, such as

$$a_1 s^4 + a_2 s^3 + a_3 s^2 + a_4 s + a_5 = 0$$

The coefficients are then set up in an array, without having assigned values to the controller settings. The array is then solved in a stepwise manner such that the algebraic sign of the terms in the left-hand column does not change; i.e., as the solution proceeds, values are assigned to the controller settings that will just prevent the signs of the left-hand column from changing. These values bring the system close to instability but will not produce instability. The procedure, however, is rather tedious for complex systems.

- *Phase and Gain Margin*

These techniques are based on the use of the open-loop Bode diagram for obtaining an approximate value for controller gain. The phase margin method uses the concept that the open-loop magnitude ratio should not exceed unity when the phase angle is in the range of −120 to −150°; i.e., when the phase margin (the approach to −180°) is 30 to 60°. A typical vaue of allowable phase margin is 45°, corresponding to a phase angle of −135°.

The gain margin method states that the reciprocal of the open-loop magnitude ratio should be in the range of 2.5 to 10 when the phase angle is −180°. A typical value of the gain margin is 5.

As an example of the application of these techniques, consider Figure 7-21, which is the Bode diagram that corresponds to the signal flow diagram in Figure 7-20, in which arbitrary values have been assigned to the gains and time constants. Only the dynamic terms are plotted, but note that the open-loop steady-state gain will be

$$K_o = (2.0)(2.0)(K_c)(5.0) = 20\,K_c$$

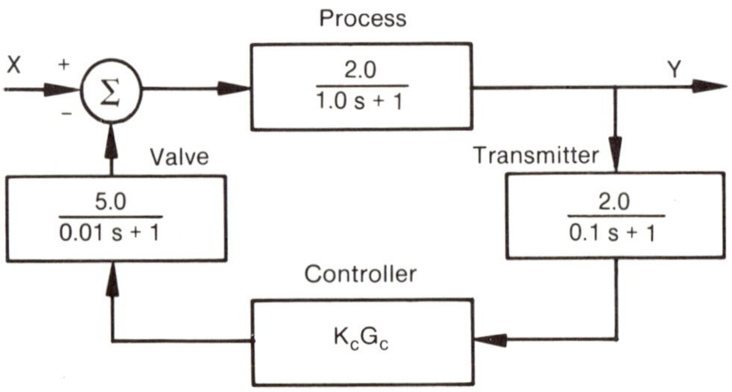

Figure 7-20. Phase and Gain Margin Signal Flow Diagram

Applying the phase margin technique, the magnitude ratio is 0.065 when the phase margin is 45°; i.e., when the phase angle is −135°. Since the allowable loop gain is unity at this phase angle, the controller gain K_c that will meet this condition can be calculated:

$$1.0 = 0.065\,(20\,K_c)$$

$$K_c = \frac{1}{(0.065)(20)} = \begin{array}{l}0.77 \text{ or } 130 \text{ percent}\\ \text{proportional band}\end{array}$$

Applying the gain margin technique, the magnitude ratio is 0.0073 when the phase angle is −180°. If the gain margin is 5 (i.e., the reciprocal of the open-loop magnitude ratio = 5), the controller gain K_c that meets this condition can be calculated:

$$5.0 = \frac{1}{(0.0073)(20\,K_c)}$$

$$K_c = \frac{1}{(5.0)(0.0073)(20)} = \begin{array}{l}1.37 \text{ or } 73 \text{ percent}\\ \text{proportional band}\end{array}$$

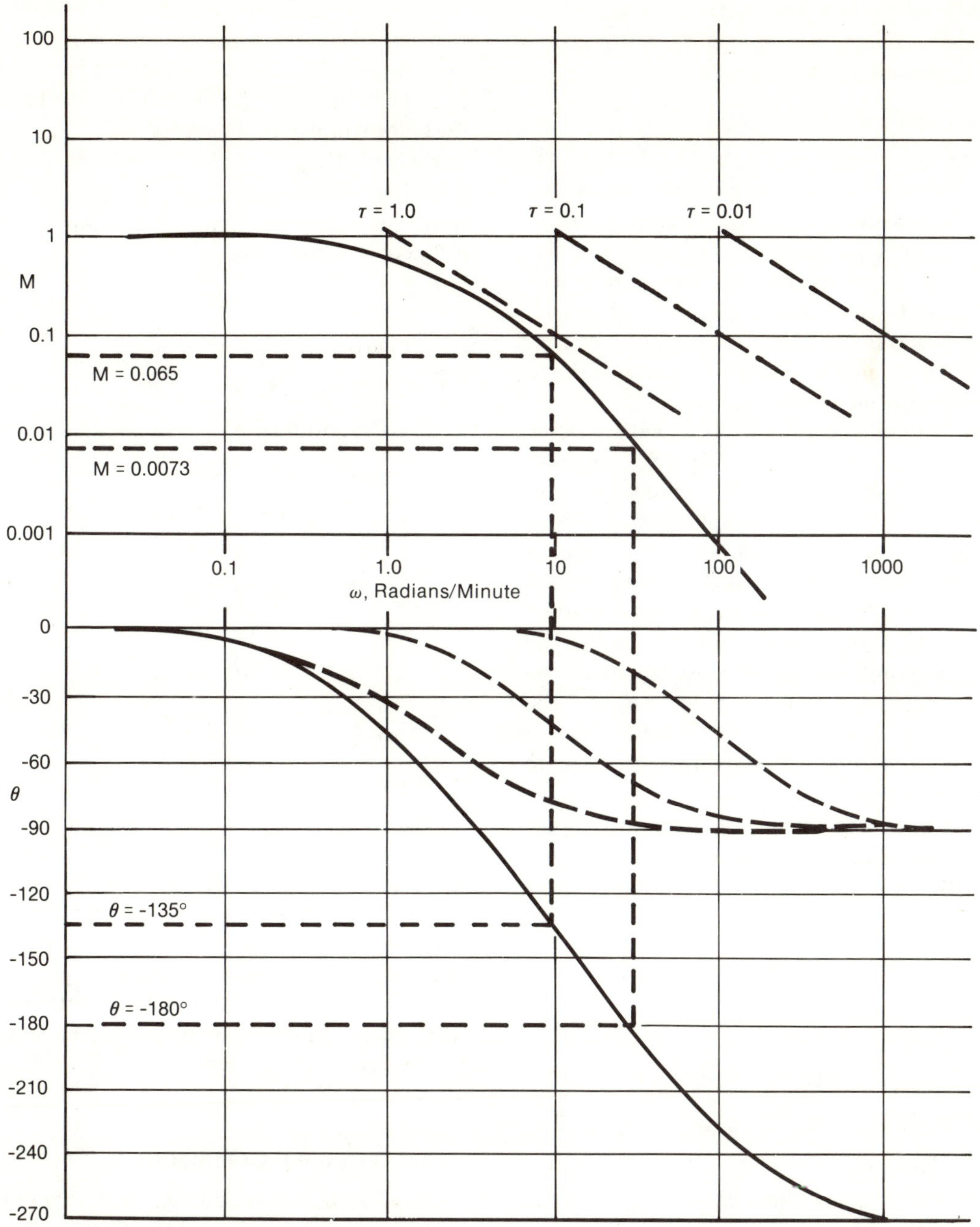

Figure 7-21. Bode Diagram for Phase and Gain Margin

Note that the values of K_c determined by the phase and gain margins differ appreciably although not greatly. Had a more conservative value of 10 been taken for the gain margin, K_c would have been 0.69. Or, had the less conservative phase margin of 30° been taken, K_c would have become 1.35.

Preliminary values for reset and derivative time constants can be estimated by setting the reset (integral) time constant equal to the largest time constant in the loop, and the rate (derivative) time constant equal to the second largest time constant in the loop.

7-8-2 Final Adjustment of Controller Settings

It is often necessary to retune controllers while the process is in operation. Such retuning may result from major sustained changes in throughput, ambient conditions, or operating conditions, which in turn cause changes in the overall loop gain. Again, there are a variety of methods for approaching optimum controller settings, of which two will be discussed here.

- *Ultimate-Sensitivity Method*

This method, developed by Ziegler and Nichols,[110] requires that the process be forced into a stable, continuous oscillation under closed-loop control. For this reason, its application is limited to processes that are reasonably stable in themselves and have a reasonably prompt response. The method would be risky when applied to high inertia, underdamped systems (such as turbines, compressors, etc.) and tedious when applied to very slow processes (large stirred kettles, distillation columns, kilns, etc.).

To apply the method, the controller reset and derivative actions are eliminated as far as possible (i.e., maximum reset time and minimum rate time). With the controller on automatic, the controller gain is gradually increased (proportional band decreased) until a sustained oscillation at constant amplitude is attained. At this point, the loop gain (magnitude ratio) must be unity and the phase lag must be 180°. The controller gain setting and the period of the oscillation (i.e., the time for one cycle) are recorded. The gain setting is called the "ultimate gain" (K_u) or the "ultimate sensitivity," and the period of oscillation is called the "ultimate period" (θ_u) and is usually expressed in minutes (as are reset times).

Controller settings that will give a slightly oscillatory response to a step change in input are then calculated as follows (using K_c as controller gain = $\dfrac{100}{\text{P.B.}}$):

(1) Proportional control only

$$K_c = 0.5\ K_u$$

(2) Proportional-plus-reset

$$K_c = 0.45\ K_u$$

$$\tau_i = (\theta_u/1.2) \text{ minutes}$$

(3) Proportional-plus-reset-plus-rate

$$K_c = 0.6\ K_u$$

$$\tau_i = (\theta_u/2) \text{ minutes}$$

$$\tau_d = (\theta_u/8) \text{ minutes}$$

- *Reaction-Curve Method*

This method, also developed by Ziegler and Nichols,[110, 112] uses the open-loop response of the system to a step change as a means of calculating controller settings. It is more amenable to operating processes as the steps can be made as small as possible consistent with the ability to read the resulting response on a recorder chart.

Consider the S-curve in Figure 22, which represents the response of a multiple-lag system to a step change in valve position. In general, the response curve can be approximated by a dead time (a) and a first-order lag, as shown by the tangent drawn at the steepest part of the response curve in Figure 7-22.

Using the values of a, r, and ΔF determined from a plot of the reaction curve, controller settings are calculated as follows (K_c = controller gain):

(1) Proportional control only

$$K_c = \Delta F/ar$$

(2) Proportional-plus-reset

$$K_c = 0.9\ \Delta F/ar$$

$$\tau_i = (a/0.3) \text{ minutes}$$

(3) Proportional-plus-reset-plus-rate

$$K_c = 1.5\ \Delta F/ar$$

$$\tau_i = (a/0.5) \text{ minutes}$$

$$\tau_d = (0.5a) \text{ minutes}$$

- *Popular Method*

What might be called the "popular method" is a simplified version of the Ziegler-Nichols ultimate-sensitivity method, which generally gives satisfactory settings for proportional-plus-reset controllers. In this method, reset action is eliminated insofar as possible (longest reset time), and the controller gain is increased (proportional band decreased) until a stable oscillation is attained. The gain is then halved, and the reset time is decreased until satisfactory response is achieved.

7-8-3 General Comments

It should be recognized that the term "optimum control performance" is subjective in nature. A given amount of overshoot on recovery from an upset may be acceptable in one system and intolerable in another. For example, a certain amount of overshoot (overpeaking) on recovery from an upset may be desirable in a continuous process as a means of reducing the recovery time. The same amount of overshoot may be disastrous on start-up of a batch operation.

Similarly, it may be desirable to control critical process variables within very narrow limits despite severe upsets. On the other hand, it may be desirable in the case of noncritical variables to allow considerable deviation in these variables as a means of absorbing and attenuating the upset, rather than passing it along to the next process unit. An example is the use of averaging level controls to prevent flow transients from being transmitted through a distillation train.

Processes and load changes tend to fall into broad categories, as shown in Table 7-2.

Process characterization has been discussed in the foregoing sections. Reference to Table 7-2 provides guidelines as to the controller actions that best control the specified process.

Finally, since people as well as processes must be dealt with, be prepared to occasionally encounter individuals who disagree as to the definition of optimum control performance. Such differences can generally be settled by negotiation — or coercion.

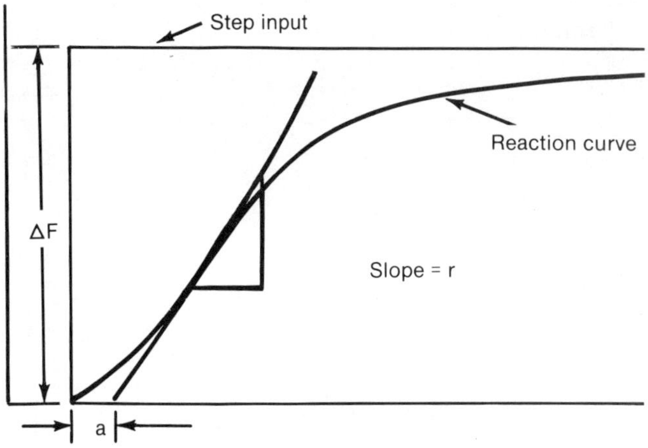

Figure 7-22. Process Reaction Curve

Table 7-2
Classification of Controller Actions for Process Characteristics[113]

Number of Process Capacities	Load Changes		Process Reaction Rate	Process Lag Times		Appropriate Control Action
	Size	Speed		Resistance-Capacity	Dead Time	
Single	Any Moderate	Any Slow	Slow	Moderate to large	Small	Two position. Two position with differential gap Multiposition. Time proportioning
Single (self-regulating)	Any	Slow Moderate	Fast	Small	Small	Floating actions: Single speed; multispeed Integral
Multiple	Small	Moderate	Slow to moderate	Moderate	Small	Proportional
Multiple	Small	Any	Moderate	Any	Small	Proportional plus derivative
Multiple	Large	Slow to moderate	Any	Any	Small to moderate	Proportional plus integral
Multiple	Large	Fast	Any	Any	Small	Proportional plus integral plus derivative
Any	Any	Any	Faster than that of the control system	Small or nearly zero	Small to moderate	Wideband proportional plus fast integral

(From Considine, Douglas M., 1957, *Process Instruments and Control Handbook*, McGraw-Hill Book Company, NY.)

Problem

1. Consider the process shown in the figure below.

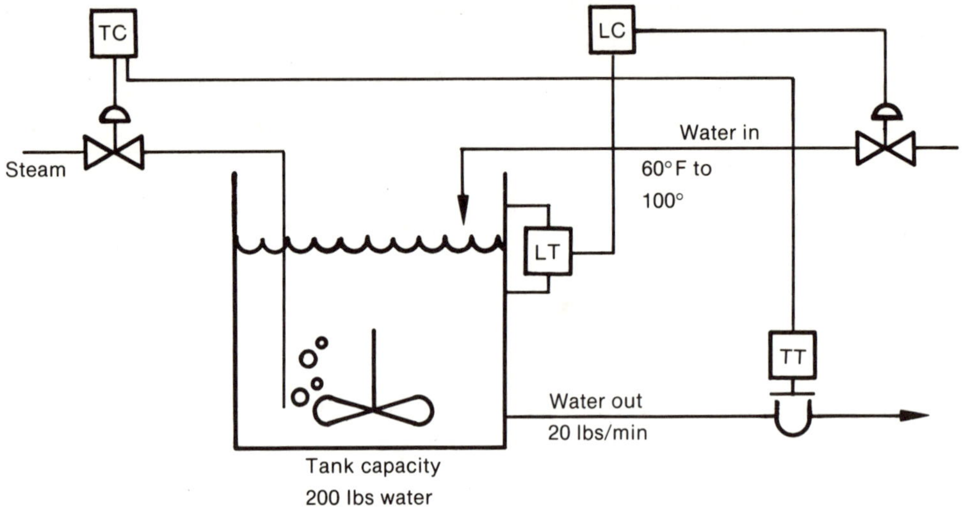

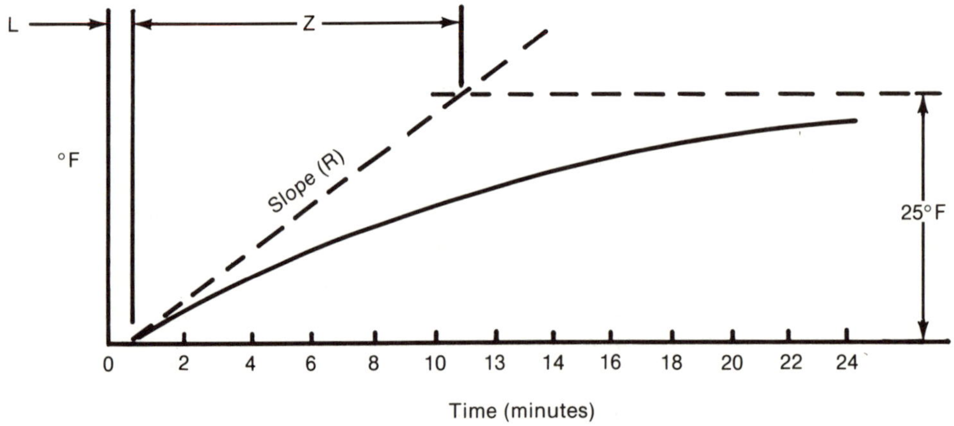

Cold water flowing into a tank is heated by steam injection. Water flows in at a constant rate and leaves through a pipeline to the sensor of a temperature controller located some distance away. The principal load change to the process comes from variations in the temperature of the incoming water. Assume the following data:

(a) Water in tank, 200 lb

(b) Water in line between tank and temperature sensor, 12 lb

(c) Water flow, 20 lb/min

(d) Steam flow (maximum), 6 lb/min

(e) Incoming water temperature, 60°F (minimum)
100°F (maximum)

(f) Outlet water temperature, 160°F

Calculate the temperature controller settings.

STUDENT SUMMARY NOTES and QUESTIONS FOR INSTRUCTOR

SECTION 8

DYNAMIC RESPONSE DETERMINATION BY EXPERIMENTAL TESTING

8-1 INTRODUCTION

Experimental testing is required (1) to verify the mathematical analysis in the foregoing sections; (2) to determine the transfer function of processes where the physical relationships are uncertain; and (3) where mathematical complexity makes testing more economical than analysis. In general, instrument dynamics are determined by testing, since they can be taken out of service and removed to a location where testing is convenient. Every effort is made to describe a process analytically, since it is a production unit that cannot be moved, and thus testing is extremely expensive in lost production and manpower to set up and record test data. The criterion for process testing is to select a procedure that minimizes curtailment of process operation and interference with normal production. The time spent in the production facility should be a minimum. This requires a wise selection of test procedure, extensive planning, and well versed participants in the test. Thus plant testing is quite expensive.[36]

On the other hand, the gains and time constants used in dynamic analysis have little requirement for precision. In fact, the transfer function in Figure 8-1 will describe just about any process to the degree of fidelity required for analysis.[35]

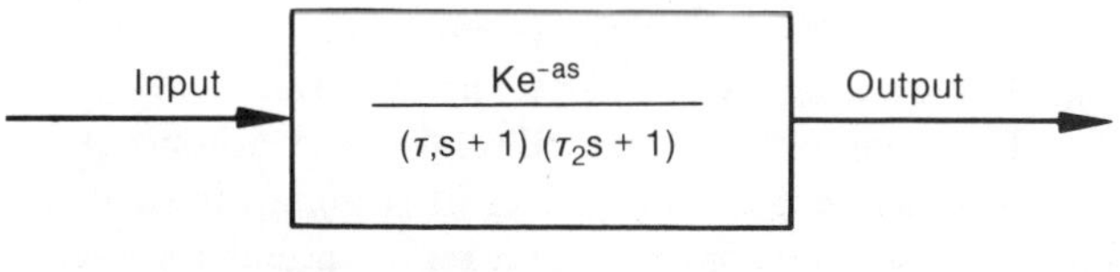

Figure 8-1. Process Transfer Function

Appendix C shows that the response can be underdamped and that the denominator (characteristic equation) cannot be factored as shown in the figure. However, a transfer function of the order shown in Figure 8-1 will almost always be adequate for the control of processes. Figure 8-2 gives the schematic representation.

Figure 8-2. General Transfer Function

Consider the equations that characterize a system's dynamics, as developed in the foregoing sections. A knowledge of $R(s)$ and $C(s)$ should allow determination of $G(s)$. This being true, the system's differential equation can be obtained. Thus an experimental test procedure is based on a measurement of the system's response to an input or forcing function, as shown in Figure 8-3.

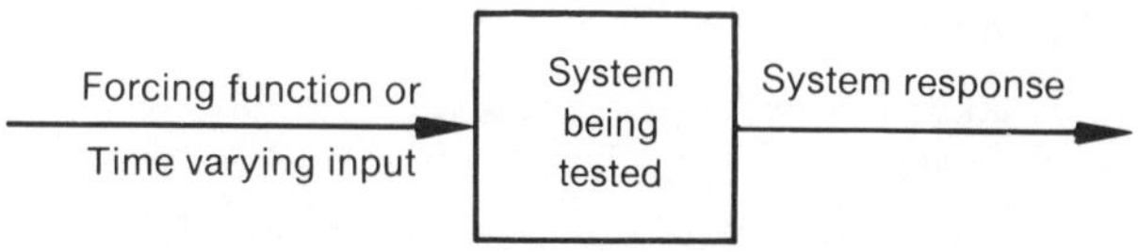

Figure 8-3. Testing Configuration[37]

All techniques for determining system dynamics may be conveniently subdivided into three general categories. These are based on the use of periodic inputs, nonperiodic (but forced) inputs, and random inputs. A number of techniques have been developed and can be used to evaluate the system dynamics, including:

- Frequency response
- Step testing
- Pulse testing
- Correlation
- Parameter evaluation

Correlation is a method of establishing the coherence or similarity of two signals as a function of the time delay between them. Frequently, correlations can be performed using only normal inputs and outputs of the process,[38] but a computer is required to analyze the results. Parameter evaluation will be discussed in Section 14 along with self-tuning regulators.

The other techniques are amenable to paper and pencil analysis. Each has advantages and disadvantages. In process analysis, the frequency response technique requires infinite patience; the pulse technique, great cleverness; and the step response technique, substantial courage. The best method in any given instance is dictated by the objective of the test, the equipment available, and the availability of the process unit.

8-2 FREQUENCY RESPONSE TESTING

Frequency response techniques call for the application of sinusoidal test signals and the observation of the gain-phase characteristics. Frequency response has two major uses:

(1) For the analysis, synthesis, design, and compensation of systems and components, as described in Section 7.

(2) For the experimental determination of characteristics and transfer functions to describe systems without relying entirely on the theoretical development of analytic equations.

Section 7-2 describes the use of an experimental Bode plot to determine the system time constant for a single-capacity (first-order) system. The following example, using experimental numbers, illustrates the success of this technique.[39]

Example

To make a frequency response analysis, it is necessary to study both magnitude ratio $\frac{(\Delta P_O)}{(\Delta P_I)}$ and phase shift for a number of different frequencies. Hence, a special piece of equipment is required to produce a sinusoidal pressure change. This unit is known as a pneumatic sine wave generator. The sinusoidal output pressure from the generator is obtained by using a synchronous motor and the gear train to drive a baffle through a cam. The moving baffle changes the nozzle-baffle relationship of a follow-up unit. The chart speed increases as the frequency increases so that one cycle always covers 3 inches on the chart. It can be adjusted through quick-change gears from 0.00229 to 44.35 cycles/min. The arrangement of the sine wave generator in the system for frequency response testing is shown in Figure 8-4. Phase shift and magnitude ratios for several frequencies are plotted to form a Bode diagram. This consists of plotting magnitude ratio versus frequency on log-log paper, and phase shift in degrees versus frequency in cycles/min on the semi-log paper (Figure 8-5).

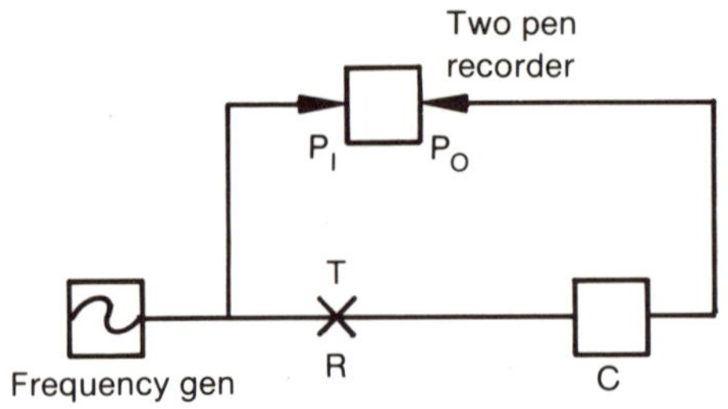

Figure 8-4. Single-Capacity System[39]

The other points on the curve (Figure 8-5) were determined experimentally. It can be noted that curve *A* of magnitude ratio versus frequency asymptotically approaches two lines: (1) a horizontal line and (2) a line with a slope of minus 1. These lines intersect at a corner frequency of 8.75 cycles/min. Note that at the corner frequency the phase lag is 45°, while at low frequencies it approaches 0° and at high frequencies it approaches 90°. This is a typical Bode plot of a single time constant.

There is a definite relationship between the value of the time constant and the corner frequency. The value of the resistance (R_2) for the curve equals .0595 psi/in.^{3}/min. at 9 psi. The capacity (C_2) equals .272 in.^{3}/psi. The time constant (τ_2) equals $R_2 C_2$ equals .0162 min. The relationship between corner frequency and the time constant is $f = \frac{1}{2\pi\tau}$. Hence, f = 9.84 cpm. This is located at X just to the right of the corner for the curve and checks well with the experimentally determined corner frequency of 8.75 cycles/min.

Frequency response testing can be extended to a multi-time constant (high-order) systems. Let the system be tested over a wide range of frequencies, as shown in Figure 8-7. The Bode plot is prepared, and the magnitude ratio plot is approximated by a series of straight lines having integer slopes that conform to the known attenuation rates of first- and second-order systems (Section 7).

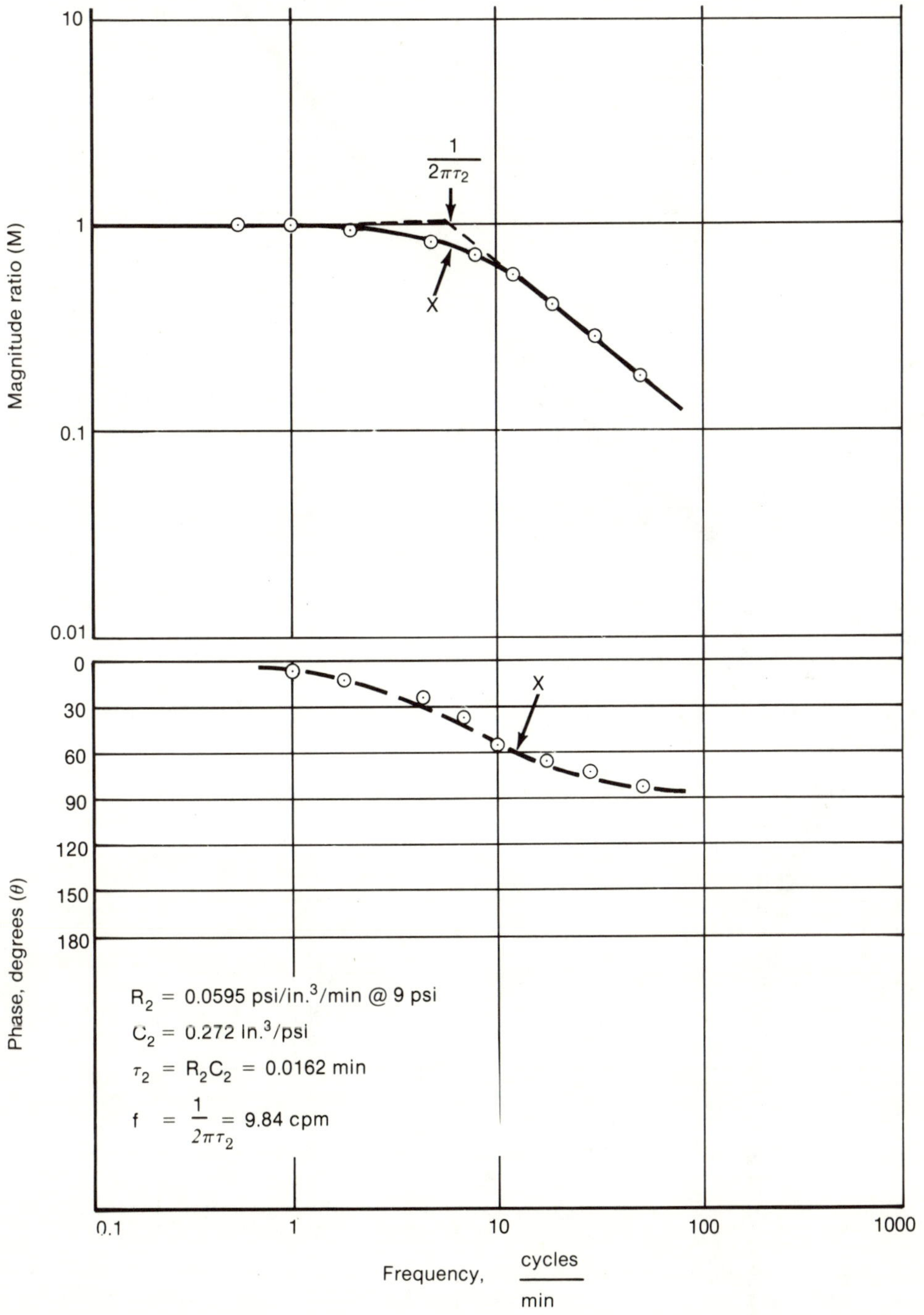

Figure 8-5. Frequency Response Analysis
(Courtesy Taylor Instrument Division)

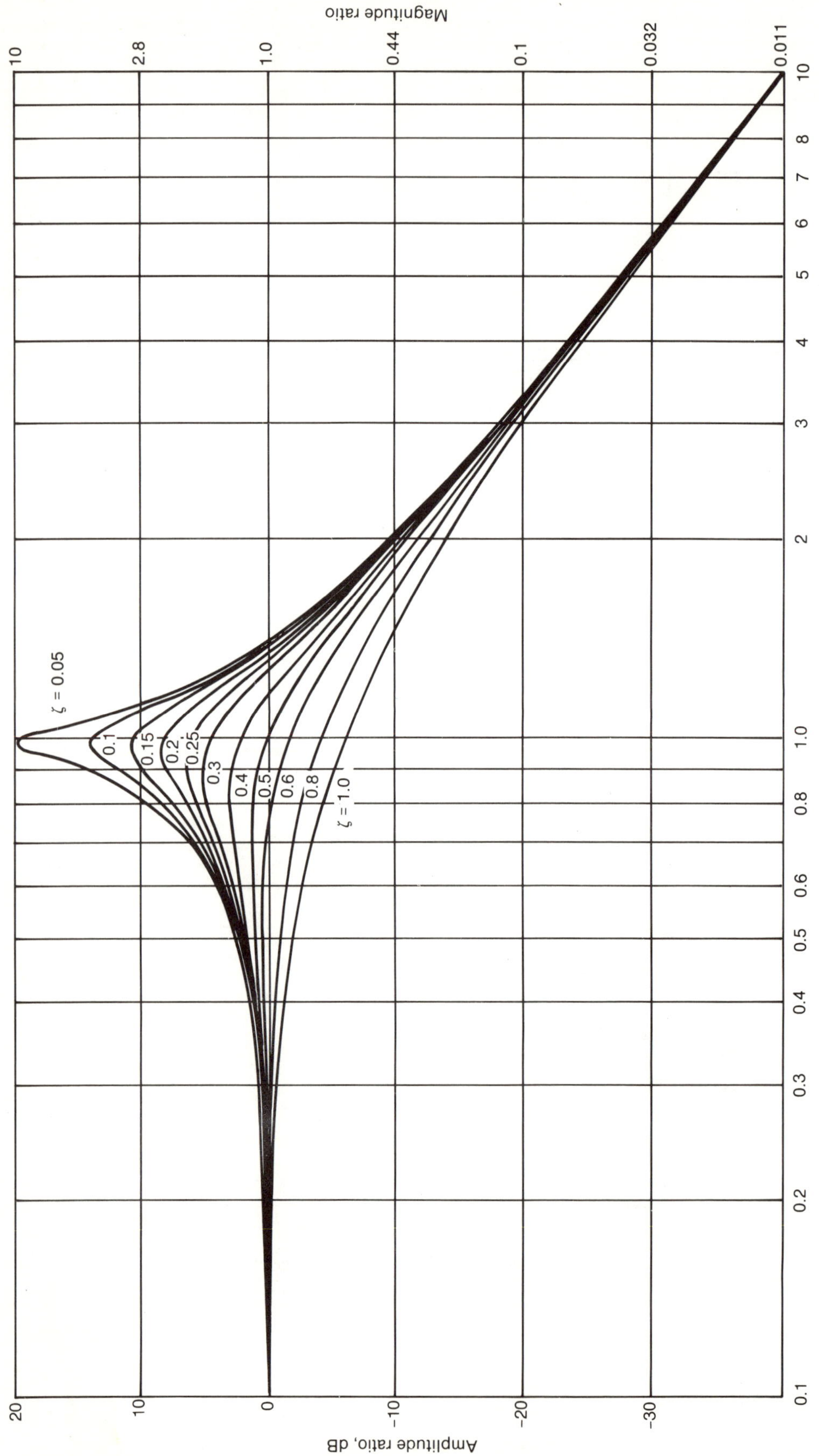

Correction to straight line asymptotes for magnitude ratio and phase on Bode plot for values of damping coefficient ζ of less than 1.

(a)

Magnitude ratio of $\frac{1}{1 + \zeta 2\tau s + \tau^2 s^2}$ for various values of $\zeta \leq 1$.

Figure 8-6. Amplitude Ratio

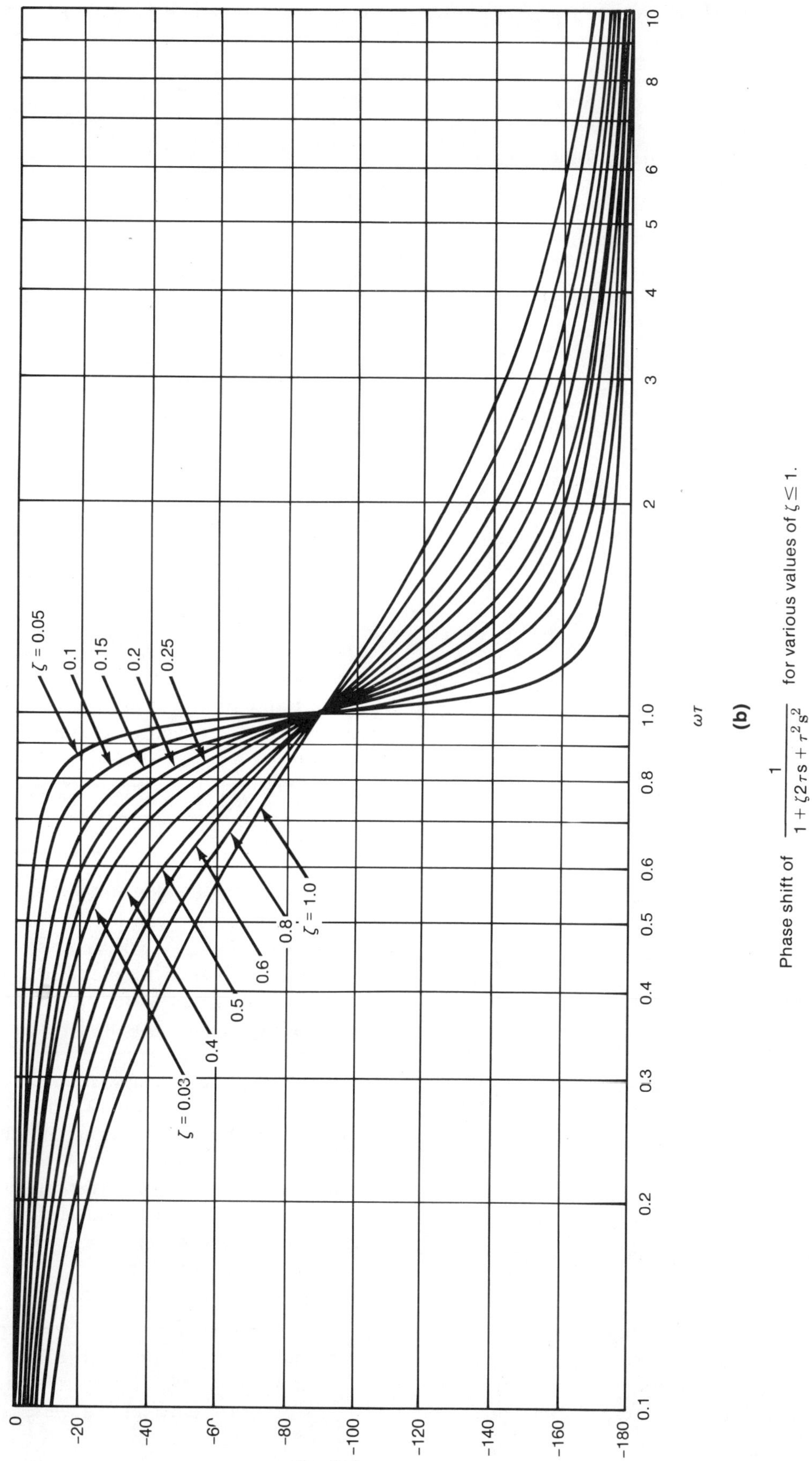

Figure 8-6. Amplitude Ratio (continued)

Starting from the low frequency end, the first slope is an integrator, described by the transfer function

$$\frac{1}{s}$$

Then, the point where slope changes from −1 to 0 must be a lead (derivative) term, with the frequency:

$\omega_1 = 0.1$ cps

$\tau_1 = 1.6$ sec

$(1.6s + 1)$

The slope changes from 0 to −2 at the next break point, typical of a second-order log term:

$\omega_2 = 1$ cps

$\tau_2 = 0.16$ sec

$\tau_2^2 = 0.025$

$$\left[\frac{1}{0.025s^2 + 0.16\zeta s + 1}\right]$$

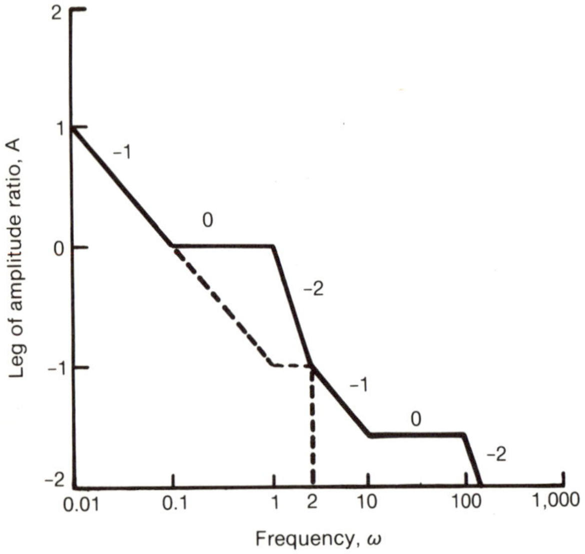

Figure 8-7. Straight Lines Approximate Data[40]

The factor ζ is a measure of system stability and is called a damping factor or damping coefficient. It is extensively discussed in Section 9, and the time response for several values of the coefficient are shown in Figure 9-5. That figure shows that systems with a damping coefficient of less than 1 become increasingly oscillatory. An oscillatory system will amplify at its natural frequency. This phenomenon is shown in terms of magnitude and phase on Figure 8-6a and b.

Thus, experimental points that appear above the asymptotic straight lines indicate underdamping ($\zeta < 1$), and the value of ζ for use in the analytical expression is determined by sketching a line through the experimental points and comparing the resulting curve to Figure 8-4. The phase relationship (Figure 8-6b) serves to support the magnitude ratio curve, since experimental points are noisy and scattered.

The next change in slope is from −2 to −1, indicating another derivative term:

$\omega_3 = 2$ cps

$\tau_3 = 0.08$ seconds

$(0.08s + 1)$

Change from −1 to 0 slope,

$\omega_4 = 20$ cps

$\tau_4 = 0.016$ sec.

$(0.016s + 1)$

Finally, change from 0 to −2,

$\omega_5 = 100$ cps

$\tau_5 = 0.0016$

$\tau_5^2 = 0.0000025$

$$\left[\frac{1}{0.0000025\, s^2 + 0.0016\, \zeta s + 1}\right]$$

where ζ can again be determined by experimental data and Figure 8-6.

The transfer function of the generalized system shown on Figure 8-7 after being tested, recorded, and analyzed is therefore

$$\text{input} \longrightarrow \boxed{\frac{K(1.6s + 1)\,(0.08s + 1)\,(0.016s + 1)}{(0.025s^2 + 0.16\zeta s + 1)(0.0000025s^2 + 0.0016\zeta s + 1)}} \longrightarrow \text{output}$$

The gain (K) is as yet unknown and might be more easily calculated analytically from known system parameters. The value of the gain has no effect on the foregoing result but merely shifts the plot vertically.

The foregoing transfer function could be simplified for use in the dynamic analysis of a typical process by ignoring the smaller time constants, as discussed in Section 7.

While frequency response testing is a valuable tool in the frequency range of 0.1 to 10 cycles per second (hertz), a number of problems arise when the technique is applied to a much slower reacting process. As an example, consider a process with a time constant of one hour. A sine wave with a period of 8 hours (a workman's shift) would have to be applied to find the magnitude ratio of 1 point on the Bode plot. The problems then become:

- The acquisition or design of a sine wave generator that will generate the required frequency.
- The necessity of protecting the process from extraneous disturbances for several shifts while Bode plots are taken at different frequencies.

- The unavailability of a production unit for several shifts.
- The cost of high level manpower to run the test.

For these reasons, frequency response is widely used to characterize the dynamics of instruments and rapidly responding processes but is seldom used for the typical large processing unit.

8-3 TRANSIENT RESPONSE TESTING — STEP ANALYSIS

For a less accurate test, but one requiring far less time, a step input may be used. As part of the background for the frequency response method, it was stated that any complicated wave form (provided it is single valued and periodic) may be written as a Fourier series of sinusoids of various amplitudes, frequencies, and phase angles. Moreover, the sharp corners of, for example, a square wave are represented by higher frequency harmonics. While a step change is not apparently periodic, it may be considered as half a square wave and therefore reducible to a Fourier series. Used as an input, a step essentially applies to the system a whole spectrum of frequencies simultaneously. Thus the response to a step, if properly interpreted, will give the complete Bode diagram for the system at a single shot.

The step input to a system can be a sudden increase or decrease in flow, pressure, firing rate, concentration, or any of a number of input variables that can be changed suddenly. The gain can be calculated immediately from a knowledge of the steady-state conditions both before and after the test, as in the following example.

Example

The static gain K of an element in a system is found by taking the ratio of the steady-state output to the input change, in the proper units. Steady-state is the condition achieved when all transient responses to the input step change have disappeared.

Note that this procedure is straightforward for any part of a system that reaches a constant steady-state value. In the case of an integrating or differentiating element, some care is necessary with units.

The response of a first-order system to a step input was shown previously to be

$$\theta = (1 - e^{-t/\tau})\, w$$

The signal flow diagram and the response curve are as shown in Figure 8-8.

Figure 8-9 shows the step response method to be extremely simple in application. Ideally, a step change is imposed by changing a controller set point suddenly, and then the time constant is read from the recorded response. In practice, it's not that simple. Consider the process discussed earlier in which a response was measured in hours. At least two variables (input and output) will be out of specification for several hours while the test is completed. This could quite possibly mean that off-specification product is being made during this period. The courage to run the test previously mentioned has to do with suggesting to an operating plant that it make off-specification material for several hours in order to do testing. With deference to this reservation on the part of operating personnel, it is

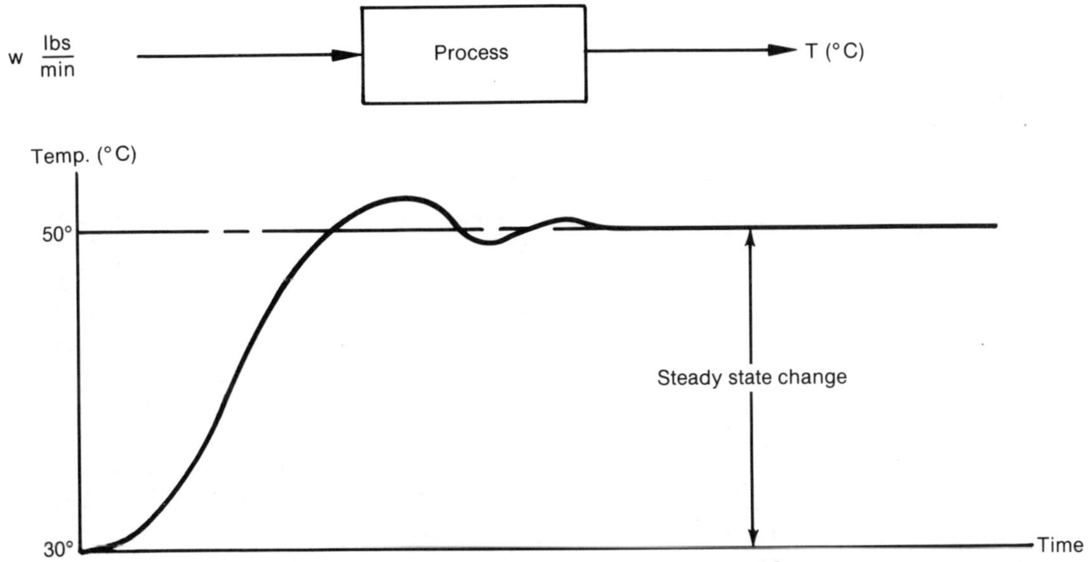

$$K = \frac{\Delta\ \text{output}}{\Delta\ \text{input}}\ SS = \frac{50-30}{6.2-5.0}\ \frac{°C}{\text{lbs/min}} = \frac{20}{1.2} = 16.7\ \frac{°C}{\text{lbs/min}}$$

Figure 8-8. Static Gain

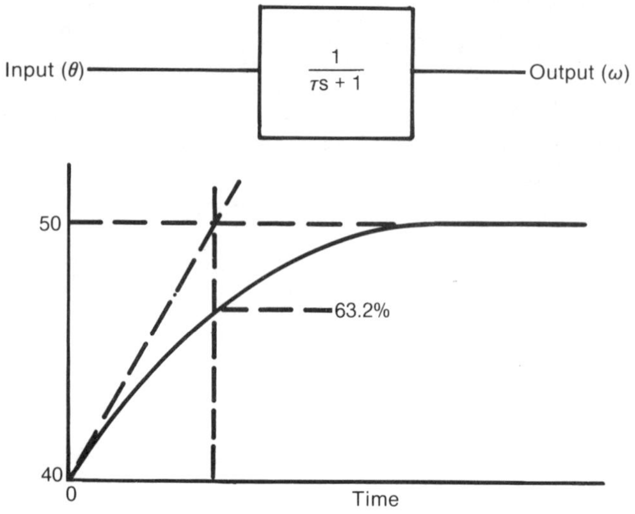

1. Curve is pure exponential.
2. Response starts immediately at time zero.
3. Initial slope is finite.

Figure 8-9. Step Response, Single-Capacity System

necessary that (1) the step disturbance be as small as possible, and (2) the test duration be as short as possible.

The desirability of small disturbances and short tests is thwarted by the noisy response of the output variable. Several step disturbances of increasing magnitude will almost always be tried before a discernable response is attained, thus increasing the length of the test. The noise further contributes to the length of the test by introducing uncertainty as to when the new steady state has been reached, thus necessitating a longer test to discern the trend.

The uncertainty of the response recording will sometimes require additional analysis in order to obtain maximum information from the data. The "percent incomplete" response plot on semi-logarithmic paper is useful.[42] Such a curve represents the percent of total rise that has not taken place at each point in time on Figure 8-9. Thus the line starts at 100 percent (zero response) and decreases downscale. The "incomplete" feature means that the following equation is being plotted:

$$f(t) = e^{-t/\tau}$$

This pure exponential plots to a straight line on the semi-logarithmic plot shown as Figure 8-10(c). Thus a straight line drawn among the noisy experimental points better defines the time constant than does the raw data. Since Figure 8-10 is a percent incomplete curve, the time constant is picked off the curve at the 37 percent (36.8 percent) point, as shown.

In order to determine the first time constant (τ_1), the curve is drawn by extending the linear portion of the reaction curve (the experimental points) back to the zero time axis. The intercept of the straight line at zero time is the first intercept, which can be called I. The time constant is the time it takes the straight line curve to reach 37 percent of I. The value of time constant τ_1 is found by determining point 0.37 I on the percent incomplete axis, drawing a horizontal straight line to the tangent curve in Figure 8-10(c), and reading the value on the time scale.

The percent incomplete technique also provides information as to whether the system being tested is first-order, in which case the experimental points will cluster around the straight line across the entire time span, or higher-order, which is indicated by the reaction curve (experimental points) bending away from the straight line as shown in Figure 8-10(d).

The second time constant (τ_2) is evaluated from the differential curve on Figure 8-10(d). This curve represents the numerical difference between the reaction curve and the tangent curve; at any time the ordinate of the differential curve is the difference between the ordinate of the tangent curve and the ordinate of the reaction curve. Extending the straight line that represents the differential curve to the zero time axis defines a second intercept. The second time constant is found at the 37 percent point on the differential curve, as shown in Figure 8-10(d).

The percent incomplete method provides a technique for evaluating the two largest time constants, which is an adequate description for the dynamic analysis of most processes, as shown in Figure 8-1. An alternative method for evaluating the time constants in Figure 8-1 is the Oldenbourg-Sartorius method,[43] which uses the response of a system to a unit step input in order to obtain the two non-interacting time constants.

The response of the second-order system to the step input is shown in Figure 8-11. The two terms T_A and T_c are obtained by drawing a tangent to the response curve through the point of inflection. A relationship between the two time-related parameters T_A and T_c and time constants τ_1 and τ_2 can be developed and plotted as Figure 8-12. The values of the ratio T_c/T_A are parallel straight lines that intersect the axis at 45° angles, where the value of the ratio is determined from the response curve as shown in the example on Figure 8-12. Note that the numerical value of T_c/T_A is the point on each axis that defines the ends of the 45° line. The method is limited to values of T_c/T_A of between 0.735 and 1.0. At a ratio of 1.0, one time constant is zero, and the system is first-order. At 0.735, the two time constants are equal. Below 0.735, the system is underdamped or interacting ($\zeta \leq 1$) and the method does not apply. The most difficult step in applying this method is the initial step of determining the tangent to the response curve at the point of inflection.

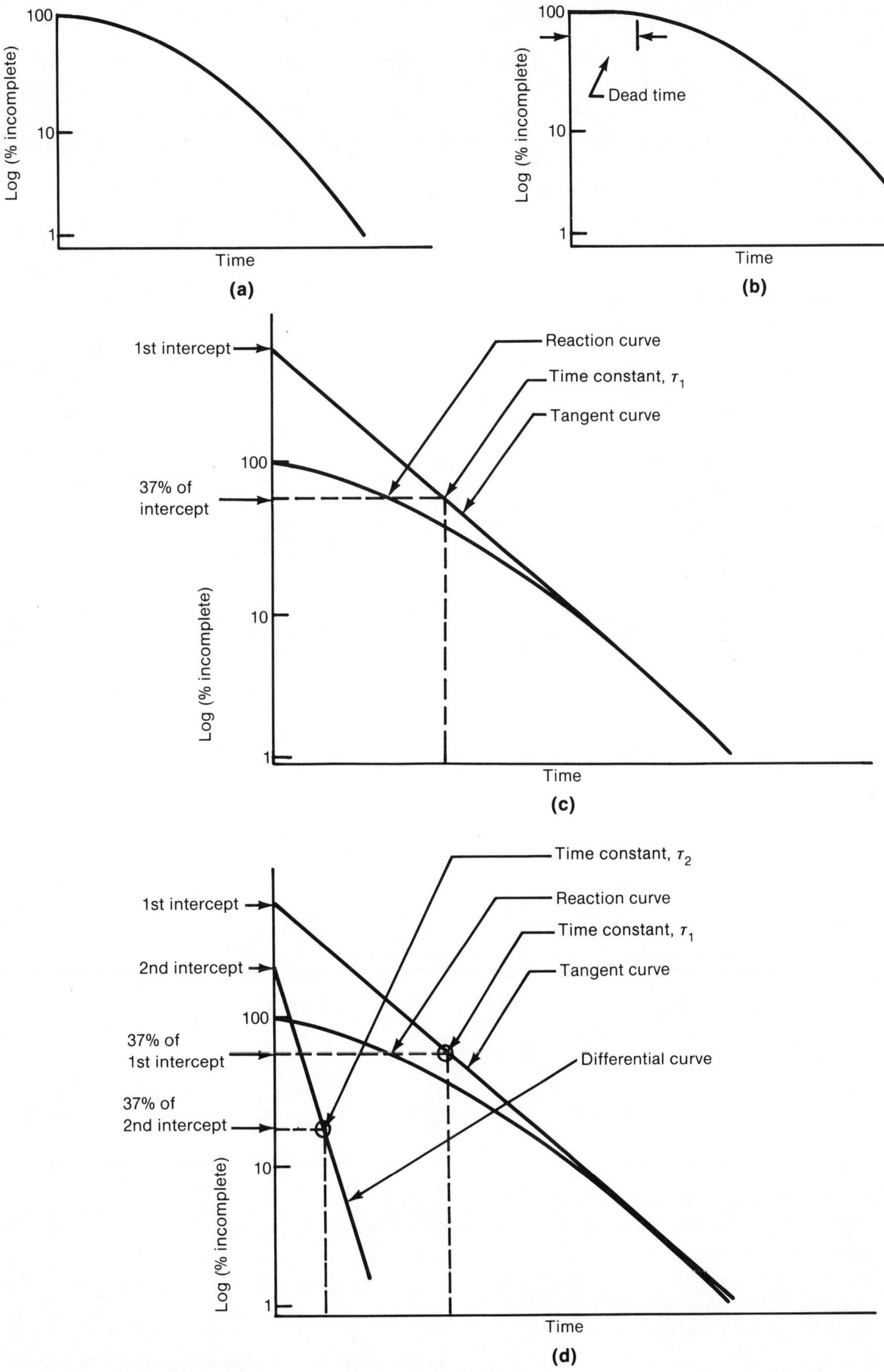

Figure 8-10. Non-Interacting Time Lags — Percent Incomplete Plot[41]

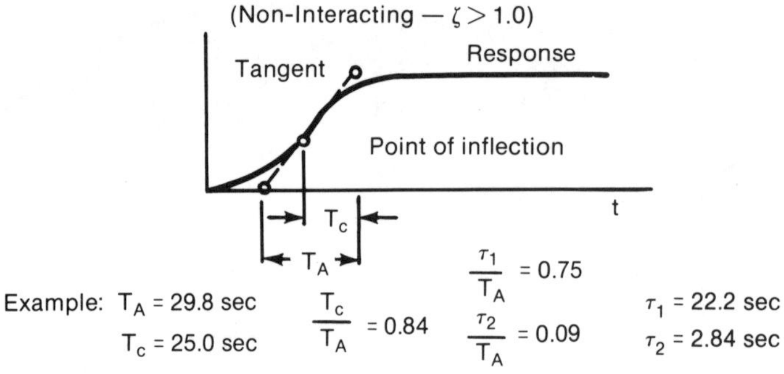

Figure 8-11. Response of Second-Order System

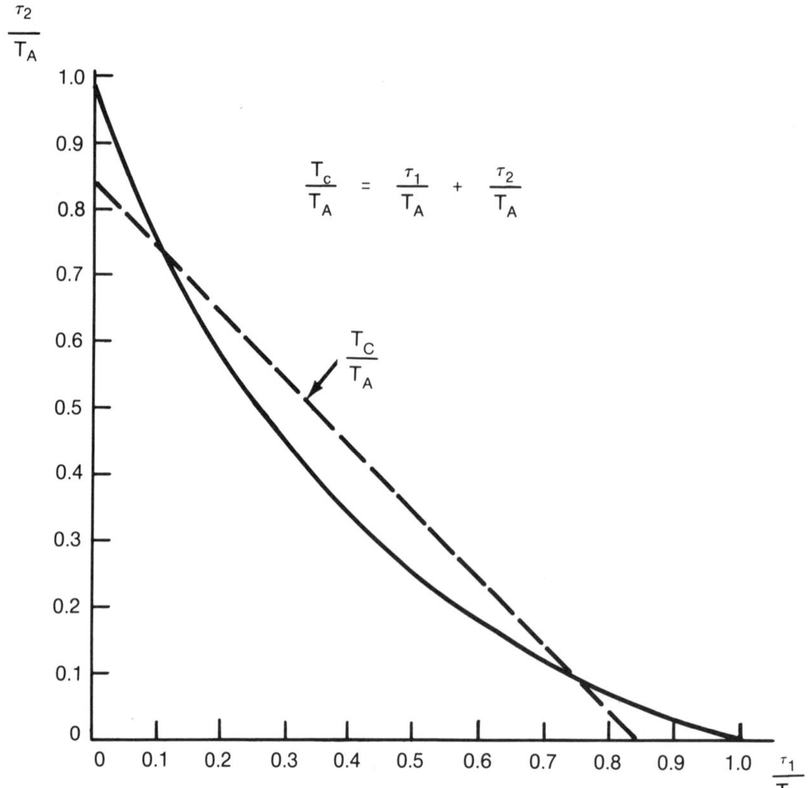

Figure 8-12. Determination of Two Equivalent Time Constants

In many applications, a very rough approximation of system dynamics is fully adequate for a dynamic analysis of the control system. Thus the simplest possible model is preferable, as shown in Figure 8-13.

Figure 8-9 shows that two simple methods exist for determining the first-order time constant (τ). One way is to take τ as the time at which the response has reached 63.2 percent of its final value. The other is to take the initial slope and extrapolate it to the final value of the response

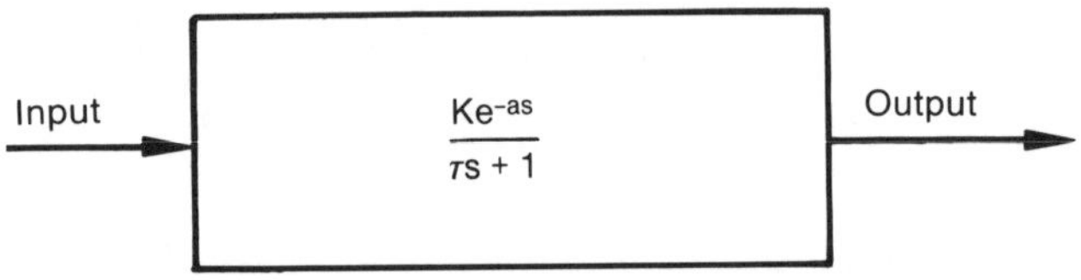

Figure 8-13. Simplified Process Dynamics Signal Flow Diagram

curve. The time at which the slope intercepts the final value is τ. Both methods are quite easy to apply. The first method usually gives better results since it avoids the difficulty of finding an initial slope. However, given the noisy data formerly discussed, the two methods can be used to verify each other.[44]

Figure 8-14 shows the retardation of the initial response as additional equal time constants are added in series. With a bit of imagination, it can be observed that the retardation is something like a dead time, with the first-order curve being moved to the right and superimposed on the curve for the correct order of the system. A remarkably good fit will be found for the experimental data from many higher-order systems. Figure 8-15 shows the technique for evaluating the dead time and time constant.

Following the technique of Figure 8-7, the slope is taken as a tangent at the inflection point of the "S-shaped" response curve. The intercept with the initial and final steady-state conditions serves as an approximation of the time constant. Finally, the intercept with the initial condition determines the value of the equivalent dead time.

The alternative technique is illustrated for a higher order curve in Figure 8-16. The tangent at the inflection point is again drawn, and the initial value intercept used to approximate the equivalent dead time. Then the 63.2 percent point of the total response is found, and the time constant is the time value of that point minus the dead time.[44]

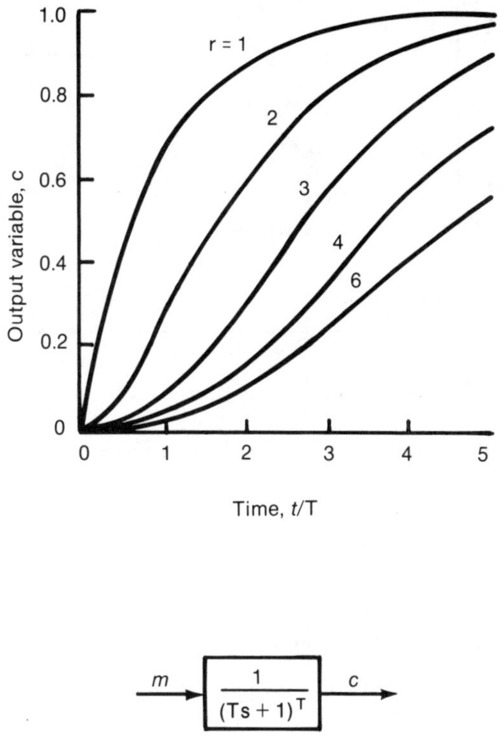

Figure 8-14. Response of Processes with Several Equal Time Constants in Series[45]

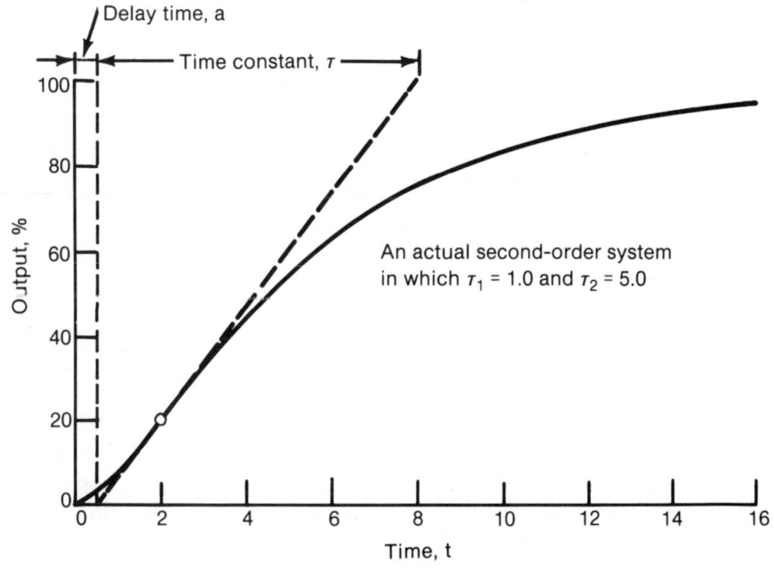

Maximum slope establishes the system delay time and the time constant from response curve.

Figure 8-15. Multi-Time Constant Approximation Method[44]

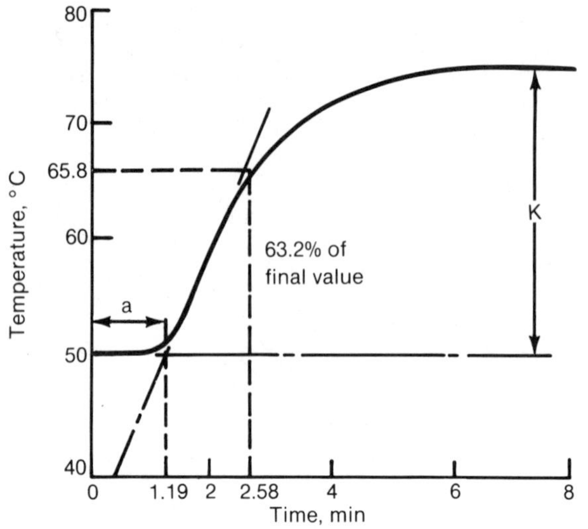

Unit step change in the output of the controller produces typical response curve for the temperature-control loop of a process.

Figure 8-16. Higher-Order System Approximation[44]

8-4 PULSE RESPONSE TESTING[46]

Pulse testing entails the imposition of a nonperiodic input of any general mathematical shape on the system. The nonperiodic input returns to its original value at the conclusion of the disturbance. The testing technique has the advantages of returning the process to its design condition and of being of short duration (less interruption of production). The difficulty is in the arbitrary nature of the pulse. It can be of any mathematical shape and of "appropriate" magnitude for a "proper" duration.[47] A poor choice of shape, magnitude, or duration can result in a need for additional tests. Thus a certain amount of cleverness is inherent in pulse testing.

The pulse testing method can be used to approximate the dead time-time constant model of the system as shown in Figure 8-13, or to extract the frequency response characteristics of the system. Both the input and output responses have frequency spectra that can be characterized by Fourier transforms, as discussed previously. The key to successfully applying the pulse test method to determine frequency response characteristics is the proper design of the pulse to excite the system. The general requirement is that the forcing pulse must contain all the frequencies at which frequency response measurement is desired, and it must contain sufficient energy to produce a measurable response in process output that is readily discernable from system noise. On the other hand, the pulse amplitude and its derivatives must not be so great as to drive the system into saturation or into a nonlinear region.[36,47,48,49] Triangular pulses are widely used in pulse testing to develop frequency response characteristics because they contain a relatively flat amplitude spectrum at low frequencies, while the phase angles vary linearly with frequency.

The Fourier transform is similar in form to the Laplace transform described in Section 3.

$$c(t) = G(p)r(t)$$

$$c(s) = G(s)f(s) \qquad (8\text{-}1)$$

$$G(s) = \frac{c(s)}{r(s)}$$

$$G(s) = \frac{\int_0^\infty c(t)e^{-st}\,dt}{\int_0^\infty r(t)e^{-st}\,dt} \qquad (8\text{-}2)$$

As previously,

$$s \longrightarrow j\omega$$

Since s may be replaced by $j\omega$, Equation 8-2 can be rewritten

$$G(j\omega) = \frac{\int_0^\infty c(t)\,e^{-j\omega t}\,dt}{\int_0^\infty r(t)\,e^{-j\omega t}\,dt} \qquad (8\text{-}3)$$

Equation 8-3 is one form of the Fourier integral and is a Fourier transform. The transform can be determined analytically if the functions $c(t)$ and $r(t)$ are known and are sufficiently simple. This is seldom the case. The transform is typically evaluated using numerical techniques on a computer. Each integration provides one point for the "gain versus frequency" and "phase versus frequency" plots. Another disadvantage is thus the voluminous data storage, manipulation, and computation that are needed to develop the Bode plot. The principal asset is the small number of brief tests necessary to complete a study.

Another type of pulse testing forcing function is the unit pulse, shown in Figure 8-7.

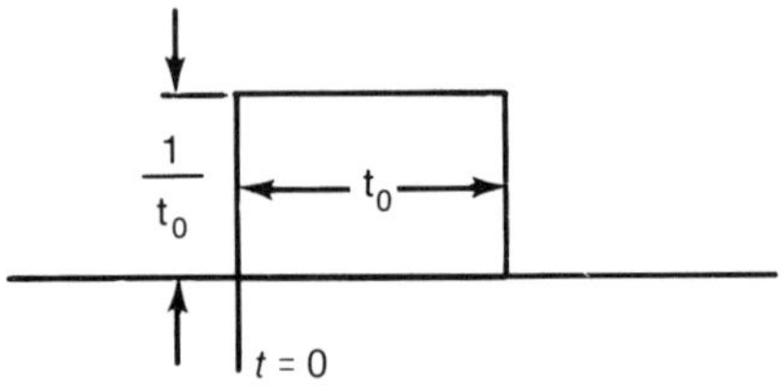

Figure 8-17. Unit Pulse[44]

The impulse function is a unit pulse in which the time t_0 has been reduced to zero. Thus, theoretically, it is zero units wide and infinity units high. The impulse has the advantage of having a Laplace transform of 1. In Equation 8-1, $r(s) = 1$ for an impulse,

$$c(s) = G(s) \qquad (8\text{-}4)$$

Equation 8-4 shows that the system response to the impulse is the system transfer function itself. This provides a very simple test to determine the equivalent dead time and time constant in Figure 8-13.

Admittedly, the theoretical impulse defies physical implementation. However, adequate approximations can be made in many testing situations. Consider the testing for the degree of mixing in an agitated vessel. Fluorescent dye is typically used for the testing, as shown in Figure 8-18.

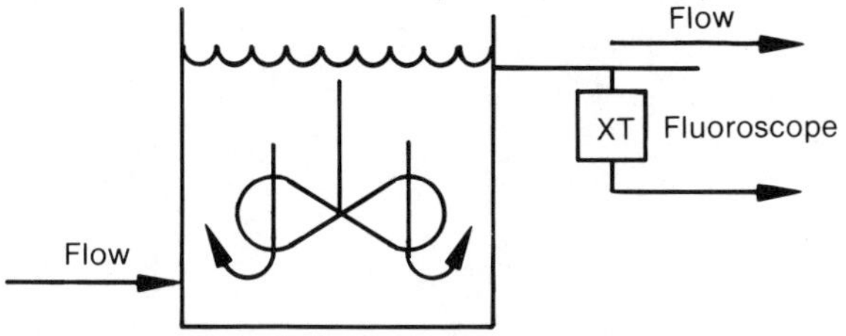

Figure 8-18. Tank Agitation Testing

A bottle of dye is dumped totally and as rapidly as possible down the shaft of the agitator. The fluoroscope is capable of analyzing a sample every few seconds, while the residence time of the tank is in minutes. The plotted chart of the experimental points is as shown in Figure 8-19.

Physically, the dye has been drawn down through the impeller, goes back up the sides of the tank in an umbrella effect, and reaches the overflow weir at time equals A in Figure 8-19. Thus the amount of time a is the apparent dead time of the process, or the dispersion time for the dye to go from entrance point to sensor. The dye then follows the internal circulation of the tank and mixes throughout the tank at point B. Then the sensor "sees" an exponentially decreasing concentration as the volumes containing dye leave the tank and are replaced by volumes of pure liquid.

The points from point B onward in time can then be plotted on semi-logarithmic graph paper with a result much like the percent incomplete plots in Figure 8-10. Since the exponential response is known to plot as a straight line on the semi-logarithmic graph, a straight line can be drawn that best represents the experimental data and can be extended back to the B time point as the first intercept. The time constant is then evaluated as the time for 36.8 percent of the response, as shown in Figure 8-20.

The impulse response test represents a simple and quick technique for obtaining a dead time-time constant model of the system.

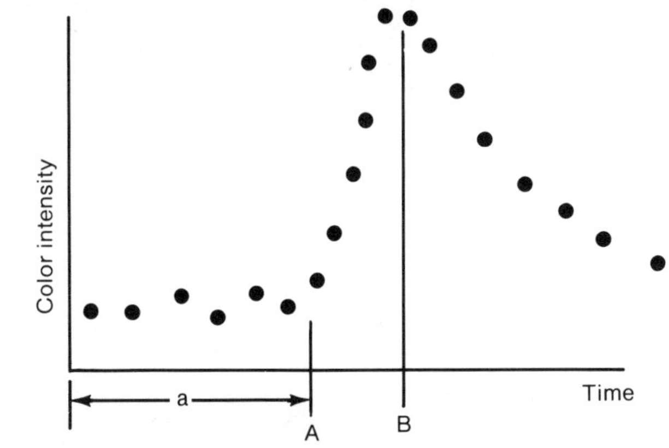

Figure 8-19. Impulse Response

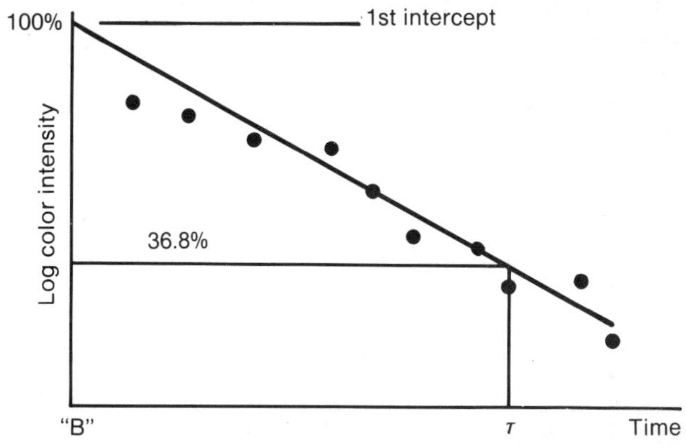

Figure 8-20. Time Constant From Impulse Test

Problems

1. Here's a set of experimental frequency response data.

Frequency, cps	0.3	0.45	0.6	0.75	0.9	1.2	1.5	2.3	3.	4.5
Input amplitude	1.	1.	1.	1.	1.	1.	1.	1.	1.	1.
Output amplitude	1.01	1.1	1.2	1.3	1.6	2.6	3.	1.1	0.45	0.1

Evaluate the time constant (τ) and damping coefficient (ζ).

2. Here's a set of experimental pulse test data.

Time, sec	5	10	15	20	25	30	35	40	45	50
Amplitude	10.	10.	18.	20.	19.	17.5	14.8	13.7	12.9	11.8

Evaluate the dead time (a) and the time constant (τ).

SECTION 9

SELECTION OF CONTROLLER MODES

9-1 INTRODUCTION

Dynamic analysis uses frequency response and Bode plots to design the control system, to choose system parameters, and to tune the controller. The design is largely an analytical procedure. The Bode plot will show the effectiveness that can be expected from each compensating mode that has been purchased in the controller. However, synthesis is required to determine which modes should be purchased in the controller.

The two-mode PI controller is purchased for an overwhelming majority of industrial applications. It is the optimum commercially available controller in a number of applications and is adequate for a large majority of others. However, other readily available controller modes are necessary in some 2 to 5 percent of the applications and are advantageous in perhaps 10 percent of other applications. For example, the PI controller is too slow for turbine governor speed control, inappropriate for level control, and inadequate for temperature control.

9-2 SYSTEM ANALYSIS

Feedback control concepts remain the same whether the control instrumentation be mechanical, hydraulic, pneumatic, electronic analog, or electronic digital. Each type of control "hardware" has its own advantages and disadvantages, and an analysis of the system will provide a reasonable choice of the most advantageous type of instruments. As a corollary, the same analysis will provide guidance to the desired control modes and an estimate of the adjustment values of those modes.

A "system" includes both the process and the instrumentation. The process can be anything from a locomotive (with its fuel injection servomechanism) to a home (with its heating system). In the petrochemical industries, processes are generally unit operations such as distillation columns, boilers, and reactors, even though the process can be as simple as a pipeline (for a flow control system) or a tank (for level control). An exothermic chemical reactor, shown in Figure 9-2, has been chosen as a specific example for illustration in these notes.

System analysis requires that each element of the system be reduced to a common mathematical form and be presented in a configuration that expedites combination and simplification. The block diagram fills this need. The block diagram is made up of lines (which represent directional information flow) and blocks (which represent static and dynamic mathematical operations on the flowing information). A generalized block can be represented as shown in Figure 9-1.

Figure 9-1. Generalized Block Diagram

The Law of Behaviour is made up of a static gain or sensitivity and a dynamic (time-varying) response characteristic in terms of Laplacian operators. It can be developed either analytically or experimentally. While transfer functions (laws of behaviour) for instruments can be developed analytically,[59] they are generally found experimentally and are also available from the manufacturers. Process transfer functions can also be determined experimentally,[46,60] but only with great difficulty. One takes his life in his hands when walking into the control room of an operating plant and announcing that he wants to upset the plant to see how it responds. Conversely, if the analysis concerns a plant that is being designed and is not yet built, the transfer function must be established analytically. The tools used for analysis are generally mass balances and heat balances. For example, the heat balance for the reactor shown in Figure 9-2 can be written:

Heat accumulated = Heat in − Heat out

$$Mc \frac{dT_2}{dt} = w_1cT_1 - w_2cT_2 + KT_2 - Q$$

where M = mass of material in reactor, lb

c = specific heat of reactant, Btu/lb °F

T = temperature, °F

t = time, minutes

w = mass flow rate, lb/min

K = reaction coefficient, Btu/min °F

Q = heat transfer, Btu/min

and where the heat transfer to the jacket water can be defined as:

$$Q = w_W C_W (T_{4W} - T_{3W})$$

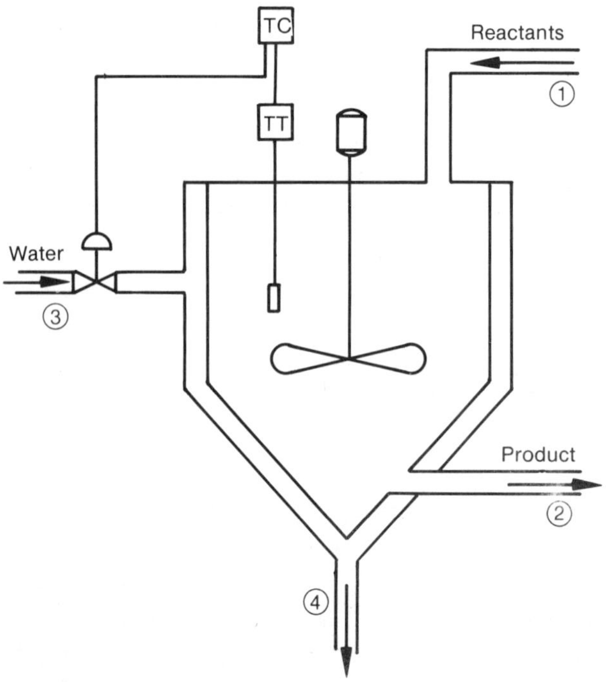

Figure 9-2. Water Jacketed Chemical Reactor with Temperature Control

The subscripted numbers refer to the line numbers on Figure 9-2, while the subscripted W refers to water.

9-3 POSITIVE FEEDBACK

Figure 9-3 shows the evolution of the block diagram for the reactor process. Simplification has been made by

$$\tau = \text{process time constant} = \frac{Mc}{w_2c - K} \text{ minutes}$$

$$K_p = \text{process gain} = \frac{1}{w_2c - K} \text{ min °F/Btu}$$

The minus sign in the process gain and time constant is worrisome, since it indicates a possibility that these terms will go negative and the system will go unstable. The tendency to instability is due to the exothermic nature of the chemical reaction. Exothermic reactions represent positive feedback, as shown in Figure 9-3(a). The positive feedback makes regulation very difficult when the reaction coefficient (K) is equal to or larger than the heat flow coefficient (w_2c). The positive feedback overcomes the self-regulation of the process and introduces the observed tendency for exothermic process to "run away." The strength of the positive feedback and its inherent control difficulties can be readily assessed from the block diagram gains. Also from Figure 9-3(a), should positive feedback predominate the process, all of the negative feedback for controllability must be provided by the control system modulating heat removal (Q).

9-4 REACTOR CONTROL SYSTEM

Figure 9-3(c) shows a complete block diagram for the reactor system shown in Figure 9-2. The process temperature (T_2) is measured by a temperature transmitter and transduced into a signal that can be transmitted to the controller. The controller algebraically calculates the difference (error) between measured variable and set point and operates to reduce the error through the gain and control modes. Controller output is a transmitted signal to the control valve that calls for a water flow (w_W). Water flow sets the heat removal (Q), which causes the process temperature to change as required for control.

The above reference to process temperature is indicative of the extreme simplification being introduced. Actually, "the" process temperature does not exist, but different temperatures are found in different parts of the reactor. The span of temperature difference is a function of the mixing in the reactor. Techniques exist for dividing the reactor into a number of segments, then calculating the temperature of each segment by individual heat balances. This technique requires a computer to monitor all of the calculations, however, while surprisingly reliable results can be obtained from the block diagram shown in Figure 9-3.

The dynamics of the temperature transmitter must also be developed by the same heat balance techniques already discussed applied to the temperature sensing bulb and thermowell.[61] The resulting time constant (τ_T) is shown in Appendix C to be:

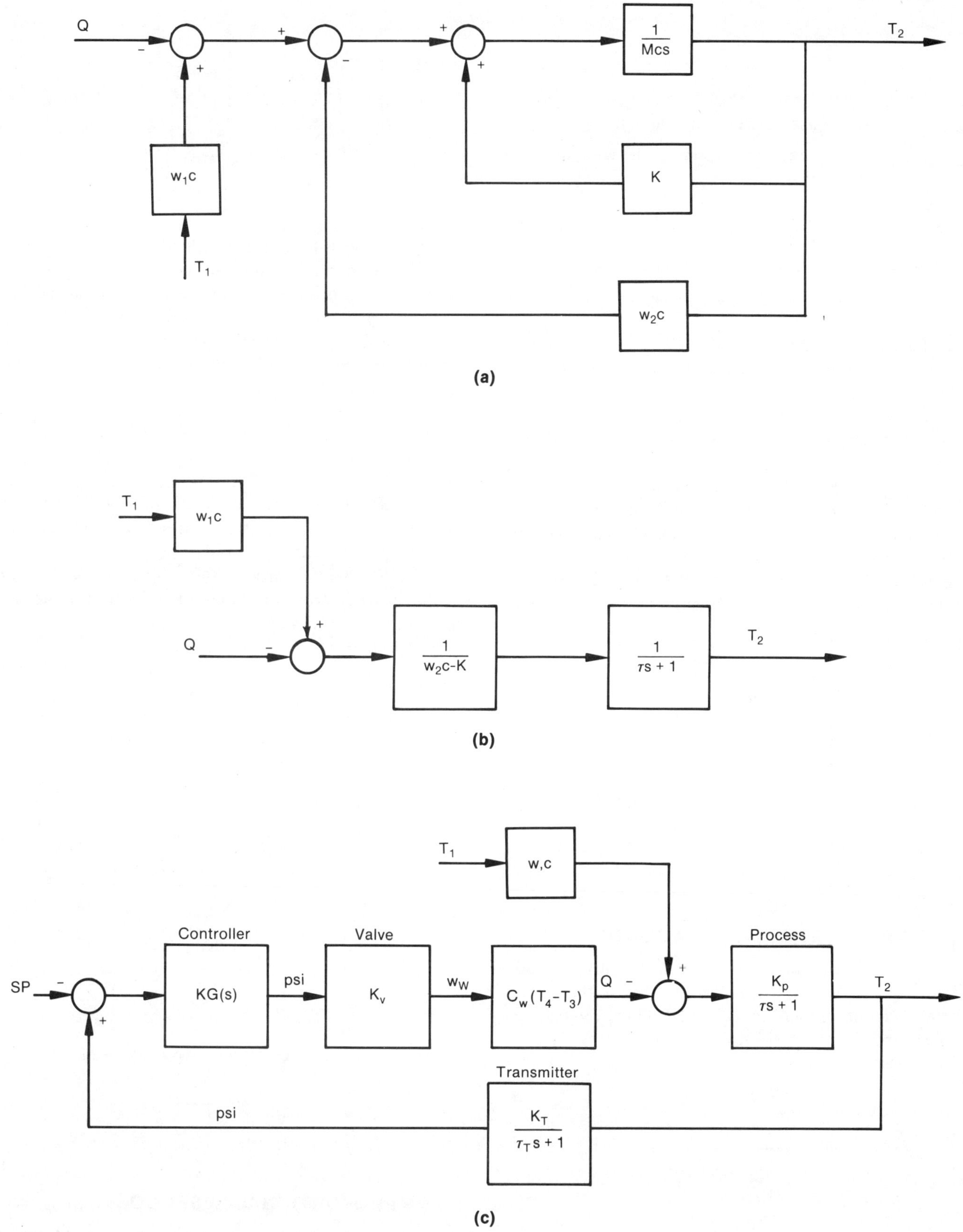

Figure 9-3. Combining the Block Diagram for a Reactor

$$\tau_T = \frac{Mc}{hA}$$

where M and c have been previously defined, and

A = surface area of the bulb, ft^2

h = film heat transfer coefficient, Btu/min °F ft^2

and

$$h = K_h (v)^{0.6}$$

where K_h = constant

v = fluid flow velocity past bulb, ft/min

Thus the bulb must be placed not only at a point of representative temperature, but also at a point of good velocity to obtain the temperature measurement.

Figure 9-4 shows the block diagram of Figure 9-3(c) entirely in terms of gains (K) and time constants (τ). The disturbance to the system has been chosen as incoming process temperature (T_1), while obviously any other variable in the foregoing equations could have been chosen. Block diagrams of the four different controller mode configurations in common usage in the petrochemical industry have also been shown in Figure 9-4. The following rule exists for the convenient simplification of the block diagram:

$$\Delta T_2 = \frac{\text{Feedforward portion}}{1 + \text{complete loop}}$$

Applying this to Figure 9-4:

$$\Delta T_2 = \frac{\dfrac{K_1 K_p}{\tau s + 1} \Delta T_1}{1 + \dfrac{KG(s)\, K_v K_p K_T}{(\tau s + 1)(\tau_T s + 1)}}$$

9-5 PROPORTIONAL-ONLY CONTROL

The question now arises as to which controller modes should be chosen for this application. By far, the simplest control mode is proportional-only (P in Figure 9-4). Inserting the mode in the block diagram, the foregoing equation becomes:

$$\Delta T_2 = \frac{\dfrac{K_1 K_p}{\tau s + 1} \Delta T_1}{1 + \dfrac{K K_v K_p K_T}{(\tau s + 1)(\tau_T s + 1)}}$$

Applying the final value theorem to the response to a step upset in T_1:

$$\Delta T_2 \Big|_{ss} = \lim_{s \to 0} \left[\frac{\dfrac{K_1 K_p}{\tau s + 1}}{1 + \dfrac{K K_v K_p K_T}{(\tau s + 1)(\tau_T s + 1)}} \right]$$

$$\Delta T_2 \Big|_{ss} = \frac{K_1 K_p}{1 + K K_v K_p K_T} \Delta T_1$$

Thus the process temperature will not return to set point after a change in inlet temperature (ΔT_1) but will assume an offset. The offset is undesirable in most applications, and thus proportional-only control is limited to a few special situations. It is sometimes used in temperature controllers where disturbances are small. The foregoing equation shows that the offset is small if the controller gain (K) is high or the proportional band is narrow.

This condition is generally met in pressure and temperature control. Thus, self-contained pressure regulators are also proportional-only devices. Section 2-12 discusses the self-contained pressure regulator, which has been analyzed extensively(65) to evaluate its dynamics and define the offset (or droop as it is called in a regulator).

The dynamic response of the proportional-only controller in the control loop in Figure 9-4 can be illustrated by multiplying out and comparing to a standard form that includes characteristic response time and damping:(62)

$$\Delta T_2 = \frac{K_1 K_p (\tau s + 1)(\tau_T s + 1)\, \Delta T_1}{(\tau s + 1)\left[\tau \tau_T s^2 + (\tau + \tau_T)s + (1 + K_o)\right]}$$

(where K_o = open loop gain = $K K_v K_p K_T$)

$$\Delta T_2 = \left[\frac{K_1 K_p}{1 + K_o}\right] \left[\frac{\tau_T s + 1}{\left(\dfrac{\tau \tau_T}{1 + K_o}\right) s^2 + \left(\dfrac{\tau + \tau_T}{1 + K_o}\right) s + 1}\right] \Delta T_1$$

Let

$$\tau_c^2 = \frac{\tau \tau_T}{1 + K_o} = \text{characteristic response time}$$

$$2 \tau_c \zeta = \frac{\tau + \tau_T}{1 + K_o}$$

where ζ = damping coefficient (see Figure 9-5).

Then,

$$\zeta = \left(\frac{\tau + \tau_T}{2}\right) \sqrt{\frac{1}{(1 + K_o)\tau \tau_T}}$$

9-6 PROPORTIONAL-RESET CONTROL

Integrating (reset) action is almost always required in process controllers to eliminate the offset, despite the

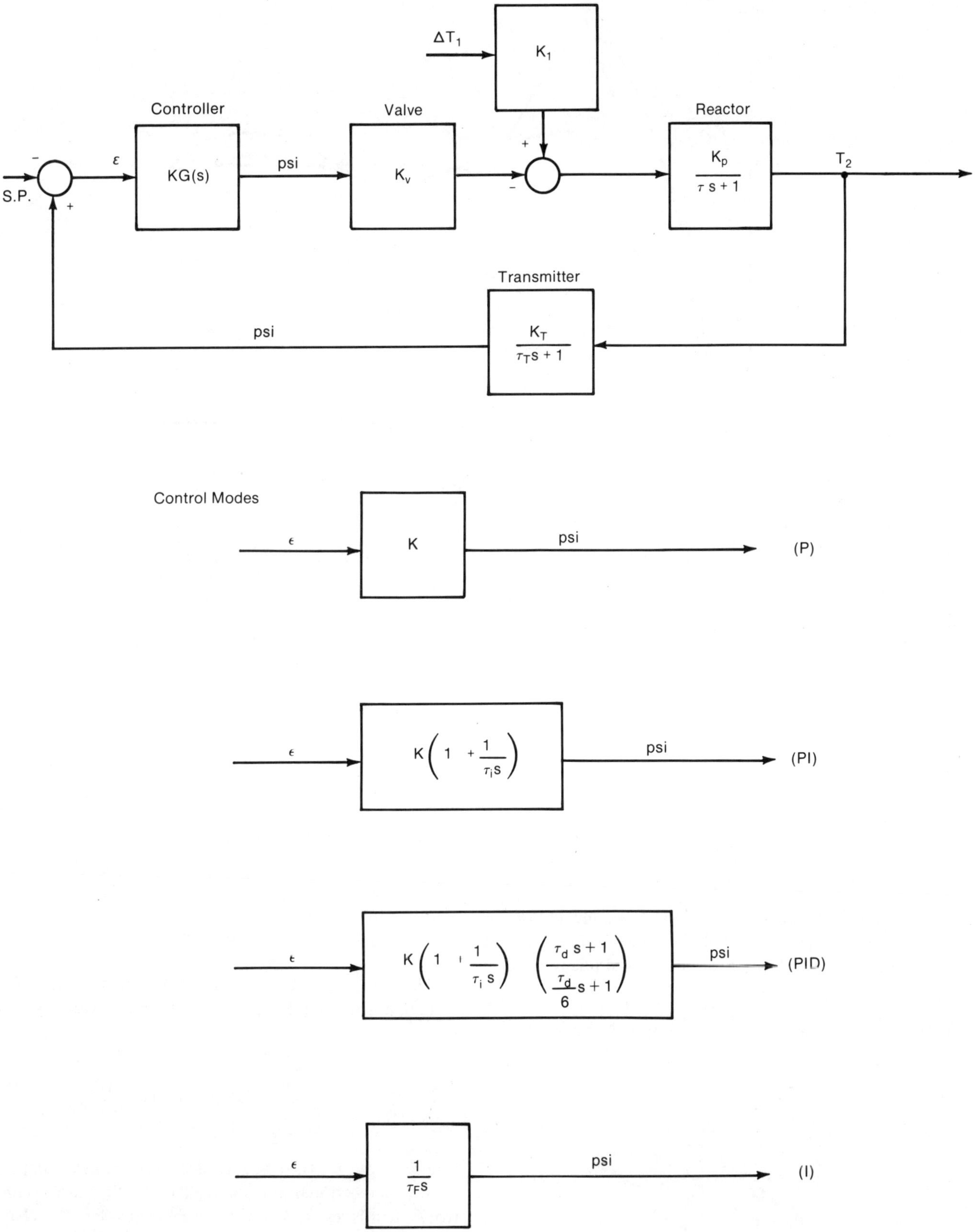

Figure 9-4. Commercially Available Controller Modes

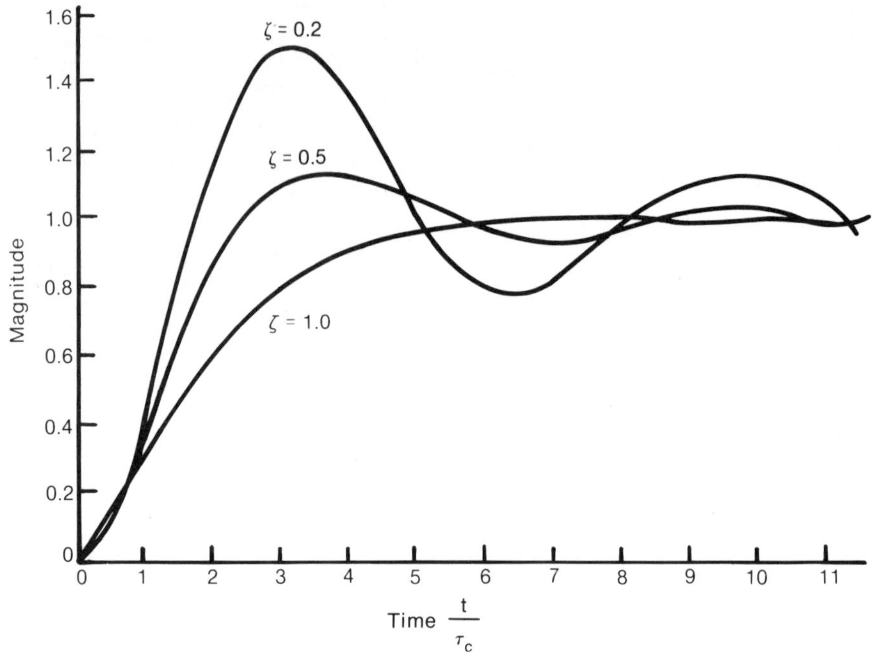

Figure 9-5. Response of a Second-Order System to a Step Input

significantly greater system instability incurred by the reset mode, as shown in Figure 9-6. Application of the final value theorem to the above closed-loop equation with proportional-reset (PI) control modes results in:

$$\Delta T_2 \Bigg|_{ss} = \lim_{s \to 0} \left[\frac{\dfrac{K_1 K_p}{\tau s + 1}}{1 + \dfrac{K K_v K_p K_T (\tau_i s + 1)}{\tau_i s (\tau s + 1)(\tau_T s + 1)}} \right]$$

$$= 0$$

The PI controller eliminates the offset, and, in addition, the controller modes are chosen to compensate for the dynamics of the control loop. Consider this compensation:

Let $\tau_i = \tau$

Then the closed-loop transfer function with the PI controller becomes

$$\Delta T_2 = \frac{\dfrac{K_1 K_p}{\tau s + 1} \Delta T_1}{1 + \dfrac{K K_v K_p K_T}{\tau s (\tau_T s + 1)}}$$

Reducing this algebraically as was done with the transfer function with the proportional-only controller:

$$\Delta T_2 = \frac{K_1 K_p (\tau s)(\tau_T s + 1)}{(\tau s + 1)(\tau \tau_T s^2 + \tau s + K K_v K_p K_T)} \Delta T_1$$

$$= \left[\frac{K_1 K_p}{K_o} \right] \frac{\tau_s (\tau_T s + 1)}{\left[\dfrac{\tau \tau_T}{K_o} s^2 + \dfrac{\tau}{K_o} s + 1 \right] (\tau s + 1)} \Delta T_1$$

where

$$\tau_c^2 = \frac{\tau \tau_T}{K_o}$$

$$\zeta = \left(\frac{\tau}{2} \right) \sqrt{\frac{1}{K_o \tau \tau_T}}$$

The damping factor (ζ) is a measure of stability of the system, as shown in Figure 9-5. The limiting values are as follows:

$\zeta = 1$: system is an oscillator

$\zeta = 1$: critically damped system

$\zeta > 1$: overdamped system

The value of ζ for the PI controller is usually less than that for the proportional-only controller, and the reactor control system is less stable with the PI controller. The characteristic response time (τ_c) is smaller for the proportional-only controller, and slower response is expected from the

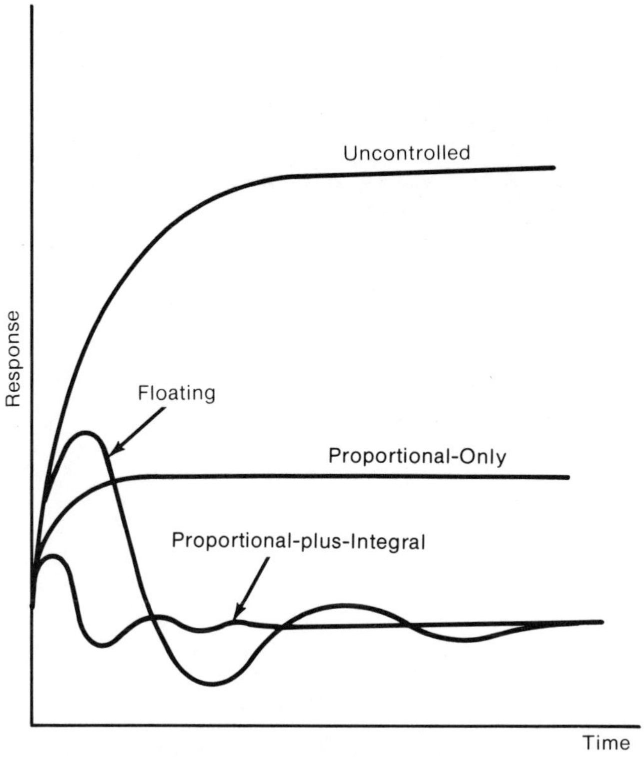

Figure 9-6. Response of a System to a Step Disturbance with Controller Modes as Shown[64]

system with the proportional-reset controller. Basically, the controller modes are chosen to compensate for process and instruments dynamics.

9-7 PROPORTIONAL-RESET-DERIVATIVE CONTROL

Apply the PID controller to the block diagram on Figure 9-4:

$$\Delta T_2 = \frac{\dfrac{K_1 K_p}{\tau s + 1} \Delta T_1}{1 + \dfrac{K K_v K_p K_T (\tau_i s + 1)(\tau_d s + 1)}{\tau_i s (\tau s + 1)(\tau_T s + 1)}}$$

and let $\tau_i = \tau$

$\tau_d = \tau_T$

Then,

$$\Delta T_2 = \frac{\dfrac{K_1 K_p}{\tau s + 1} \Delta T_1}{1 + \dfrac{K K_v K_p K_T}{\tau s}}$$

But, using the more complete transfer function for the derivative control mode:

$$\text{Derivative mode} = \frac{\tau_d \quad s + 1}{\dfrac{\tau_d}{6} \quad s + 1}$$

Then,

$$\Delta T_2 = \frac{\dfrac{K_1 K_p}{\tau s + 1} \Delta T_1}{1 + \dfrac{K_o}{\tau s \left(\dfrac{\tau_d}{6} \; s + 1\right)}}$$

Converting to polynomials algebraically,

$$\Delta T_2 = \frac{K_1 K_p (\tau s) \left(\dfrac{\tau_T}{6} \; s + 1\right)}{(\tau s + 1) \left(\dfrac{\tau \tau_T}{6} \; s^2 + \tau s + K_o\right)} \Delta T_1$$

$$\text{where} \quad \tau_c^2 = \frac{\tau \tau_T}{6 K_o}$$

$$\zeta = \frac{1}{2} \sqrt{\frac{6 \tau}{K_o \tau_T}}$$

Thus the derivative mode has decreased response time and increased stability. The PID controller is the best choice for this application. This conclusion was intuitively evident from the original closed-loop transfer function, since each dynamic term in the system was compensated for by a controller mode. Developing a block diagram not only specifies the controller modes but also provides estimates of the adjustment of the dynamic terms.

The foregoing result can be generalized further. The derivative mode is useful only when a second time constant appears in the control loop. Since most control valves respond so quickly that they contribute no significant dynamics, the second time constant will usually be in the measurement. But most measurements are also very fast (time constants of a second or less). In general, temperature is the exception because of the necessity for heat transfer to make the measurement. Many large purchasers of instruments have established a general policy of PI controllers for all applications except temperature, where PID controllers are specified.

9-8 SYSTEMS WITH DEAD TIME

It should be emphasized that the foregoing discussion pertains to measurement time constants, an exponential response to a disturbance. It does not apply to measurements with dead time (e.g., chromatographs, which are a translation of the measurement signal in time, during which there is no response). Since the derivative mode is by definition an advancing of the signal in time, it would seem to be helpful with dead time. However, derivative also contributes high frequency gain, and the increasing gain with strong derivative action is more detrimental than the advantage of advancing the signal.

9-9 FLOATING (INTEGRAL-ONLY) CONTROL

The final control mode, floating control (I), has been shown to be optimum with dead time systems[63] and also performs well with systems with very short process time constants (flow, pressure). However, adding this control mode to the block diagram in Figure 9-4 results in:

$$\Delta T_2 = \frac{\dfrac{K_1 K_p}{\tau s + 1} \Delta T_1}{1 + \dfrac{K K_v K_p K_T}{\tau_F s (\tau s + 1)(\tau_T s + 1)}}$$

The floating controller does not have the dynamic functions to compensate for the reactor and measurement dynamics. To simplify, assume that the measurement dynamics (τ_T) are negligible. The similarity to the PI controller assures that there will be no offset. However, multiplying out:

$$\Delta T_2 = \frac{K_1 K_p \tau_F s}{\tau_F s (\tau s + 1) + K_v K_p K_T} \Delta T_1$$

$$= \left[\frac{K_1 K_p}{K_v K_p K_T}\right] \left[\frac{\tau_F s}{\dfrac{\tau_F \tau}{K_v K_p K_T} s^2 + \dfrac{\tau_F}{K_v K_p K_T} s + 1}\right] \Delta T_1$$

where $\dfrac{\tau_F \tau}{K_v K_p K_T} = \tau_c^2$

$$\frac{1}{2} \sqrt{\frac{\tau_F}{\tau K_v K_p K_T}} = \zeta$$

Comparing this characteristic response and damping to that of the PID controller, the response time will be longer because of the larger value of the τ_F necessary to stabilize. Stability will be about the same, as shown in Figure 9-6.

The concepts of characteristic time (τ_c) and damping (ζ) permit estimates and comparisons of system response times and stabilities. However, the difficult question still exists as to the expected variation in the controlled variable (process temperature T_2) with a well chosen and well adjusted control system. Zoss[64] has provided the graph shown as Figure 9-7, which permits rapid estimation of the time to reach the new set point and the amount of overshoot with the knowledge of τ_c and ζ.

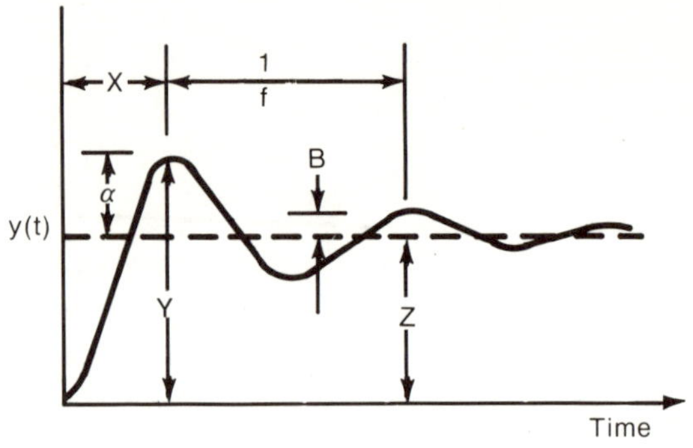

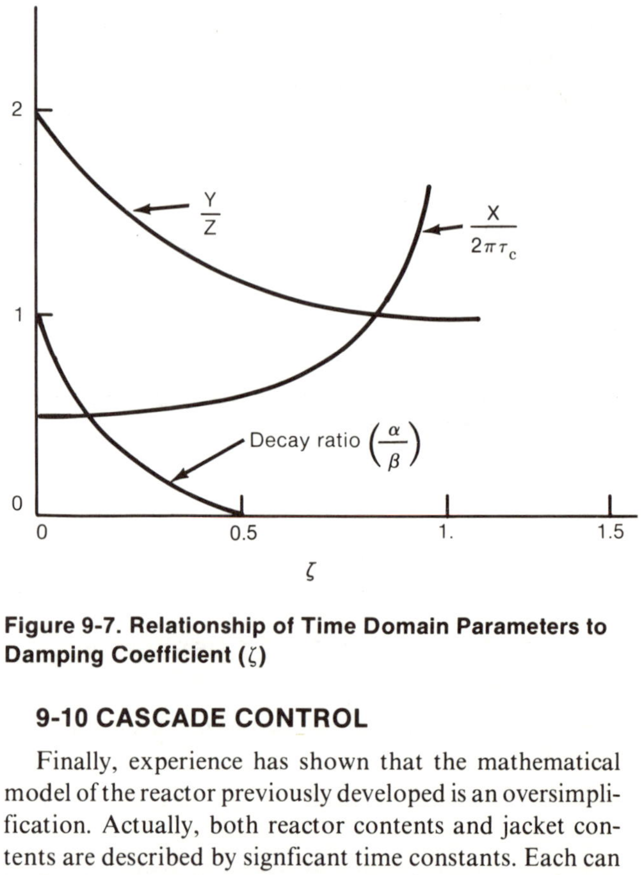

Figure 9-7. Relationship of Time Domain Parameters to Damping Coefficient (ζ)

9-10 CASCADE CONTROL

Finally, experience has shown that the mathematical model of the reactor previously developed is an oversimplification. Actually, both reactor contents and jacket contents are described by signficant time constants. Each can be developed by a heat balance, as previously shown for the reactor contents. Referring to Figure 9-2, the inadequacy of the single temperature control loop becomes apparent. To overcome a change in temperature T_1, the control system must change the jacket conditions. Since the jacket has an appreciable time constant, it changes slowly, causing a large overshoot in reactor temperature. Thus the system is characterized by slow response and cycling, since the jacket temperature will change far too much before the temperature sensor in the reactor finds out about it.

Performance improved immensely by the cascade control system shown in Figure 9-8. The block diagram of the cascade system shows that two controllers are now required, and the choice of controller modes is far more complicated.

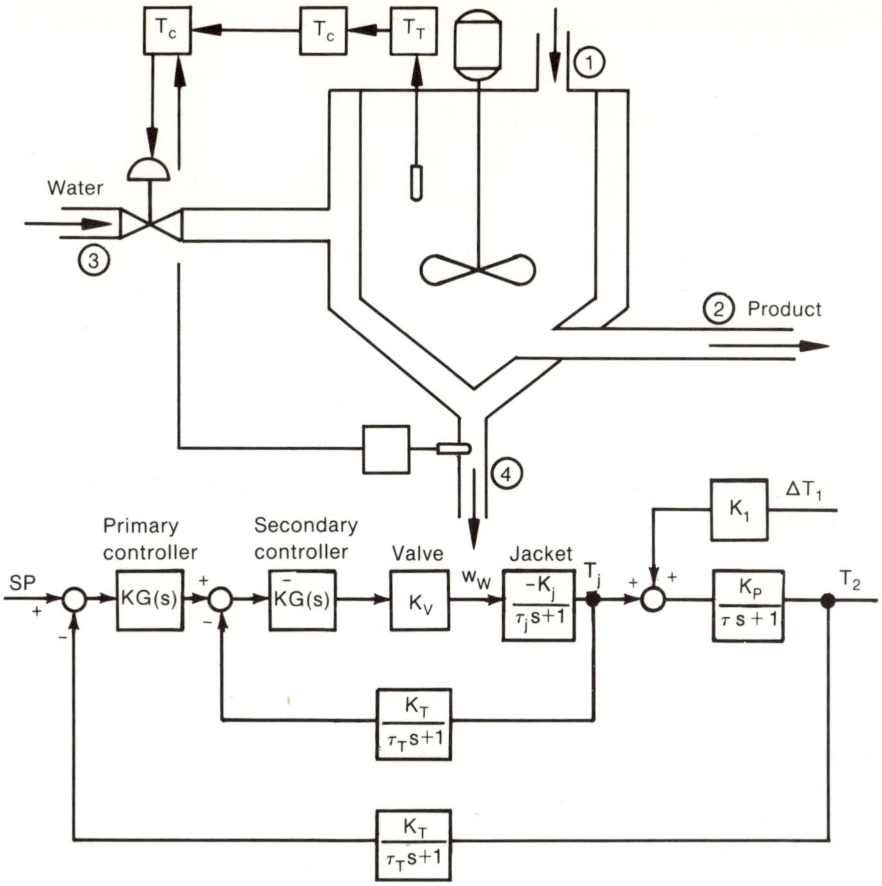

Figure 9-8. Chemical Reactor with Cascade Control System

However, note that each process time constant now has a control loop around it. The loop on the reactor temperature (the primary loop) is essentially the loop previously discussed, and it would use the PID controller.

However, the jacket temperature control loop (the secondary loop) has no need to eliminate offset, since the value of the jacket water temperature really doesn't matter. Further, it will happen that the primary loop will drive the secondary controller past the end of its span. The controller will then go into a condition called reset windup, or saturation of the reset mechanism. Thus a proportional-only controller serves extremely well in the secondary loop for a temperature process, which is characterized by narrow proportional bands (high controller gain) and slow reset. Flow as the secondary variable will require reset in the controller, since flow controller proportional bands are typically wide and permit offsets that are too great for controllability.

Problems

1. How would the equation below change if the process were endothermic (required heat to maintain reaction)?

$$M_c \frac{dT_2}{dt} = w_1 c T_1 - w_2 c T_2 + K T_2 - Q$$

2. Does the endothermic process represent a stabilizing or destabilizing effect on controllability?

3. How can positive and negative feedback be identified from the block diagram, Figure 9-3(a)?

4. With regard to the heat balance from which the time constant of the temperature sensor was developed, what would be the effect of a large heat transfer through the metal of the sensor and thermowell to the atmosphere?

5. What is the effect on the time constant if temperature sensor is in a stagnant location?

6. Could flow w in the equation in question 1 above be a disturbance?

7. What is the limitation on proportional-only control?

8. Does higher controller gain (K_c) increase or decrease the damping coefficient (ζ)? Does this make the process more or less stable?

9. If one type of controller were to be chosen for general usage, what would the control modes be?

10. With reference to Figure 9-7, what happens to response time X as damping coefficient ζ decreases? What happens to overshoot Y?

STUDENT SUMMARY NOTES and QUESTIONS FOR INSTRUCTOR

SECTION **10**

FLOW CONTROL

10-1 INTRODUCTION

Flow is the predominant controlled variable in the fluid processing industries. In the days of large hold-up volumes, flow control was used to bring the process to steady state because the process was designed on steady-state principles. The modern high performance processes (high throughput, minimum hold-up) use flow control to set production rates or as a cascade or slave variable. These requirements dictate that the flow control system not only satisfactorily attenuate disturbances but also respond well to set point changes.

Flow control systems are widely applied; they are simply tuned using the concepts to be discussed in this section; yet the investigation of flow control systems is one of the most vexing problems in the field of automatic control.[50] The anomaly involves the word "satisfactorily". There is no doubt that instability is unsatisfactory, but the control system is easily stabilized with a wide proportional band and fast reset. The questions arise concerning satisfactory response to set point changes, satisfactory rangeability, and satisfactory attenuation of disturbances.

In practice, the flow control system cannot attenuate disturbances. Appendix E shows the time constants in the control system to be in the instruments, while the process reacts instantaneously. Thus the process is faster than the instruments, and flow will deviate as it must in response to a disturbance but will be brought back to set point very quickly by the instrument system. Luckily, panel board indicators and recorders do not have a sufficiently high frequency response to make the deviation visible, and the recovery is so quick that downstream equipment absorbs the "spike" with no visible effects.

The load changes that require corrective action by the flow control system are essentially limited to the upstream and downstream pressures. Appendix E shows that the effects of disturbances are decreased by increasing total pressure drop ($p_1 - p_3$). But increasing the pressure drop is costly in energy, changes the valve gain, and changes the valve flow characteristic. Thus the flow system variables interact strongly, and computer design packages have been developed that optimize valve size, valve flow characteristics, pipeline resistance, flow transmitter gain, and controller functions for adequate regulation at minimum energy costs.[29]

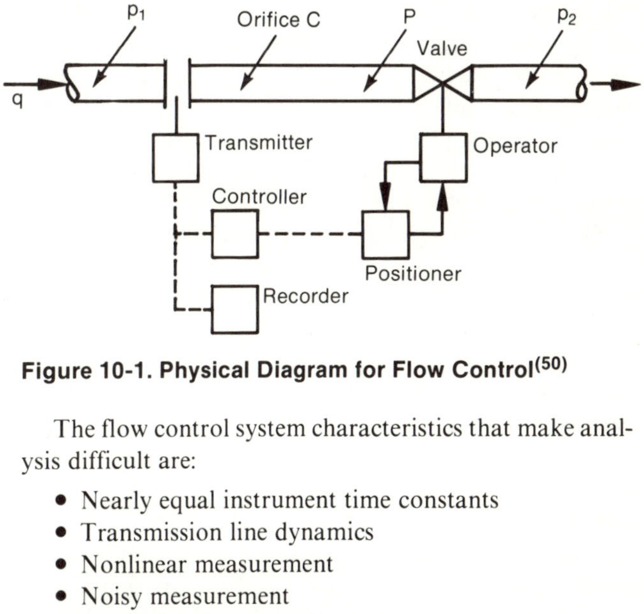

Figure 10-1. Physical Diagram for Flow Control[50]

The flow control system characteristics that make analysis difficult are:

- Nearly equal instrument time constants
- Transmission line dynamics
- Nonlinear measurement
- Noisy measurement
- Lack of process capacitance
- Control valve gain nonlinearity

10-2 PROCESS CAPACITANCE

The small length of pipe between flow transmitter and valve shown in Figure 10-1 represents the volume that will determine process capacitance. Following the development for a polymer in Section 3, the capacitance for a water-like liquid is

$$c = V\frac{d\rho}{dp} = V(10^{-8})\ \text{in.}^2$$

and for air as a typical gas,

$$c = \frac{mV}{n\bar{R}\theta} = \frac{(29)\ V}{(1.0)\ (19.3)\ (303)}$$

$$= V(5)\ (10^{-3})\ \text{in.}^2$$

Section 3 shows turbulent resistance to have values around 1 sec/in.^{2} Thus the values of the process time constants are going to be numerically equal to the capacitance. But the instrument time constants will be between 0.1 and 1.0, thus making the process dynamics negligible. The analysis shown in Appendix E is performed using algebraic relationships with the knowledge of the negligible process time constant.

10-3 INSTRUMENT TIME CONSTANTS

Ziegler and Nichols[(51)] have shown that open-loop gain reaches a minimum value when the two major time constants in the system are equal. Their analysis concerns the second-order system described by Figures 8-6, 9-5, and 9-7. With only two time constants, the system cannot become unstable in the sense of the Nyquist criterion where phase lag can exceed 180°. Rather, stability is described by the damping factor ζ. As zeta approaches zero, the system response approaches a constant oscillation (a sine wave of constant amplitude). While the second-order model can be applied to many physical control systems for analysis, it describes flow control systems very well because they have few dynamic elements.

The transmitter and control valve actuator time constants can vary widely, but both are around 0.5 second in modern instruments. Thus their ratio is around 1, forcing the minimum loop gain shown on Figure 10-2. Because of the noisy process, the time constant of the transmitter is sometimes made larger by the use of a damping adjustment found on the transmitter. Noise is the random changes in the flow pattern through piping, fittings, and the flow sensor. The small erratic variations are picked up by the flow transmitter and displayed on the chart record. In reality, the noise is simply a series of small load changes occurring at intervals less than the natural period of the control system so that the controller is unable to correct for them. In order to obtain a more legible record, the transmitter time constant is increased.

Referring to Figure 10-2, suppose that the measurement time constant is 0.2 second and the valve time constant 0.5 second. Increasing the measurement time constant by damping would (surprisingly) decrease stability and require widening the controller proportional band. Further damping would reverse the condition and allow the use of higher gain (narrower proportional band). This paradox is explained by Figure 10-2, which shows that for a two-capacity process the allowable loop gain is a minimum when both time constants are equal. Increasing or decreasing either one allows higher gain settings. Thus Figure 10-2 shows that open-loop gain can be 0.3 for stability if the two time constants are equal, but can be 2.7 if one time constant is 10 times the other.

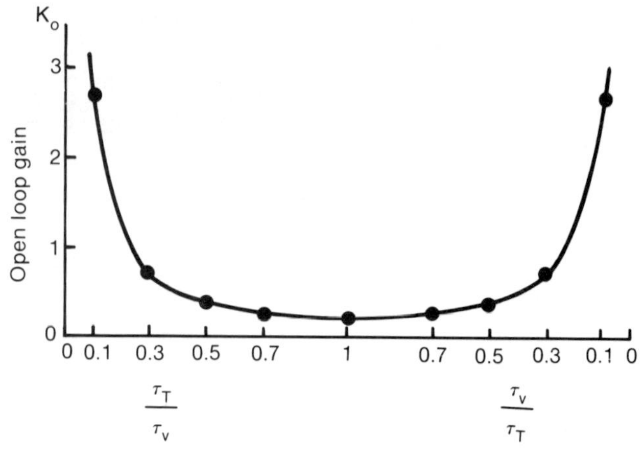

Figure 10-2. Proportional Control with Two Time Constant System

It would appear from Figure 10-2 that reducing one or the other of the time constants would enhance control. However, reducing the time constant of the transmitter will result in a noisier signal, while speeding up the control valve will cause water hammer in liquid service. Likewise, increasing one time constant or the other would appear beneficial. Actually, control can only be made poorer by slowing down the instruments. A slower transmitter will improve the appearance of the recorder chart but will not improve the actual flow regulation. Excessive damping of the valve is even worse since the valve can then no longer respond as quickly in moving to a new position.

10-4 EFFECT OF PNEUMATIC TRANSMISSION LINES

The pneumatic tubing used to transmit signals in a pneumatic control system can quite easily represent the major dynamics in a flow control system, since the system components have such small time constants. Electronic systems, of course, do not have the tubing dynamics but will typically be so noisy that capacitors will be placed on the instrument outputs to slow them down. Thus, in the final analysis, pneumatic and electronic flow control systems will respond similarly.

The transient flow in pipes is described by two partial differential equations:[(52)]

$$\frac{\partial v}{\partial l} = -\frac{1}{\rho c^2}\frac{\partial p}{\partial t}$$

$$\frac{\partial p}{\partial t} = -\rho\frac{\partial v}{\partial t} - Rv$$

where c = acoustic velocity.

These partial differential equations indicate that standing pressure waves exist in the tubing and introduce questions as to the validity of linear (lumped) analysis. The following equation can be used to determine whether the lumped parameter approach is valid or not:[53]

$$\frac{\omega_e}{\omega_d} = \frac{2nM}{\pi} \tag{10-1}$$

where M = Mach number

ω_e = lumped parameter frequency

ω_d = distributed system frequency

n = ratio of specific heats

The lumped parameter approach is valid well past the break frequency if the ratio of frequencies in Equation 10-1 is small. If the ratio is large, however, then distributed parameter effects will introduce resonances and excessive phase shift at the break frequency, interfering with control. The flow through pneumatic tubing is small so the Mach number is small and the frequency ratio is small. Equation 10-1 shows that the lumped parameter (linear) approach can be used in describing pneumatic tubing. The "bumps" indicative of the resonances will appear at higher frequencies of the Bode plot, as shown in Figure 10-3.[54]

The resonances shown in Figure 10-3 are at frequencies higher than the natural frequency of the control system and, thus, have little effect. Warnock[55] has correlated a large amount of experimental work into Figure 10-4, which provides time constants that characterize lengths of tubing.

Using the block diagram Figure E-3 of Appendix E, Eckman[50] has analyzed a close-coupled flow control system for which the dynamics are:

Flow transmitter time constant, τ_T = 0.16 sec

Control valve time constant, τ_v = 1.6 sec

resulting in an open-loop gain from the Bode plot

$$K_o = K_c K_v K_p K_T = 2.0$$

and the typical flow controller adjustments are:

$$\text{P.B.} = 100\%$$
$$\text{Reset} = 0.02 \text{ minute} \tag{10-2}$$

These tuning parameters are calculated with a ratio of τ_T/τ_v of 0.1. In practice, a factor of safety is added for time constants that are closer together, as per Figure 10-2. Thus, upon start-up, the typical flow controller will be adjusted to

$$\text{P.B.} = 150\%$$
$$\text{Reset} = 0.1 \text{ minute} \tag{10-3}$$

and will seldom be readjusted. Flow control has gotten very little attention over the years because it will respond quickly even though poorly tuned. That is changing. Cascade control and computer-generated set points demand better performance, which in turn will demand a more optimum tuning.

The effect of the dynamics of pneumatic tubing on flow control system tuning and performance has been investigated. Warnock[55] used a control loop with

Flow transmitter dynamics, τ_T = 0.12 sec

Control valve dynamics, τ_v = 0.5 sec

and found that system recovery time increased with the length of pneumatic tubing as shown in Figure 10-5. The 3/8-inch tubing showed no advantage for small transmission distances.

Buckley[56] has combined experimental data and a computer analysis to consider the effect of pnuematic line length and line diameter on controller gain and reset time. Using lagless instruments, line length has little effect on open-loop gain, as shown in Figure 10-6.

The analysis on which Figure 10-6 is based shows that an open-loop gain value of 2 is a good guideline in setting the proportional band on a flow controller. The graph also shows that the transmission line volume provides capacitance, which permits a higher open-loop gain and a narrower proportional band. Figure 10-7 shows that the transmission line capacitance necessitates a slower reset adjustment. The slower reset setting causes more deterioration in control performance than the increase in gain provides improvement, and recovery time grows longer as shown in Figure 10-5.

10-5 EFFECT OF NONLINEARITIES

Consider the partial signal flow diagram for the flow process and the control valve shown in Figure 10-8.

Flow control systems typically use a head-type measuring device and an equal percentage valve flow characteristic. Thus gains K_v and K_p are partial derivatives of nonlinear functions, which are as shown in Figure 10-9.

Note that the gain of both the equal percentage and square law functions increase as the valve is opened (x increases). With the open-loop gain a constant value, the controller gain must decrease to compensate for the increases in valve gain (K_v) and process gain (K_p). The increase is as shown in Table 10-1.

The foregoing requirement for controller retuning has largely been ignored by tuning with a wide proportional band at the set point, and, since the set point was constant, the tuning was adequate. More and more flow control systems are being cascaded in modern plants, and consequently the nonlinearities represent a difficulty. The orifice flowmeter is gradually being replaced with linear sensors, however, for better accuracy and rangeability as well as linearity. Further, energy savings dictate that less pressure

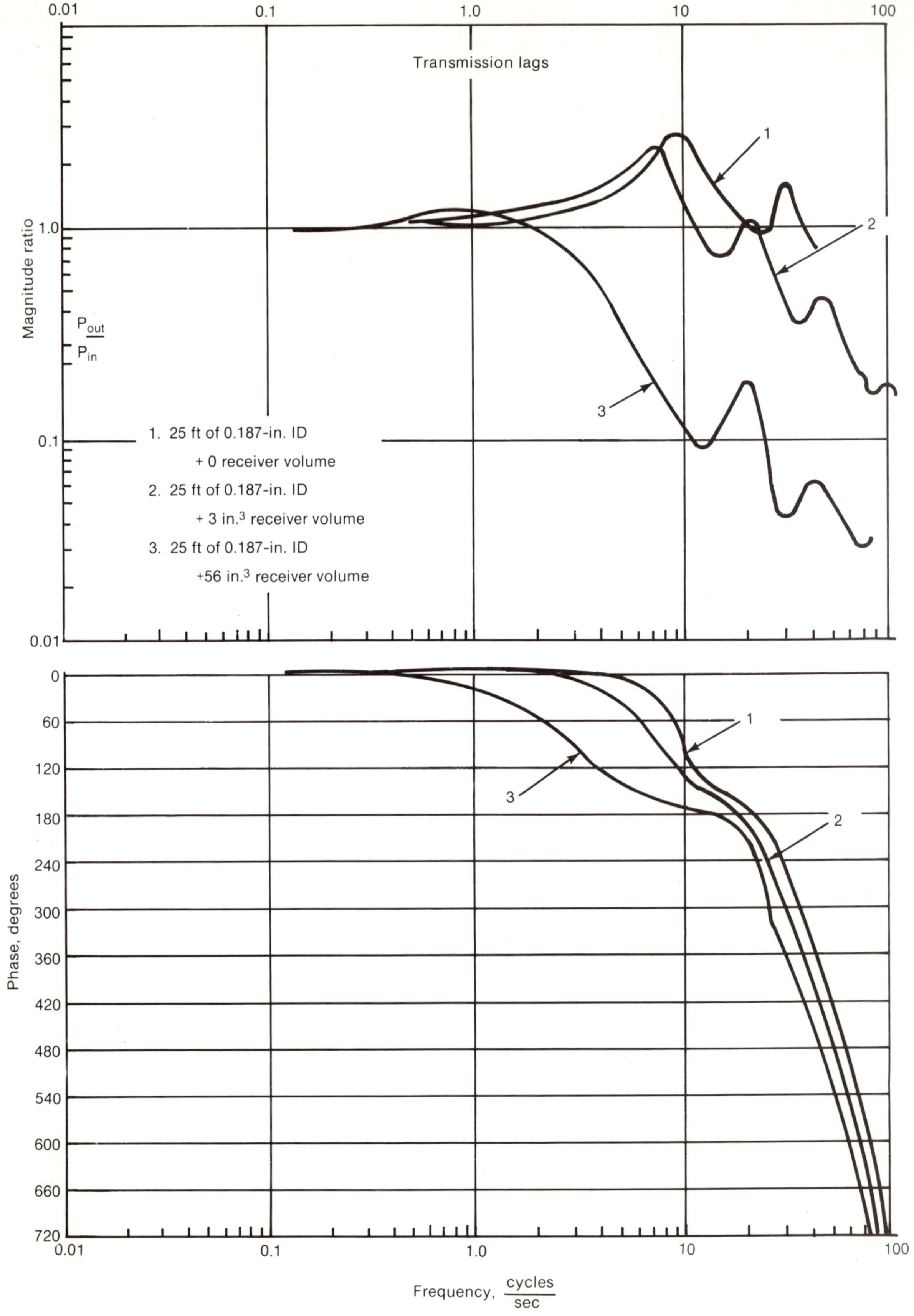

Figure 10-3. Frequency Response of Pneumatic Transmission Lines[54]

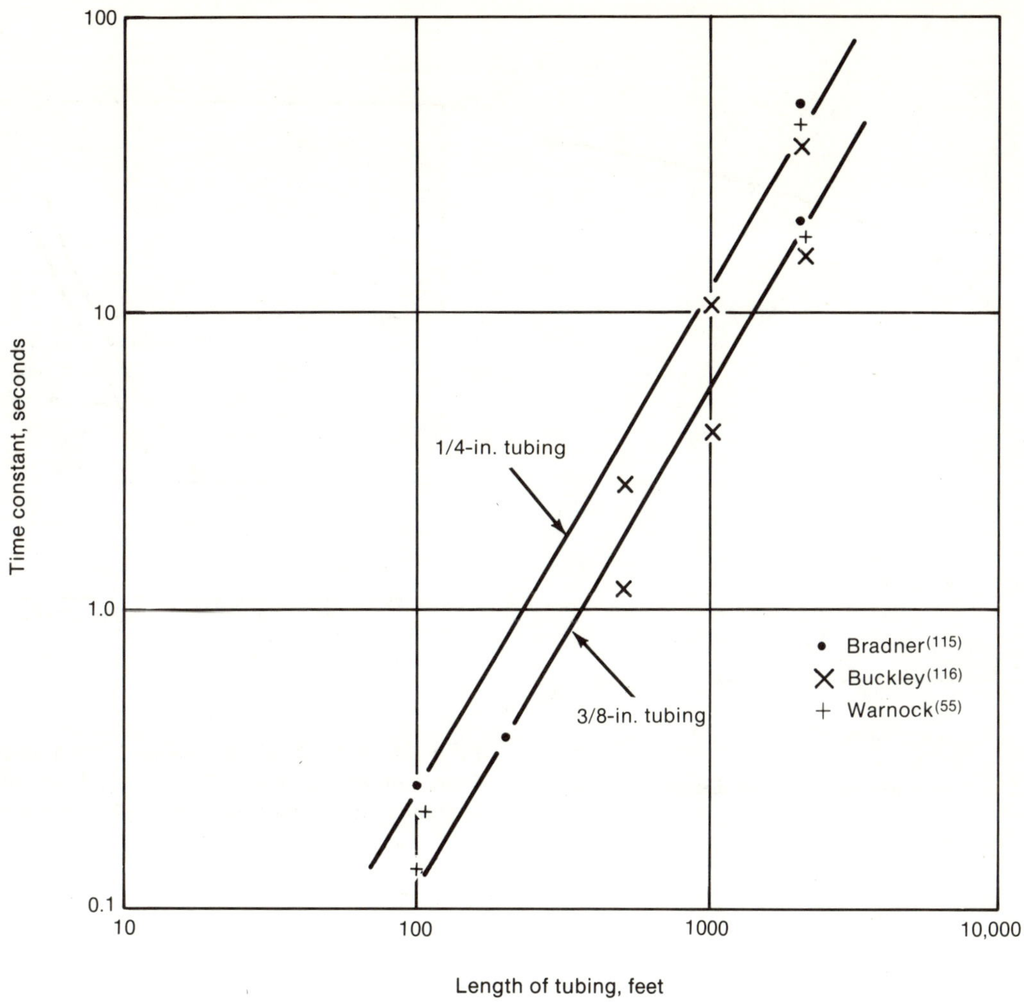

Figure 10-4. Time Constants of Pneumatic Tubing[55,115,116]

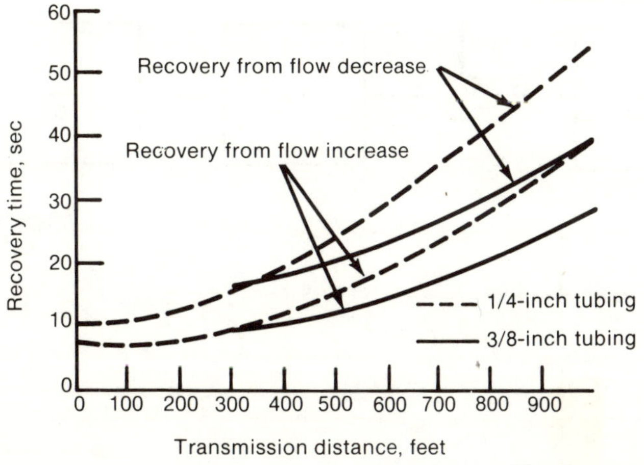

Flow loop recovery times increase linearly for distances over 250 feet.

Figure 10-5. Flow Loop Recovery

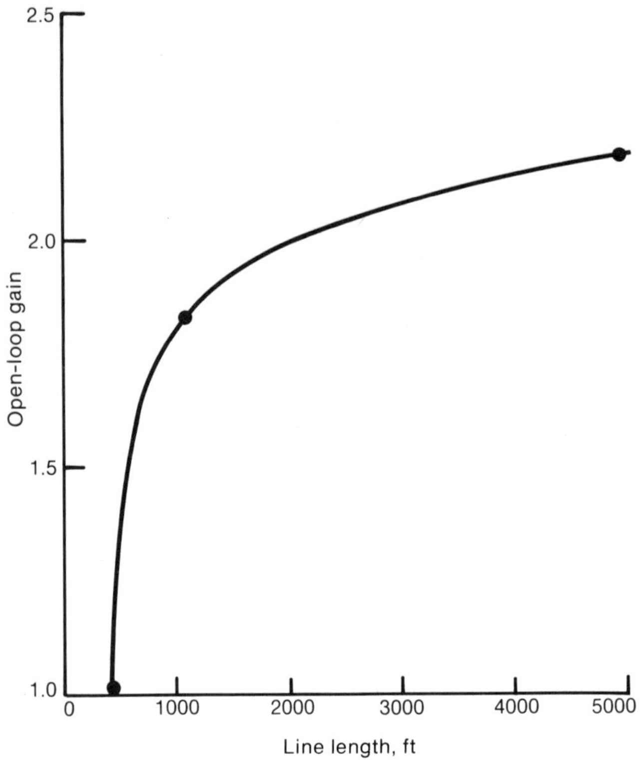

Figure 10-6. Effect of Transmission Line Length on Open-Loop Gain(56)

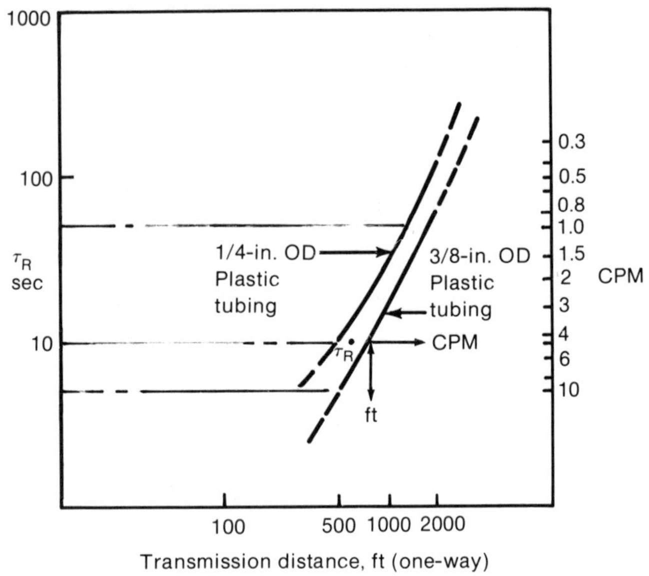

Figure 10-7. PI Controller Reset Time and Closed-Loop Resonant Frequency (CPM) versus Transmission Distance and Tubing Diameter — 2-Pipe Configuration(56)

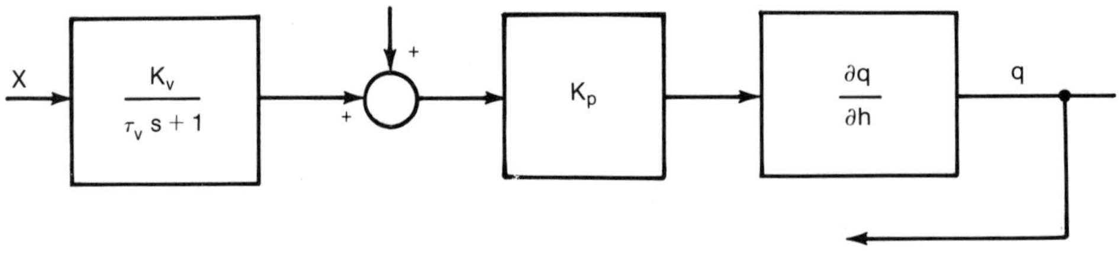

Figure 10-8. Block Diagram of Nonlinearities

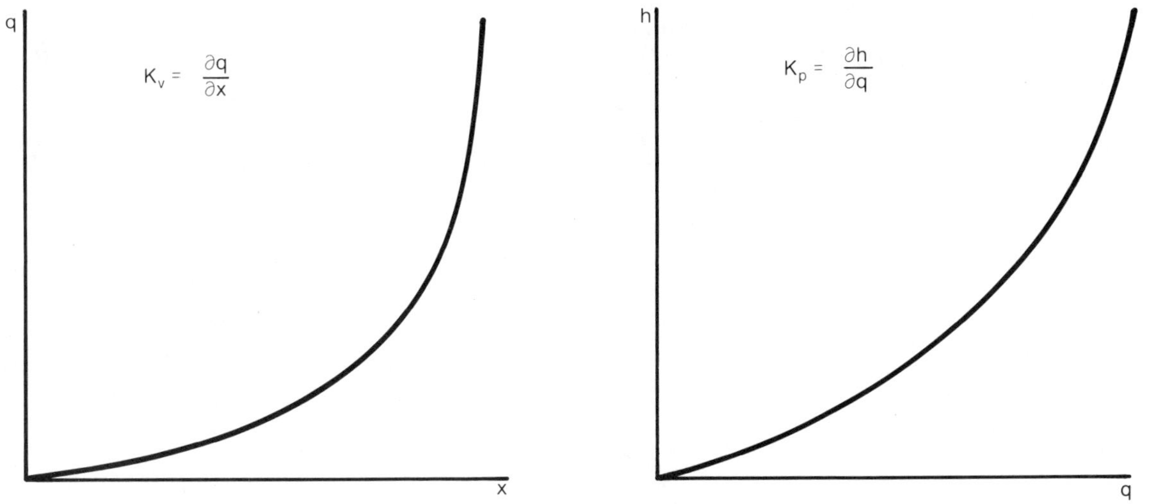

Figure 10-9. Flow System Nonlinearities

Table 10-1
Variation in Controller Gain with Flow System Nonlinearities[57]

	Equal percentage installed characteristic	
q/q max	K_c	P.B., %
0.2	6.4	16
0.4	1.03	97
0.6	0.71	141
0.8	0.40	250
1.0	0.26	391

drop be allocated to the control valve, causing the popular inherent equal percentage characteristic to become a linear installed characteristic, as seen in Figure 5-2.

10-6 CONTROLLER FUNCTIONS

While the PI controller is widely used for flow control, the floating controller also finds application for this process. It offers the advantages of economy, simplicity, easier tuning (one knob versus two), and less noisy control. It has the disadvantage, especially severe in cascade control, of having a closed-loop natural frequency that is 3 to 4 times slower than that of a similar system with a PI controller.

Cross-country pipelines are designed to move a maximum amount of fluid at a minimum energy cost or at a minimum control valve pressure drop. Valves are typically chosen the same size as the pipeline to minimize pressure drop when wide open. With a long pipeline and an oversized control valve, the ratio of valve pressure drop to system pressure drop is very small. The small ratio is shown in Figure 5-2 to convert an equal percentage inherent characteristic into an installed quick-opening characteristic. While an installed characteristic between linear and quick-opening (sometimes called a square root characteristic) compensates for the square law flow sensor nonlinearity, the extreme case of a quick-opening characteristic represents such a high gain at low flows that the controller proportional band is not wide enough to stabilize the system.

The floating controller serves very well in the cross-country pipeline application since it has no gain and needs only to be set to integrate slowly enough to be stable.

10-7 CONCLUSIONS

The dynamic analysis of flow control systems can produce puzzling results. Simply calculated controller settings turn out to be entirely inadequate. The settings will invariably be too high, leading to ridicule in the field, or will inexplicably work well one day and not the next. Perfectly good reasons exist for the inconsistent behavior exhibited by flow control systems, but they lie in the realm of the secondary effects of automatic control phenomena, nonlinearities, small capacitances, and relationships among parameters. While these phenomena must be acknowledged as better performance is required, most flow controllers will continue to be adjusted in accordance with Equation 10-3.

Problems

1. Using the block diagram of Figure E-3 in Appendix E, develop the curve plotted as Figure 10-2 using $\zeta = 0.9$.

2. Section 10 says that flow control systems cannot attenuate disturbances. Explain the reason for this from Figure E-3.

3. Going back to the polymer system of Figure 3-12, assume that the liquid was water with $d\rho / dp = 3 \times 10^{-8}$. What would be the rate of rise of pressure?

4. What will the controller tuning of Equation 10-2 change to if $\tau_T = 0.8$ sec and $\tau_v = 1.6$ sec?

SECTION **11**

MASS BALANCE CONTROL

11-1 INTRODUCTION

The control of pressure and the control of liquid level appear to be quite unrelated. Actually, they are very similar in that each attempts to maintain a constant quantity of fluid in the vessel. The pressure control system maintains a constant quantity of compressible (gases, vapors) fluid in the tank, while the level control system maintains a constant amount of liquid in the vessel. Thus, both are known as "mass balance" control systems. Every vessel requires a mass balance control to prevent it from running over (rupturing) or running dry. Some unvented vessels require both pressure and level control: pressure to regulate the compressible mass balance and level to regulate the liquid mass balance. An example of both level and pressure control is the steam boiler, where the two controls interact with each other since the decrease in the liquid mass balance due to boiling results in an increase in the vapor mass balance.

11-2 APPLICATION OF MASS BALANCE CONTROLS

The assessment of controllability is determined by combining the mathematical expressions for the controller, valve, and transmitter with the process, which is described by the mass balance equation. Essentially, this equation states that

Accumulation = Input − Output

11-3 COMPRESSIBLE FLUID MASS BALANCE

For pressure, as shown in Example 11-1, the equation becomes

$$V \frac{d\rho}{dt} = w_i - w_o$$

where the symbols are identified in Example 11-1. However, since pressure (p) is more commonly measured than density (ρ), the equation is written[22]

$$V \frac{\partial \rho}{\partial p} \frac{dp}{dt} = w_i - w_o$$

The partial derivative of density to pressure is evaluated by thermodynamic relationships for gases and tabular data for vapors, e.g., steam tables for steam.

The mass balance equation is combined with the equation describing the control valve to develop an equation denoting the sensitivity of vessel pressure (p) to inlet flow (w_i) and valve stem position (x). This is then combined with the transmitter and controller functions into a block diagram. Information contained in the block diagram permits proper choice of controller modes and valve characteristic.

A proportional-reset controller is required with, as shown by the block diagram, the reset time (τ_i, minutes) set equal to the process time constant (τ). These two terms will then cancel in the block diagram, leaving only the controller integrating term as the system dynamics and resulting in a very stable control system. The process time constant will typically be a very small number (a few seconds), and thus the controller should be purchased with a very fast reset adjustment range (a very short reset time, or a high number of repeats/minute).

Assuming the typical very small process time constant (τ), the valve flow characteristic[28] can be chosen from consideration of the open-loop gain concept, where open-loop gain is the product of all the gains in the feedback control loop. Example 11-1 shows the relationship for open-loop gain to have the flow through the control valve in the denominator. A control valve with linear characteristic is shown to have a constant valve gain (K_v), but an equal percentage valve has a gain that is proportional to throughput (w_o). Substituting the equal percentage valve gain in the open-loop gain results in a cancellation of the flow terms and makes open-loop gain independent of throughput. Thus, an equal percentage valve is required. However, in those few applications where the vessel

Example 11-1

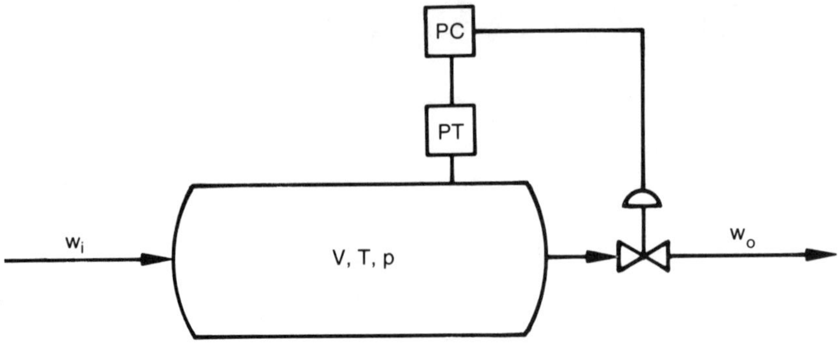

$$\frac{V}{mRT}\frac{dp}{dt} = w_i - w_o$$

$$w_o = Kx\sqrt{p - p_o}$$

$$\frac{V}{mRT}\frac{dp}{dt} = w_i - \frac{w_o}{2(p - p_o)}\,p - \frac{w_o}{x}\,x$$

$$(\tau s + 1)p = \frac{2(p - p_1)}{w_o}\,w_i - \left(\frac{2(p - p_o)}{w_o}\right)\left(\frac{w_m}{x}\right)x$$

where w_m = maximum flow through valve; $\frac{w_m}{x} = K_v$

where
V = volume, ft^3
m = polytropic constant
R = gas constant
T = temperature
p = pressure, psi
t = time, sec
w = flow, lb/sec
x = valve stem position

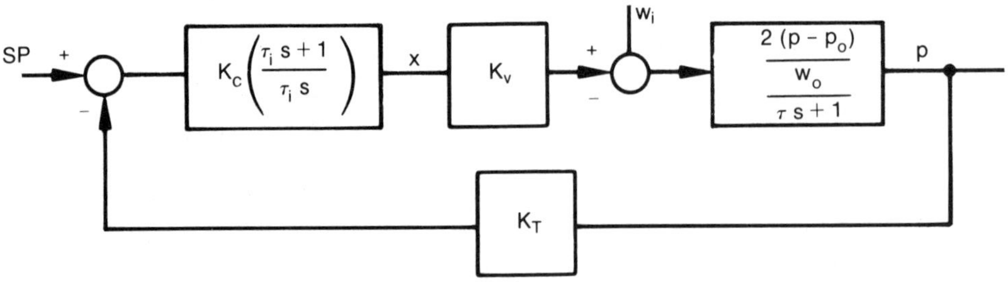

Let K_o = open-loop gain

$$= K_c K_v \left(\frac{2(p - p_o)}{w_o}\right) K_T$$

Control valve characteristic:

Linear – K_v = constant
Eq. % – $K_v = K_1 w_o$

volume is large enough to result in a significant time constant ($\tau \geq 15$ seconds), a linear valve characteristic is preferable.[102]

11-4 LIQUID MASS BALANCE

The liquid mass balance is shown in Example 11-2 and is actually seen to be a volume balance with the volume in the tank measured by level. The requirements of the vessel operation should be discerned early in the design of the level control system, since control strategy will vary with operational requirements. For example, some vessels require that level be held very closely to a set point. A boiler drum is such a vessel, since high level will result in carryover of water into the steam system, while low level will result in the burning out of a boiler tube. Other vessels have no requirement that level be held in one place but operate satisfactorily as long as level is retained within the vessel.

Most industrial flows leaving tanks are pumped, as shown in Example 11-2. Thus, level in the tank has only a negligible effect on the flow out of the tank, the level control valve has essentially a constant upstream pressure, and the process has no self-regulation. With no self-regulation, the level process becomes an integrator, as shown in the block diagram in Example 11-2. This immediately leads to two conclusions:

- Throughput flow does not appear in the denominator of the open-loop gain, and a linear installed valve flow characteristic is preferred.[28]
- The system is represented by two integrators (the other being the controller reset) in a feedback loop, which is by definition an oscillator. The system is therefore difficult to tune.

Going back to Example 11-1, some compressible fluid processes (e.g., a steam boiler) have such a large volume (V) that the gas process time constant (τ) is much larger ($\tau >>> 1$) than the self-regulation term. The self-regulation can then be assumed insignificant and can be ignored.[102] Thus, this process (see block diagram of Example 11-1) has two integrators in the feedback loop, similar to the level control system, and represents the same tuning challenge.

An elegant solution to the problem of the feedback loop being an oscillator is to remove the reset mode from the controller and specify a proportional-only model. Simplifying even further, the controller can be omitted altogether, and the open-loop gain established by the level transmitter span. Then, looking at the process diagram on Example 11-2, the control valve will be closed when level is at the bottom of the transmitter span and full open when level is at the top of the span. In between, level will reach a value dictated by the flow through the tank. The optimum surge tank (provides maximum attenuation of flow changes) has no controller and as wide a level transmitter span as the tank will allow. Again referring to Example 11-2, notice that the characteristic equation (denominator of the transfer function) becomes first-order with this choice of controller, eliminating all possibility of instability.

However, either by personal preference or by process requirement, the offset that appears when reset is omitted is unacceptable in many applications. Following Buckley[63] and Shunta,[103] the transfer function at the bottom of Example 11-2 can be converted to

τ_c = characteristic time or time constant

$$\tau_c = \sqrt{\frac{7.48 A \tau_i}{K_c K_v K_T}}$$

where K_v = valve gain = maximum flow = q_{max}

K_T = LT gain = 1/level span = $1/L$

$$\tau_c = \sqrt{\frac{7.48 A L \tau_i}{K_c\, q_{max}}} \qquad (11\text{-}1)$$

where $AL = V\ \text{ft}^3$ = tank volume

$$\text{and } \zeta = \frac{\tau_i}{2\tau_c} = \frac{1}{2}\sqrt{\frac{\tau_i K_c\, q_{max}}{7.48\, V}} \qquad (11\text{-}2)$$

$$(2\zeta)^2 = \frac{\tau_i K_c\, q_{max}}{7.48\, V}$$

For stability $0.5 \leq \zeta \leq 1.0$.

Finally,

$$\frac{7.48\, V}{q_{max}} \leq \tau_i K_c \leq \frac{30\, V}{q_{max}} \qquad (11\text{-}3)$$

The values of V, K_c, and τ_i can then be determined rapidly by judicious trial and error experimentation of Equations 11-1, 11-2, and 11-3.

The foregoing equations show the level control system to have the oddity of becoming more stable as the proportional band is narrowed (gain K_c is increased), since damping (ζ) is proportional to gain (K_c). Thus, level system can be tuned very tightly and be made to draw a straight line on the recorder. In many systems, however, it is desirable to have the level control perform less well in order to reduce variations in the outgoing flow, since those variations upset the next unit operation. This is particularly true in a distillation train where the bottom of one column feeds the next column. It is desirable to loosely tune the level controllers, let bottom levels vary (within reason), and stabilize the feed to the next column. Elimination of the controller, as discussed previously, removes the temptation to tune tightly.

11-5 LEVEL CONTROL — HIGH PRESSURE SEPARATOR

Some processes generate a liquid product that must be separated from its vapor at a very high pressure (1500 to

Example 11-2

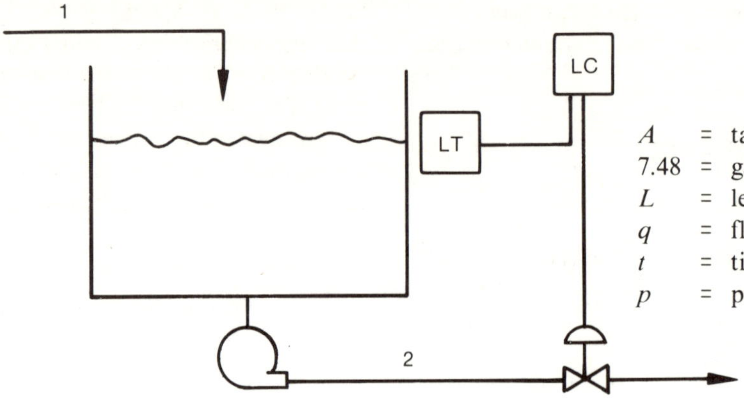

A = tank area, ft^2
7.48 = gal/ft^3
L = level, ft
q = flow, gpm
t = time, min
p = pressure, psig

The mass balance:

$$7.48\,A\,\frac{dL}{dt} = q_1 - q_2$$

With liquid being pumped from the tank:

$$p_2 \approx \text{constant, valve } \Delta p = \text{constant}$$

The control valve: $q_2 = \dfrac{q_{max}}{x}\, x = K_v\, x$

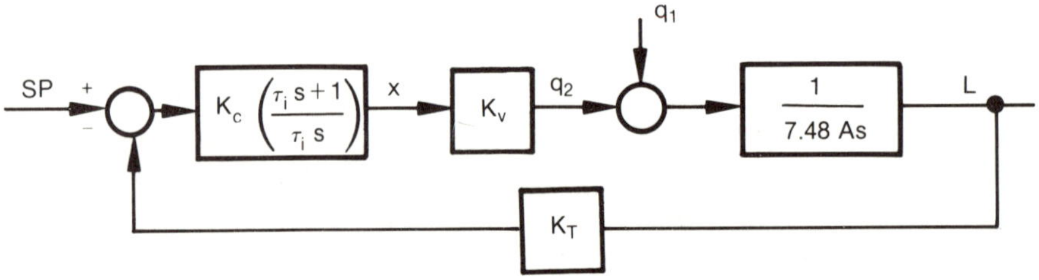

Open-loop gain $= K_o = K_c\, K_v \left(\dfrac{1}{7.48\, A}\right) K_T$

Closed-loop equation:

$$L = \frac{\dfrac{1}{7.48\, As}\, q_1}{1 + \dfrac{K_c K_v K_T (\tau_i s + 1)}{7.48\, A \tau_i s^2}}$$

$$= \frac{\tau_i s q_1}{7.48\, A \tau_i s^2 + K_c K_v K_T \tau_i s + K_c K_v K_T}$$

5000 psi) (see Figure 11-1). The separation vessel is very expensive, and one such vessel was quoted at an additional $20,000 in cost for every inch added to the diameter. Thus V in Equation 11-3 must be minimized. Yet set point level control that will restrict level deviations to small values is required, since high level permits liquid carryover into a compressor and low level permits liquid blow-by into the next vessel. Separators are oriented both horizontally and vertically, as shown in Figure 11-1. Quite obviously, the volume V is independent of orientation. In operation, however, the vertical orientation results in wide swings in level, which make good control performance as seen on the recorder chart quite difficult to attain. The horizontal orientation suffers from level gradients and "sloshing," because of the liquid traffic in the small vessel, and thus requires a well designed vessel in terms of baffling and nozzle orientation.

In addition to the necessary smallness of the vessel, other factors detrimental to controllability are:

- The displacer type level transmitter typically used in this service has a natural frequency as shown in Example 11-3. This frequency can be troublesome if it is the same as the natural frequency of the sloshing or the same natural frequency as the level control system.
- The level control valve is also used as a pressure letdown valve, and may have several thousand psi pressure drop across it, resulting in a very high valve gain.

While most applications receive little attention, the high pressure separator and its level control system must be very well designed to attain satisfactory regulation. The combination of small volume dictated by economy, very high liquid traffic, tremendous pressure drop across the control valve, and the necessity for close regulation combine for a difficult application. Secondary problems abound also, such as the variation in cross sectional area (A) when the separator is oriented horizontally. Equation 11-3 shows the controller tuning to be a function of area (A), and thus the desired tuning varies with level (L).

The major difficulties, however, will be in the small volume of the process, e.g., high liquid velocities into the level transmitter impulse line, causing erroneous level measurements. The difficulty of testing high pressure vessels to identify fluid flow phenomena and the difficulty in revising such heavy equipment will invariably lead to the conclusion that the difficulty "must be" the instruments. Thus will the instrument engineer's fortitude be tested.

11-6 LEVEL CONTROL — DISTILLATION COLUMN BOTTOMS

The bottom of a distillation column, with vapor separating from a boiling liquid, is a level control application similar to the separator. Many years ago, the residence time of the liquid in the bottom of the column could be as

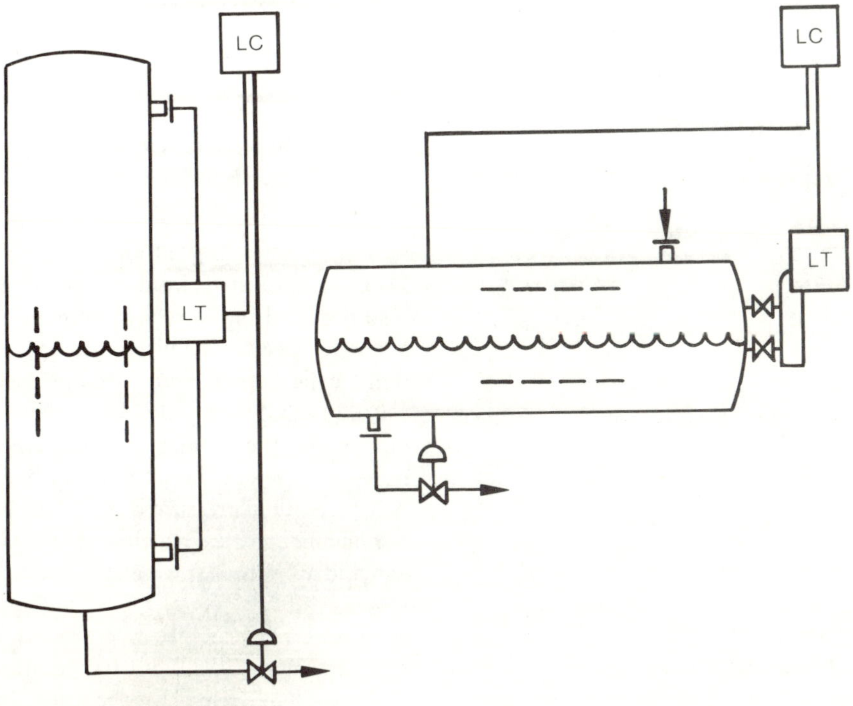

Figure 11-1. Separator Level Control

Example 11-3

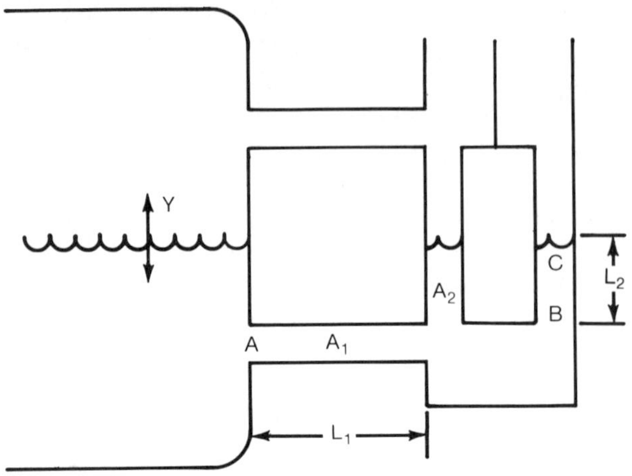

Float Cage Nomenclature(102)

To find the undamped natural frequency, neglect friction and apply Newton's Second Law to the fluid mass in Sections 1 and 2.

$$P_a A_1 - P_b A_1 = A_1 L_1 \rho \frac{d^2U}{dt^2}$$

$$P_b A_2 - P_c A_2 = A_2 L_2 \rho \frac{d^2X}{dt^2}$$

U and X are the displacements in Sections 1 and 2, respectively. Making the substitutions:

$$\frac{d^2U}{dt^2} = \frac{d^2X}{dt^2}\frac{A_2}{A_1}, \; P_c = WX, \; P_a = WY$$

and combining the two equations gives:

$$\frac{d^2x}{dt^2} \; \frac{L_2 + L_1\left(\frac{A_2}{A_1}\right)}{g} + X = Y$$

so that the undamped natural frequncy of the cage is:

$$\omega_n = \sqrt{\frac{g}{L_2 + L_1\left(\frac{A_2}{A_1}\right)}} \text{ radians/sec}$$

where g = acceleration of gravity, ft/sec^2
L_1 = length of liquid equalizing line, ft
L_2 = wetted length of displacer, ft
A_1 = area of liquid equalizing line, ft^2
A_2 = area of annular ring between displacer and cage, ft^2

much as an hour, and the regulation of level was easily accomplished. Economics, including the cost of the column and its supports and the cost of product inventory, have required reduction of residence time. Further, the thermal degradation of some chemical products with time necessitates reduction of the residence time below the 1 to 3 minute minimum considered essential for controllability.

Heat to vaporize the bottom liquid in a distillation column will typically be applied with either a thermosyphon or a kettle reboiler, as shown by Figure 11-2. The thermosyphon reboiler will jet a liquid-vapor mixture back into the column. The return is below the bottom tray, where vapor disengages and goes up through the tray, and the liquid joins liquid from the downcomer in the bottom of the column. Level must be regulated below the lower edge of the return nozzle to avoid entrainment and deterioration of column performance.

The high velocity return, the disengagement, and the small bottoms residence time lead to high liquid traffic, level gradients, and even foaming, which combine to make level control difficult. As shown in Figure 11-2(b), baffling with internal overflows is sometimes installed to minimize the difficulties. With this particular arrangement, bottoms level is divorced from reboiler and column fluid traffic.

A kettle-type reboiler is shown in Figure 11-2(c). Liquid from the lowest downcomer and from disengagement flows to the reboiler. The difference between incoming liquid flow and returning vapor flow spills over a weir, from which it leaves the column as bottoms product. The volume in the end of the reboiler past the weir is generally very small, and thus little surge capacity exists to minimize flow variations to the next unit operation.

11-7 PRESSURE CONTROL — DISTILLATION COLUMN OVERHEAD

The mass and energy control systems for the top of a typical vacuum distillation column are shown in Figure 11-3. A saturated fluid of constant concentration is assumed, and thus a relationship exists between temperature and pressure. Typically, temperature will be used to remove the latent heat in the condenser, providing a liquid for reflux that is only slightly subcooled. Since pressure is known for the condensing temperature, the saturation pressure serves as the set point of the pressure controller. Any increase in pressure is thus due to the partial pressure of noncondensibles, which always leak into a vacuum system and are vented through the pressure (vacuum) control valve.

System performance will generally be poor. Pressure will control very well, but temperature will not stabilize. There are two reasons for the instability. First, since minimum subcooling is desired for energy conservation and

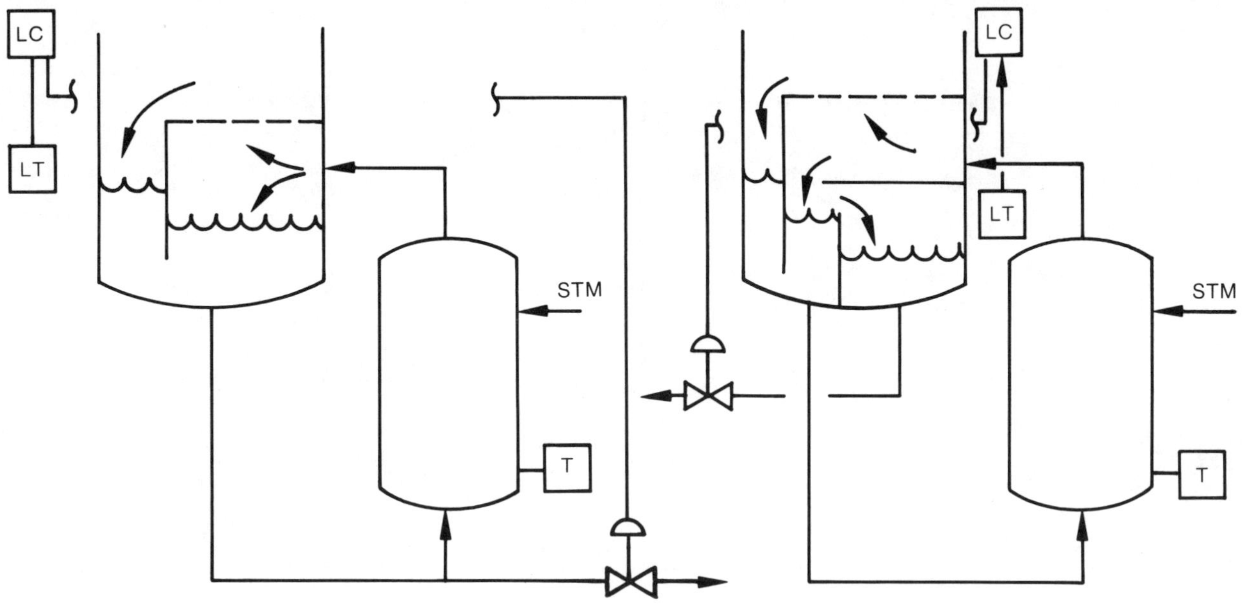

(a) Thermosyphon Reboiler

(b) Thermosyphon Reboiler

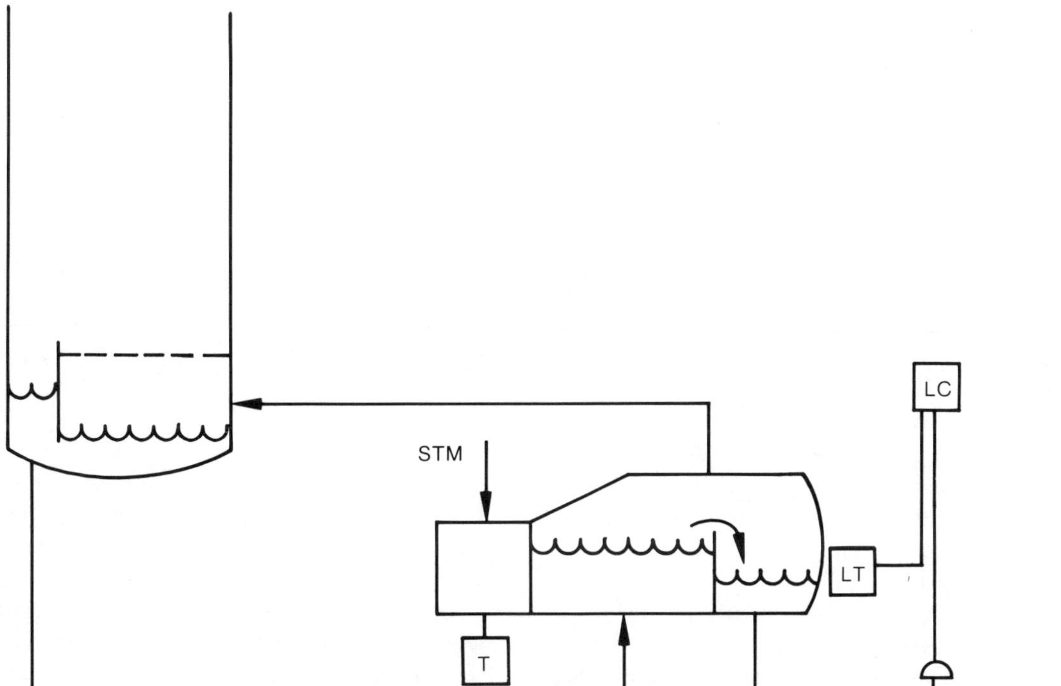

(c) Flooded Kettle Reboiler

Figure 11-2. Level Control with Reboilers

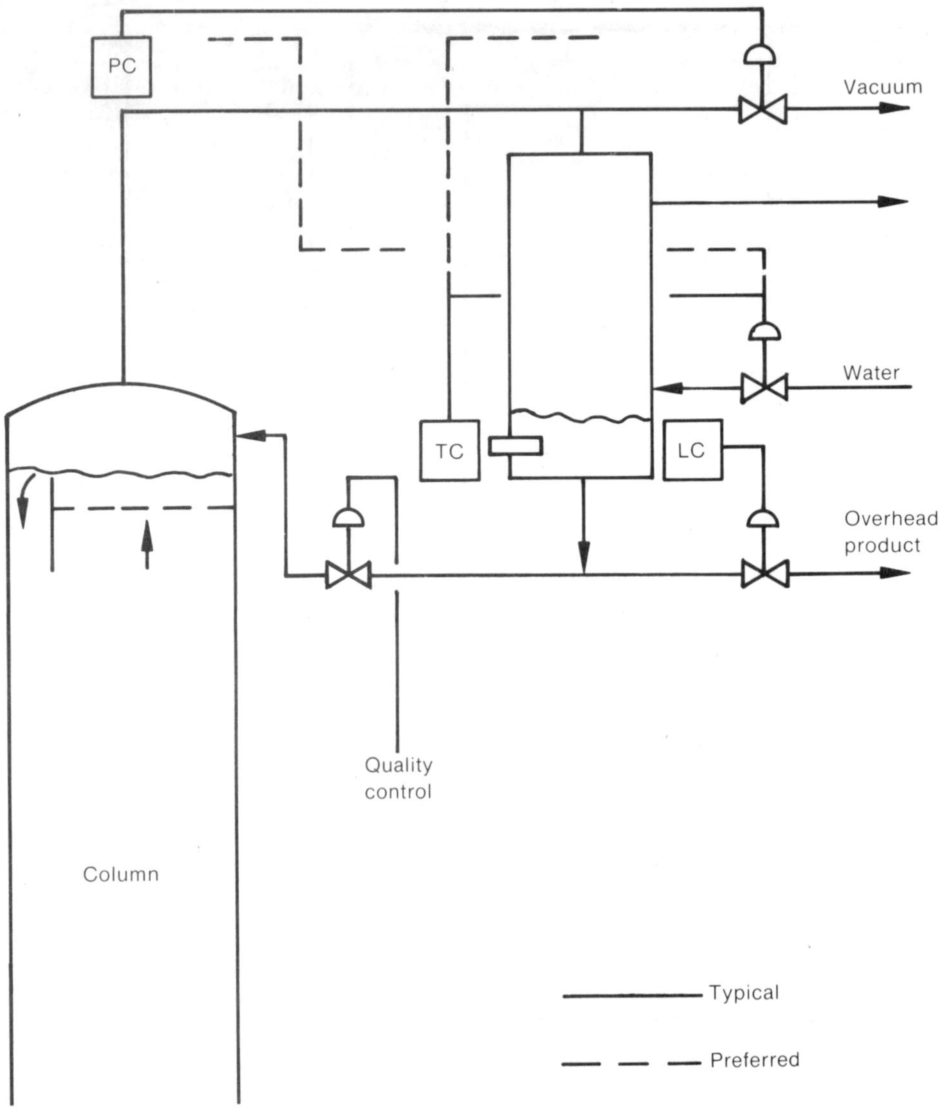

Figure 11-3. Overhead Control System for Vacuum Distillation Column

since temperature changes only with subcooling, the span of the temperature transmitter is quite narrow and represents a high gain to the control loop; second, the latent (condensing) heat is the large portion of the total heat to be removed, a large control valve is required for the large heat load, and the large valve is also "seen" as a large gain by the temperature control system, which is actuated only by the sensible heat change. Therefore, a pneumatic controller will not have a sufficiently wide proportional band to stabilize and will end up on "manual". A simple solution is to replace the standard proportional-reset controller with a floating controller, which has no gain and the proportional band adjustment problem disappears. However, many operating plants desire only "standard" proportional-reset controllers, and an alternate configuration is required.

An ingenious concept is to interchange the two control loops, as shown by the dashed lines on Figure 11-3. While ingenious, the justification for the interchange is not intuitively evident. However, remember that temperature and pressure are directly related for a saturated fluid. The pressure control is the basic system in the new configuration and maintains the vapor mass balance by condensing more or less fluid — a system that will perform very well. Now the saturation temperature is known for the condensing vapor and serves as the set point of the temperature controller. Thus a difference in temperature is indicative of inert gases in the system, causing the opening of the temperature (vacuum) control valve.

11-8 PRESSURE CONTROL — SYSTEM HEAT EXCHANGERS

The relationship of pressure and temperature in a vapor is again useful to provide remarkable improvement in the temperature control on a heat exchanger used to heat a process fluid with steam. The typical control system is shown at the top of Figure 11-4, where the temperature of the process fluid is measured, an analogous electrical or pneumatic signal transmitted to a controller, and the controller output setting the position of a control valve. A decrease in the temperature of the incoming process fluid (ultimately demanding more steam) will draw heat from the steam in the heat exchanger, the metal of the heat exchanger, and the metal of the thermowell, and thus the temperature decrease sensed by the temperature transmitter will be quite slow. With the slow measurement, steam flow will be increased slowly, resulting in sluggish temperature control.

Should process conditions warrant it, the cascade control system shown at the bottom of Figure 11-4 provides much faster control response. Steam pressure in the heat exchanger is regulated by a piloted, self-contained pressure-

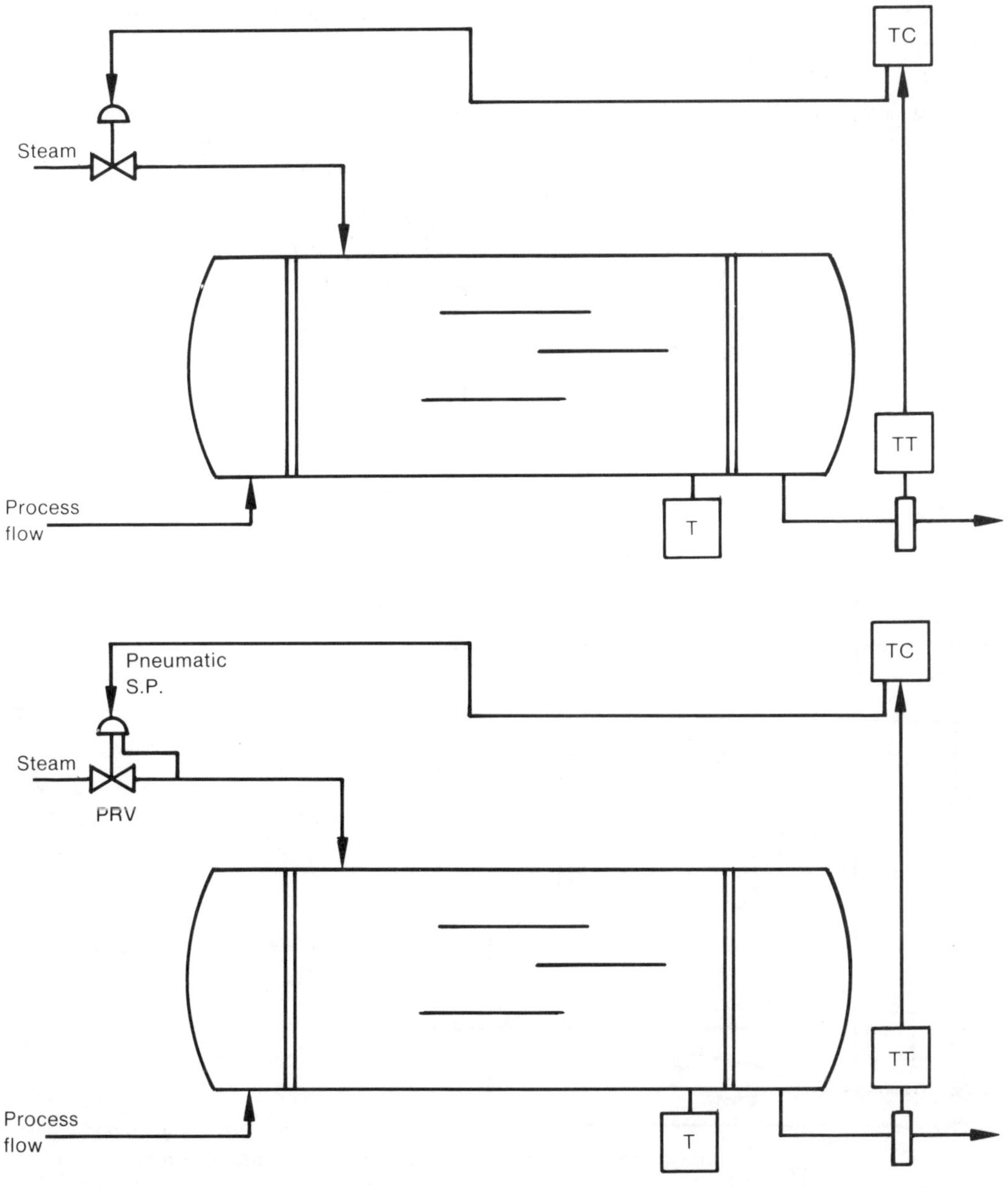

Figure 11-4. Steam Heater with Temperature Control

reducing valve (PRV) with a pneumatic set point. The PRV set point is adjusted by the output of the temperature controller. Now when the temperature of the incoming process fluid drops, heat is drawn from the steam, and the decreasing temperature results in a decreasing pressure in the saturated vapor. But the PRV "sees" the decreasing pressure and begins supplying more steam even before the temperature-measuring device senses the temperature change. Much faster response is attained, and the temperature controller slowly brings the steam pressure to the new set point for the new load condition.

11-9 PRESSURE CONTROL — FIRED VESSELS

A liquid vaporizer is shown on Figure 11-5, with liquid entering the bottom, being brought to its boiling point, vaporized, and even superheated in the tubes of the vaporizer as it leaves at the top. Exit vapor temperature is measured and used to adjust the air flow to the unit. Fuel is then ratioed to the air.

The fired side of the vaporizer is operated at a slight vacuum to assure that any leakage is "in". However, the vacuum cannot be allowed to become too great, since the differential pressure forces on the very large-area walls of the vessel could collapse them. A valve placed in the stack is used to regulate pressure.

The first anomaly to be noted is that the instrument engineer is being asked to size a valve with flow from a vacuum to atmospheric pressure. The flow is determined by a combination of the "chimney effect" that causes the hot air to rise, the various pressure drops in the tubing and the ducting, and the valve itself. In view of this complexity, a line-size butterfly valve is usually selected.

A bit of experimentation will show that the inherent flow characteristic of the butterfly valve is converted to quick-opening installed characteristic, as shown in Figure 11-5. The valve has a very high gain throughout its useful range, as does the pressure transmitter because of its very narrow span. Typically, a pneumatic controller will not

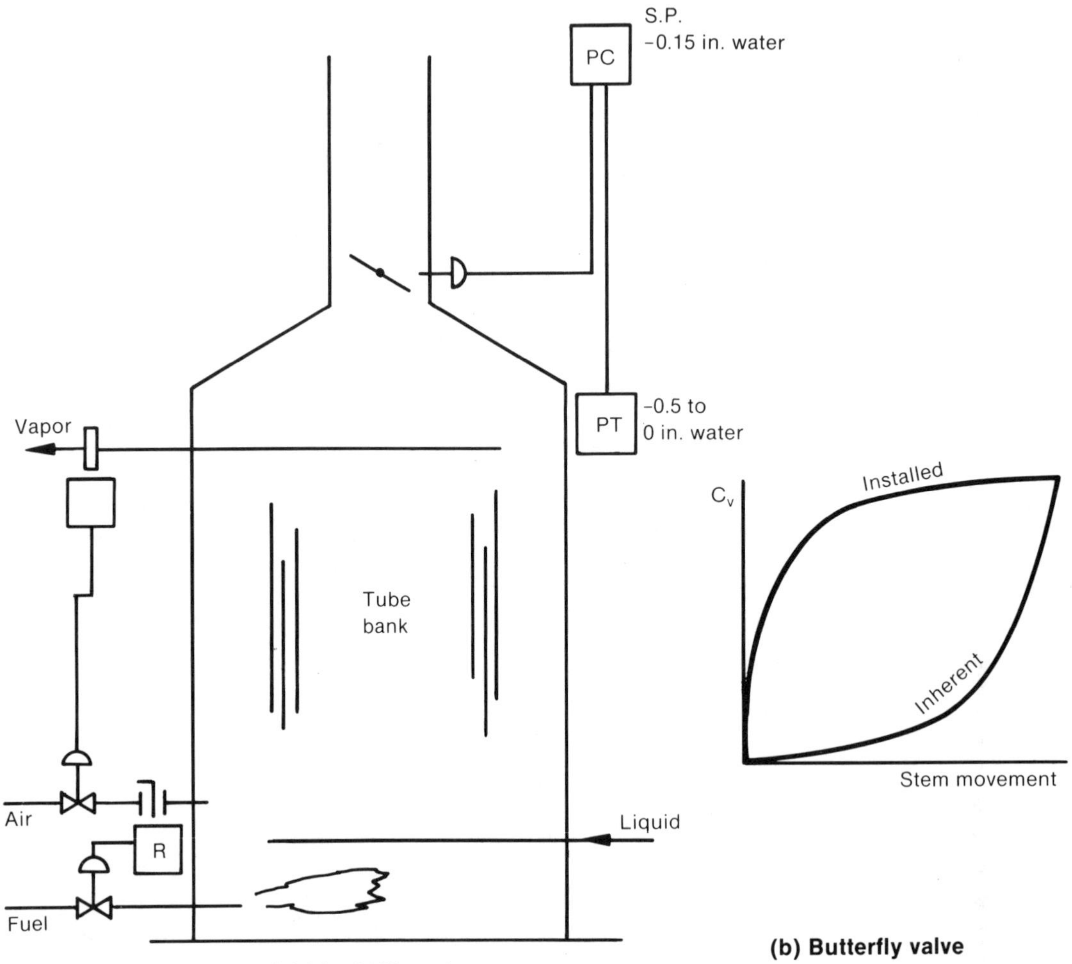

(a) Liquid Vaporizer

(b) Butterfly valve characteristic

Figure 11-5. Vaporizer Pressure Control

have a sufficiently wide proportional band to stabilize and will be operated in "manual".

The first action to be taken is to adjust the valve so that the full range of the actuating signal (3 to 15 psi) moves the valve disc only through its useful range, perhaps 40° of rotation. Thus valve gain is reduced. Second, a floating controller should be installed. The floating controller has no proportional band and accommodates the very high gain system quite well.

11-10 PRESSURE CONTROL — CENTRIFUGAL COMPRESSOR

Axial and centrifugal compressors are gaining in popularity because of their ability to economically pump large quantities of vapors and gases. The centrifugal compressor is a constant pressure ratio-variable throughput machine, as contrasted to the reciprocating compressor, which is a constant throughput-variable pressure machine.

One basic control system on the centrifugal compressor sets throughput. As shown on Figure 11-6, the system regulates the mass balance on the discharge of the compressor by forcing the fluid pumped by the compressor to match the fluid demanded by the users. Pressure is used to sense increases or decreases in mass in the discharge piping. In this typical system, the pressure controller positions a control valve in the inlet of the compressor. Since the compressor is a constant pressure ratio machine, closing the inlet valve causes inlet pressure to decrease, which in turn causes discharge pressure to decrease. Thus if the user

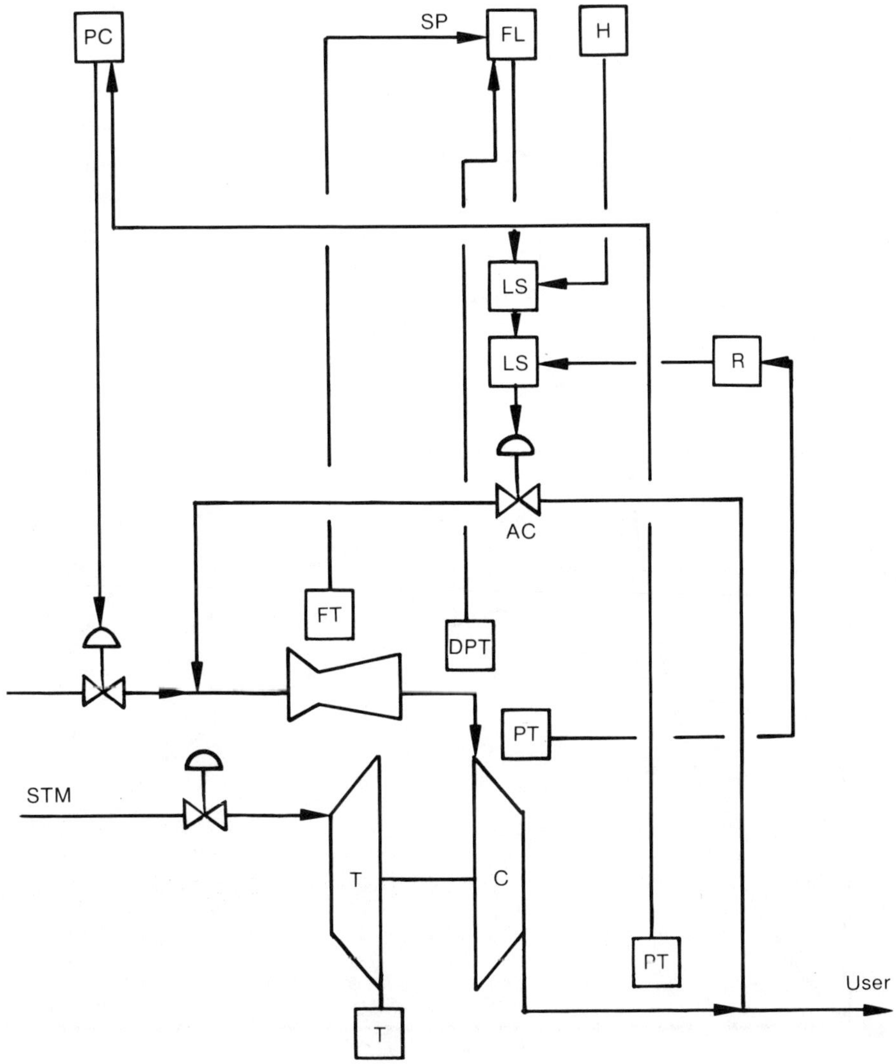

Figure 11-6. Centrifugal Compressor Controls

demand should decrease, the discharge pressure rises, the controller closes the control valve slightly to decrease the inlet pressure, and the discharge pressure is decreased toward the set point.

Energy savings can be attained by directing the output of the pressure controller to the speed of the driver, whether it be a variable speed electrical motor or a steam turbine. The energy (pressure or head) demanded by the user is then matched to the energy used by the driver, as contrasted to the energy loss of the control valve in order to make the two match. However, variable speed drivers are somewhat more expensive, and long-term savings are determined by balancing amortization costs against energy costs.

One undesirable feature of centrifugal and axial compressors is an aerodynamic instability that occurs if flow through the compressor is reduced below a certain value. This point is denoted the surge point, or the surge line, as shown on the compressor map in Figure 11-7. As flow is reduced, the compressor wheels, in pulling fluid from their inlet volumes, lower the inlet pressures until flow actually reverses, raising the inlet pressure and lowering the discharge pressure until the design pressure ratio is again established. Then flow again goes forward until the phenomenon occurs again. This surging is audible, sometimes barely discernable and sometimes violent enough to shake the whole building. Violent surge can cause the compressor wheels to rub on their casings and overload thrust bearings;

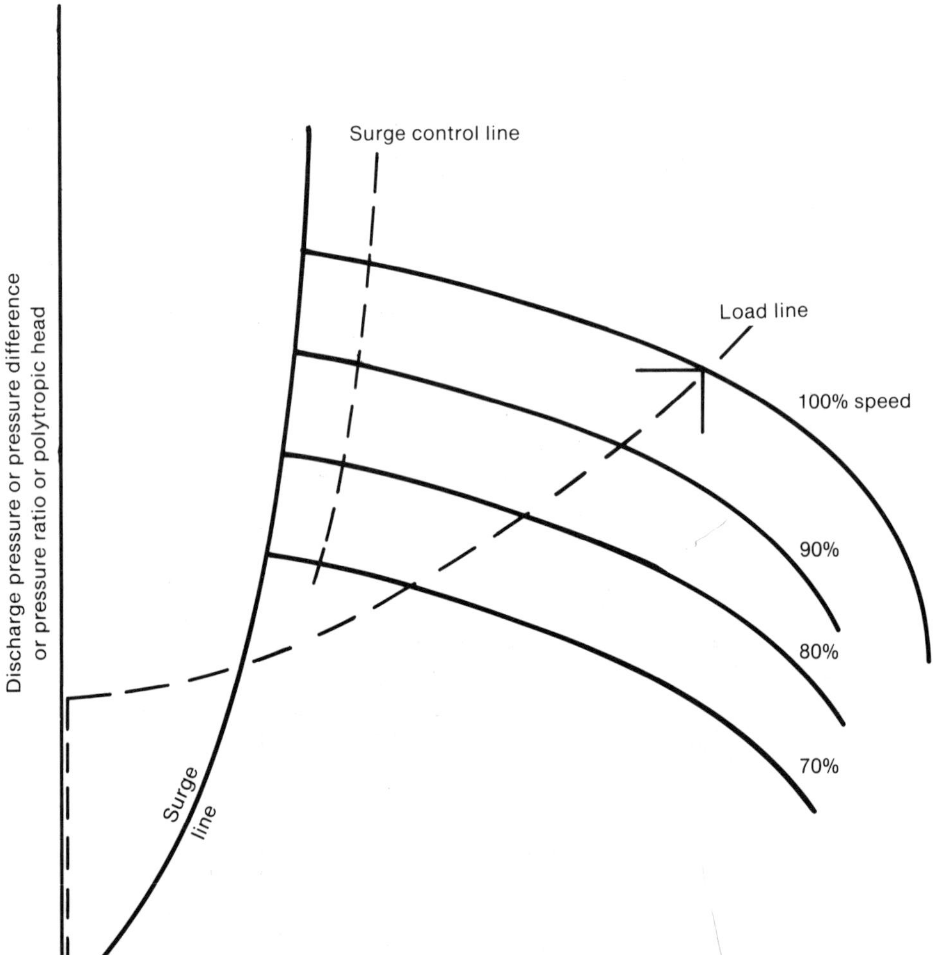

Figure 11-7. Compressor Map

it can sometimes disable the machine completely in a very few seconds. Thus surge must be prevented to protect the compressor.

The anti-surge control system shown in Figure 11-6 actually computes a line parallel to the surge curve but some conservative distance to the right. Then as process demand causes throughput to decrease to the surge control line, the anti-surge controller opens the recycle valve to maintain a minimum flow through the compressor. Since compressor flow is intimately related to pressures, a strong interaction is generally found between the pressure and anti-surge control systems. Therefore, the tuning of the two control loops must be approached cautiously.

Since protection from surging is imperative, many newer compressors include pressure transmitters that actually follow the surge cycles. Figure 11-8 shows the high frequency of the pressure variation, a frequency so high that it cannot be tracked by the typical bourdon-type process pressure transmitter. Instead, the high performance strain gage or piezoelectric sensor pressure transmitter is required to detect the incipient surge.

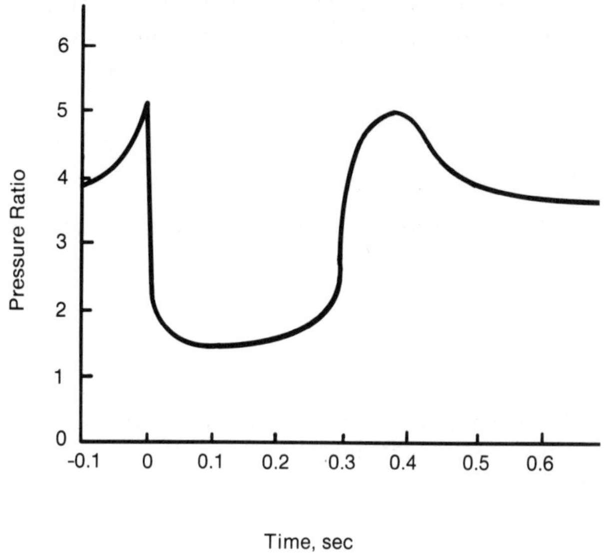

Figure 11-8. Compressor Surge[117]

Problems

1. Suppose the open-loop gain (Example 11-1) must be between 1 and 3 for good control. Flow varies between 10 and 90 lb/sec with a linear valve characteristic. Would the controller have to be retuned to maintain performance?

2. In question 1, above, what could be done to avoid controller retuning?

3. For a level control system (Example 11-2):

   $7.48\ V = 10{,}000$ gallons $\qquad q_{max} = 10$ gpm

   What are the values of controller gain (K_c) and reset (τ_i)?

4. Small boiler drums and separators (Figure 11-1) have high internal flows that impinge on the liquid LT nozzle and bias the level measurement. What can be done about this?

5. What can be done (Example 11-3) to change a troublesome natural frequency of the displacer?

6. If a kettle reboiler (Figure 11-2) has a large vapor space, what can be done to increase the level control volume (V) to provide smoother control?

7. Why does the temperature controller set point have to be set below the actual condensing temperature in a condenser (Figure 11-3)?

8. A decrease in process temperature (Figure 11-4) causes steam pressure to decrease. Why is this?

9. Where is a good point to put the upper limit of valve stem movement on the installed valve characteristic in Figure 11-5?

10. If the pressure and anti-surge controllers on a centrifugal compressor were interacting severely, which one must be detuned to attain stability?

SECTION **12**

TEMPERATURE CONTROL

12-1 INTRODUCTION

Level and pressure are mass balance control variables, flow is a production rate control variable, and temperature is a quality control variable. Should the whole chemical plant be considered the process, only one production rate (flow) system is required, while each unit operation in the plant requires a mass balance control and a quality control. Temperature control is therefore applied to many different kinds of processes of different capacitances, different dynamic relationships, different disturbances, and different accuracy specifications. These factors must be considered in selecting the proper type of automatic temperature controller for an application. For ease in controller tuning and minimum initial cost, the simplest controller that will produce the desired result should be selected.[66,68,69] In some instances, the required response will be unobtainable even with the most sophisticated controller. This situation indicates an inherent process heat balance problem that must be analyzed and revised before the controller is selected.

Process controllability can be assessed by the analytical methods of Sections 3 and 4. Several pertinent questions should be asked. Is the heat source oversized or undersized? Oversized represents a high gain, which can lead to stability problems. Undersized produces slow response. Is the thermal mass large or small? Large means slow response, while small will respond quickly. What is the distance from process to sensor? Large distances represent dead time and a less stable system. Comparing the process to the temperature control requirements will indicate the type of controller to be used.

12-2 ON-OFF CONTROL

Consider a large storage tank containing liquid with an electrical heating coil to prevent freezing, as shown in Figure 12-1.

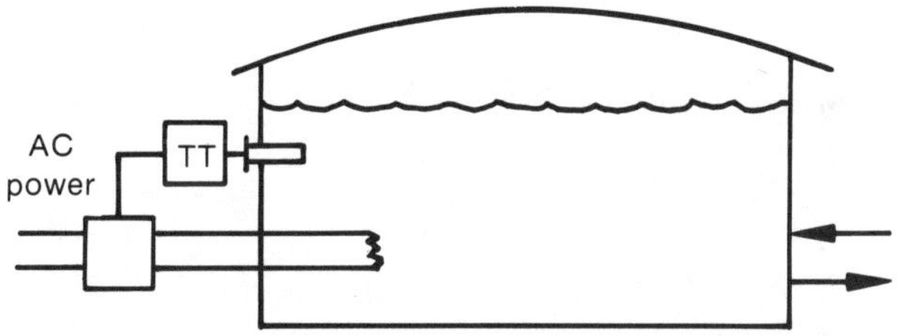

Figure 12-1. Temperature Process, Large Capacitance

Analysis indicates that the vessel is a single large capacitance with a time constant measured in hours. Temperature is relatively unimportant except that tank contents shall not freeze or boil. A secondary restriction is to minimize the use of steam for energy conservation. The process can be controlled by a simple on-off controller, shown in Figure 12-2.

In response to intelligence from the sensing element, the amplifier will open and close a relay in such a manner that the average power to maintain the set point temperature is delivered to the tank. The relay will close when the temperature falls below the set point, causing the element to heat, and will open as the temperature rises above the set point. This response is therefore cyclic, as shown in Figure 12-3.

Acceptable temperature regulation with on-off control requires that two conditions be met: (1) The cyclic peak-to-peak temperature variation equal to the controller hysteresis is acceptable, and (2) the process heating and cooling rates are long enough to prevent too rapid cycling of the final control device.

The temperature cycle magnitude can be reduced by the application of time proportioning on-off control. This is a method of on-off control in which on time and off time of the final control element are proportioned according to how far temperature has moved from the set point. As the temperature signal approaches the set point, on time is reduced proportionately. Operation at 50 percent power

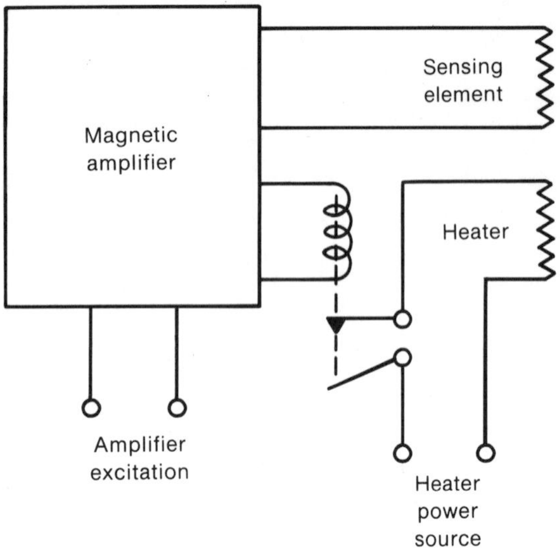

Figure 12-2. Switching System(69)

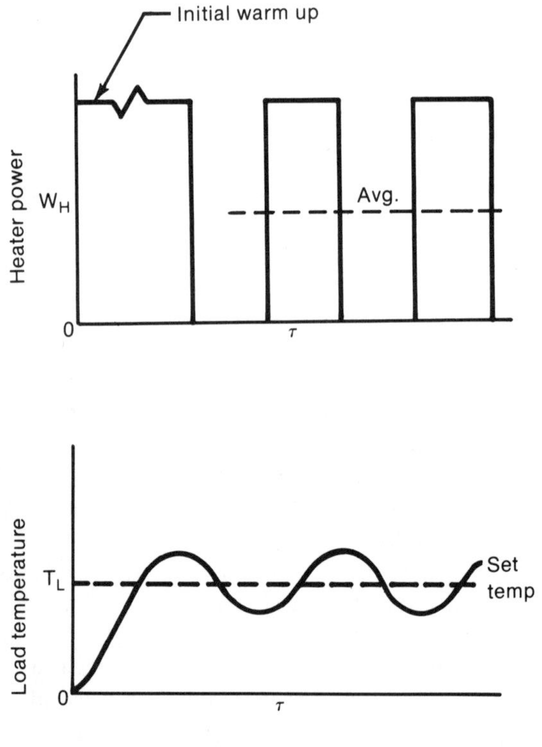

Figure 12-3. Switching System(69)

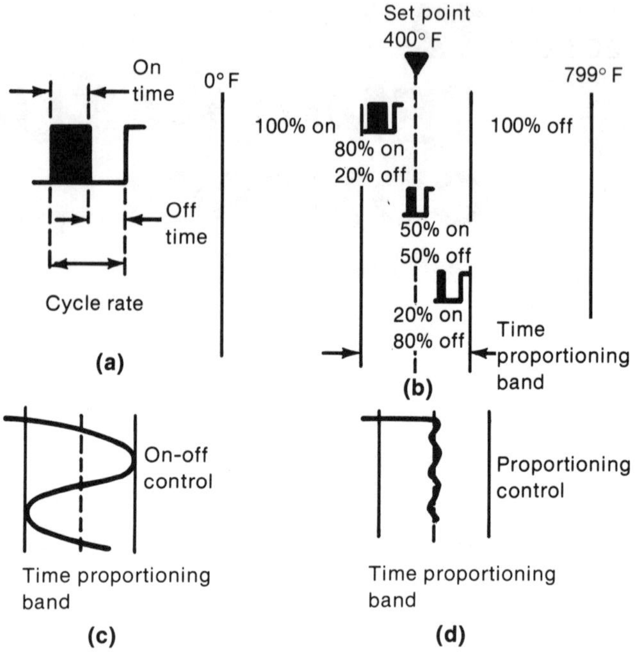

Figure 12-4. Time Proportioning Avoids Large Overshoot of Set Point Temperature(68)

means heat is applied to the load (the liquid in the tank in Figure 12-1) 50 percent of the set cycle time, and heat is not applied for the other 50 percent. Cycle time is the time for the output device (relay or solenoid valve) to change from "on" to "off" and back "on" again. The improvement in cyclic temperature variation is shown by the comparison in Figure 12-4.

Advantages of on-off control include:(69)

(1) The amplifier and relay circuits are simpler than a proportional amplifier designed for the same heater wattage.

(2) Total cost of the amplifier and relay is, in general, somewhat less than the comparable proportional amplifier.

(3) Since the amplifier and relay can be separate devices, they can be packaged into equipment already designed.

(4) Obtaining and utilizing the intelligence that the system is operating is easier with a switching system. An extra set of relay contacts can be used to interlock warm-up circuitry or to provide visual indication that temperature is within the operating band.

12-3 PROPORTIONAL CONTROL

As process capacitance becomes smaller and as control accuracy requirements become more stringent, the cycling temperature variations due to on-off control become unacceptable.

The use of a proportional-only controller is indicated. The offset of a proportional-only controller when used in a control system is the phenomenon that the temperature does not return to the set point after a load change, as discussed in Section 9. Manual reset is an adjustable controller bias that permits matching of temperature and set point at the design load condition. Offset will then appear only after a load change. Offset is minimized by adjusting the controller to a narrow proportional band (high gain). Frequency response analysis indicates that a single-capacity (first-order) process cannot become unstable (the maximum phase lag is 90°), and thus the proportional band can be set as narrow as desired. In the limit, zero proportional band is on-off control. The more typical process is multicapacity, and thus higher-order, and stability limits the minimum value to which the controller can be set. The latter condition prevails in the typical petrochemical processing unit, and proportional-only temperature controllers are seldom applied because of the large offsets.

Figure 12-5 shows the response to the tuning of a proportional-only controller.

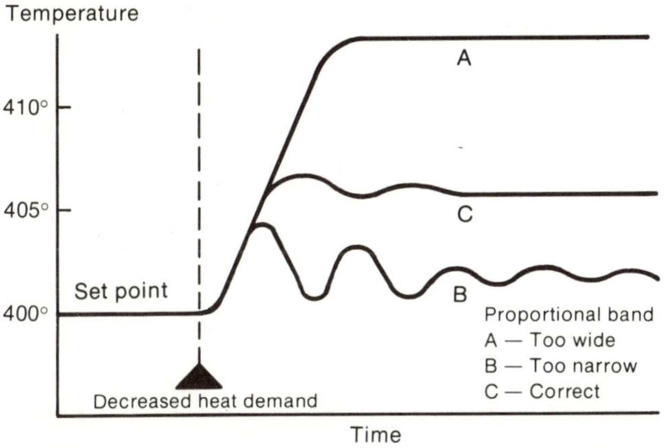

Figure 12-5. Process Characteristics for a Proportional Controller[(66)]

The response curves show the resulting temperature change after decreased process heat demand. The large offset in curve A results from the proportional band being too wide. The offset can be reduced by narrowing the proportional band. Instability results if the proportional band is made too narrow as in curve B. Acceptable response, as in curve C, is achieved with a proportional band slightly wider than that which causes oscillation. Should process parameters or operating conditions change, retuning of the proportional band will be necessary. Retuning can be minimized by using a proportional band slightly wider than optimum to avoid future instability.

Some of the advantages of proportional-only control, as compared to on-off, are as follows:

(1) As can be seen in a comparison of Figures 12-4(c) and 12-4(d), the proportional system is intrinsically more accurate and can hold the load temperature excursions within tighter tolerances.

(2) Since no relay is used, overall reliability is higher.

(3) Since no contacts are "making" and "breaking" the heater load current, no radio noise problems exist.

(4) Since no relay is used, system life is longer; in fact, it is virtually indefinite. Better resistance to environmental conditions of shock and vibration is obtained, since the relay is always the limiting factor in these considerations.

(5) Power to the heater is smoothly applied and slowly modulated in response to demand. No sudden load changes are placed on the power source such as a switching system applies.

The proportional-only controller finds application in processes that are started up frequently or in batch processes. Since controller reset is an integrating function, it will wind up or saturate during inoperative periods. Upon start-up, the reset mechanism will not come out of saturation until the process has almost reached the set point, causing a significant overshoot that may be unacceptable.

Figure 12-6 shows three possible temperature responses from cold start-up conditions using a proportional-only controller.

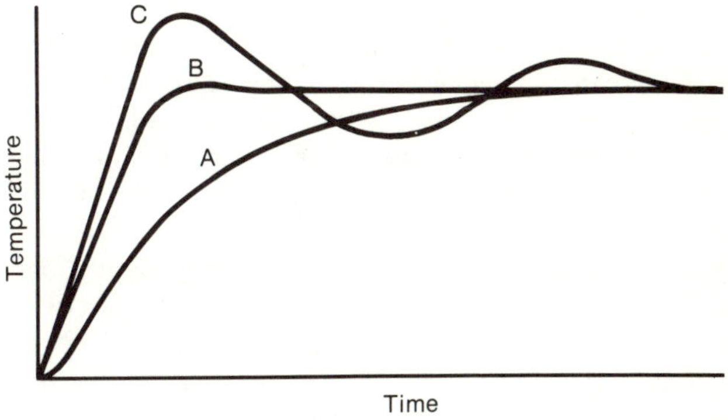

Figure 12-6. Proportional-Only Response for a Temperature System

Response curve A could be considered a good response if the material being heated required a slow, controlled heat-up. Curve B is a good response within the specification that start-up be as fast as possible without overshoot. Curve C is the fastest possible start-up. Thus "good response" cannot be defined but depends on the process and operational requirements.

12-4 PROPORTIONAL-RESET (PI) CONTROL

Processes with long or multiple time lag and fast rates of change of heat demand (e.g., a process heat exchanger) require a wide proportional band for stability, resulting in large offsets that occur with changes in load. The offsets are eliminated by the automatic reset function in a PI controller.

Section 9 discussed the practice of setting the adjustable reset time constant equal to the largest process time constant. All calculated controller tuning parameters are subject to further, preferably small, trimming in the field due to the approximations in the mathematical technique. The reset tuning can be verified from the temperature response curves shown in Figure 12-7.

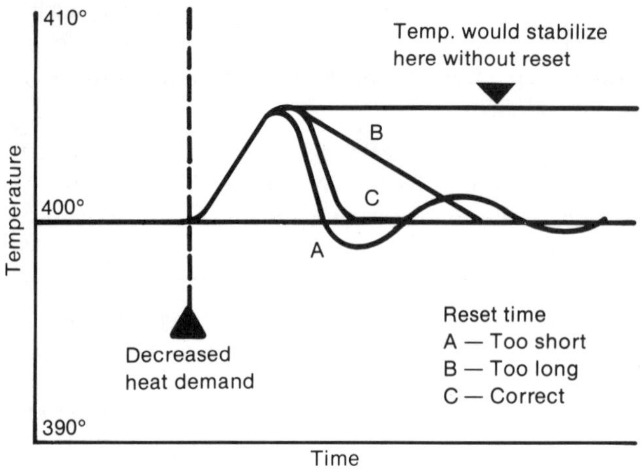

Figure 12-7. Temperature Response to Reset Adjustment(66)

Figure 12-7 shows the temperature system response for a decrease in heat demand. A reset time adjustment that is too long for the process ($\tau_i > \tau_p$) will take a long time to return the temperature to the set point, as shown by curve B. A reset adjustment that is too short will permit the controller to outrun the process, causing the temperature to cross the set point and recover as a damped oscillation, as in curve A. A reset time that is much too short (fast) will cause continuous oscillation called, appropriately enough, a reset cycle. A reset time adjustment that returns the temperature to set point as rapidly as possible without overshooting the set point is generally considered most desirable. This response is shown as curve C.

The units on the reset adjustment knob on a standard controller will usually be in terms of time (minutes) and will be the value of the reset time constant (τ_i). Sometimes the units will be in repeats/minute, or $1/\tau_i$, to accommodate the integral form of the function. Care must be exercised in reversing the reset magnitudes in the foregoing discussion if repeats/minute is used.

12-5 PROPORTIONAL-RESET-DERIVATIVE (PID) CONTROL

Derivative (rate) action is used to advantage on processes with multiple time constants to speed recovery after a process disturbance, as shown in Section 7. The derivative action provides a phase lead function that cancels some of the process system lag and allows the use of a narrower proportional band without becoming unstable. The narrower proportional band results in smaller deviation from set point upon a disturbance.

Almost all temperature sensors in fluid processing plants are in thermowells in order that they might be removed from service without shutting down and draining the pipeline. The thermowell has a significant time constant, as discussed in Appendix B. Derivative action is recommended in temperature controllers to compensate for measurement dynamics.

The choice of the control valve flow characteristic can be illustrated by considering the fluid-fluid heat exchanger of Figure 12-8. Here a process fluid in the tubes is being heated by steam condensing in the shell (outside the tubes).

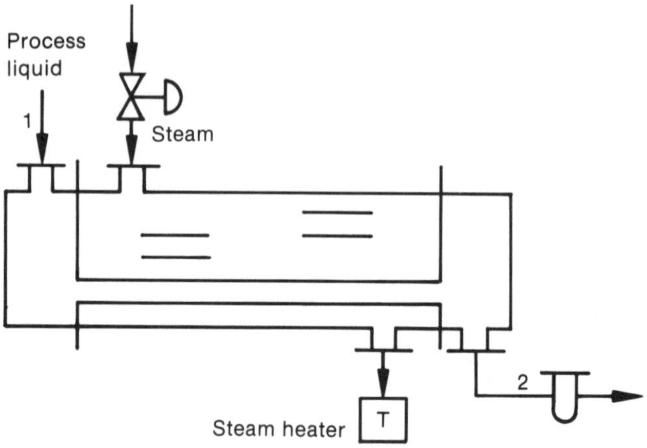

Figure 12-8. Steam Heater

The process is described by a heat balance on the process fluid:

$$Mc \frac{dT_2}{dt} = w_1 cT_1 - w_2 cT_2 + Q$$

$$Q = w_s \lambda$$

where w_s = steam flow, lb/min

λ = latent heat, Btu/lb

Transposing the self-regulation term,

$$Mc \frac{dT_2}{dt} + w_2 c\, T_2 = w_1 c\, T_1 + w_s \lambda$$

Let $w_1 = w_2$,

$$\frac{M}{w_1} \frac{dT_2}{dt} + T_2 = T_1 + \frac{\lambda}{wc} w_s$$

or

$$\left(\frac{M}{w} s + 1\right) T_2 = T_1 + \frac{\lambda}{wc} w_s$$

The process gain is

$$K_p = \frac{\lambda}{wc} \qquad (12\text{-}1)$$

A simple criterion for constant stability over a range of production rates is that open-loop gain be a constant:

$$K_o = K_c K_v K_p K_T = \text{constant} \qquad (12\text{-}2)$$

Equation 5-3 of Section 5 shows the gain of an equal percentage valve to be proportional to valve throughput, which is in turn proportional to process throughput:

$$K_v = K_s w_s = K_1 w \qquad (12\text{-}3)$$

Substituting Equations 12-1 and 12-3 into 12-2:

$$K_c K_1 w \frac{\lambda}{wc} K_T = K_c K_1 \frac{\lambda}{c} K_T = \text{constant}$$

Thus, use of the equal percentage valve characteristic results in a constant open-loop gain and is recommended for temperature control valves.

12-6 FLOATING (I) CONTROL

Thermal processes that incur a change in state require special consideration, since temperature does not change (if pressure is constant) during vaporization or condensation; yet the latent heat load in any process is typically much larger than the sensible heat load. The latent heat in the exchanger shown in Figure 12-8 is given up by the manipulated variable and is helpful in simplifying analysis by keeping shell side temperature constant throughout the exchanger. A change of state of the process fluid causes temperature control to be substantially more difficult. Consider the distillation column condenser shown in Figure 12-9.

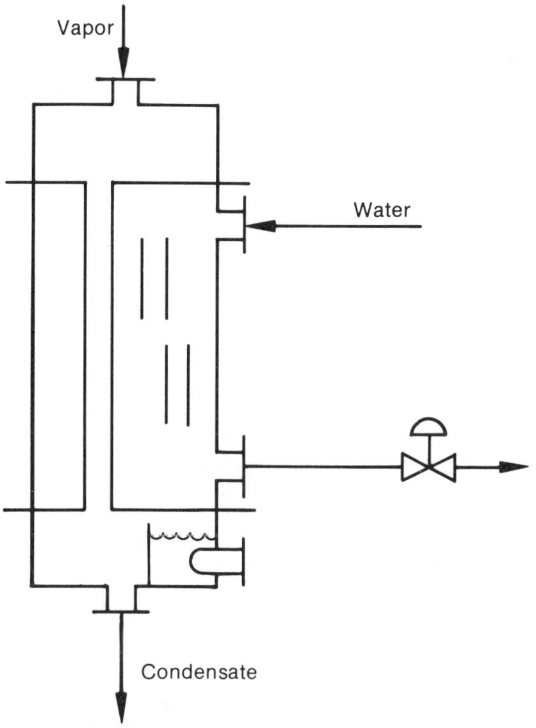

Figure 12-9. Distillation Column Condenser

The temperature cannot be controlled at the condensing or saturation temperature, since temperature does not change during condensation. Temperature is controlled a few degrees below the saturation temperature so that temperature will change with cooling water flow. The small decrease in temperature that causes the process to operate in the sensible heat span is called subcooling. Energy conservation dictates that subcooling be a minimum, because steam must be used to again bring the column contents up to boiling.

The temperature control system is characterized by a very large control valve to provide for the latent heat load and a very narrow temperature span on the sensor in order to be able to set the set point to minimum subcooling. Both the valve and the transmitter represent very high gains in the control loop. Consider the open-loop gain

$$K_c K_v K_p K_T = \text{constant}$$

Gains K_v and K_T are very large, and controller gain K_c must be made small to avoid instability. Many controllers have a maximum proportional band of 500 percent, or a minimum gain of 0.2, which is not low enough to stabilize. Therefore, many such systems are left on "manual," with too much water going to the condenser and a very inefficient use of energy.

The application of a floating controller stabilizes the process since the controller has no gain mode. Figure 12-10 shows the Bode plot characteristic of the floating controller.

The floating controller is an integrator and contributes a constant 90° phase lag. It moves the point where the phase curve crosses −180° to a much lower frequency, resulting in a much slower system response. However, as much attenuation as is necessary for stability is provided by moving the magnitude curve horizontally. Upon positioning it for adequate gain margin, the floating time constant can be read at the intersection of the magnitude curve and $M = 1$. The single adjustment knob on the controller is set to this value.

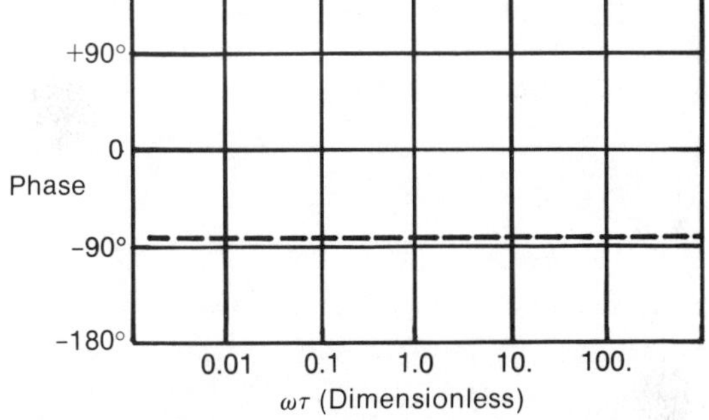

Figure 12-10. Frequency Response of a Floating Controller

Problem

1. Consider the heat exchanger in Figure 12-8 with a temperature control system including a sensor, transmitter, PID controller, and control valve. Using the following information, calculate the proportional band, reset, and derivative settings for the controller:

   Heat Exchanger

   Weight of metal in contact with process fluid, 1000 lb
   Specific heat of metal, 0.11 PCU/lb °C
   Weight of process fluid contained, 200 lb
   Specific heat of process fluid, 0.8 PCU/lb °C
   Throughput of process fluid, 200 lb/min

   Thermowell and Sensor

   Weight, 2 lb
   Specific heat, 0.11 PCU/lb °C
   Film heat transfer coefficient, 200 PCU/ft^2, °C-hr
   Area, 0.5 ft^2

   Instruments

   Transmitter span, 0 to 100 °C
   Maximum flow through control valve, 80 lb/min (water)
   Specific heat of cooling fluid, 1 PCU/lb °C
   Temperature increase of cooling fluid, 40°C
   Open-loop gain, 3.0

SECTION **13**

ADVANCED CONTROL TECHNIQUES: CASCADE AND FEEDFORWARD

13-1 INTRODUCTION

Sometimes a simple control system cannot keep a controlled variable near the set point even with the best controller tuning. Difficulty will often be experienced with processes that have multiple time constants or have a strong disturbance. Distributed hierarchical control systems, employing several superimposed loops, provide a practical solution to many industrial control problems that are difficult to resolve otherwise. Cascade control is a hierarchical system that is among the most frequently encountered advanced control techniques.[70,71,72]

The price to be paid for the improved control attained with a cascade system includes (1) the cost of more complex instrumentation, (2) the additional complexity of tuning two or more feedback loops, and (3) the added design costs in proper application, since improper application can actually cause deterioration in control performance.

13-2 MULTIPLE TIME CONSTANTS

Cascade control can be illustrated by a process made up of two time constants, as shown in Figure 13-1.

Conventionally, the output of second transfer function would be fed back to a controller, which would adjust the

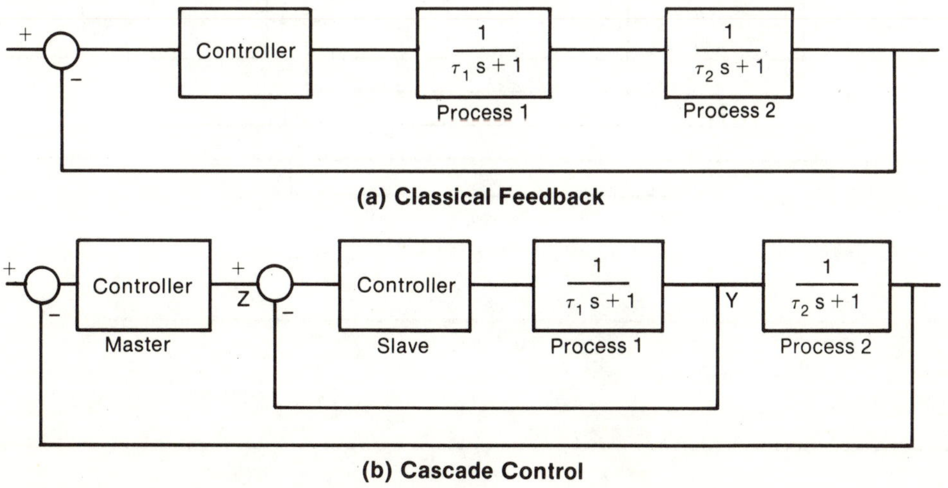

Figure 13-1. Control of a Process with Two Time Constants[72]

input to the first transfer function. In the cascade system, a secondary or slave loop is closed around the upstream transfer function, while the primary or master loop is closed around the entire process.

Improvement is attained by the reduction of the time response of the first transfer function, since closing a negative feedback loop around a dynamic element will always shorten the response time or raise the natural frequency. The closed-loop response of the slave loop must be faster than the response of the primary loop, or little improvement in control performance will be observed and satisfactory tuning of the primary controller will be impossible.(72)

With plants having a chain-like structure, partitioning of the plant into several sections, as shown in Figure 13-2, permits control of these sections by a corresponding number of superimposed loops.(71)

The controllers are designed recursively, beginning with the innermost loop. As each loop is closed, the speed of response of that part of the plant is increased. Further, the effects of disturbances in the inner loop are greatly diminished. Finally, the effects of nonlinear elements are suppressed, making possible a wider range of operation without a change in performance. Conversely, inner loops that function well are necessary for the functioning of the outer loops, each one of which is progressively slower. It is imperative that the innermost loop be made as fast as possible, since it sets the time scale for the complete system.

Figure 13-3 shows a position control system with multiple cascades.

The chemical reactor shown in Figures 9-2 and 9-8 serves to illustrate a multi-time constant process such as that shown in block diagram form in Figure 13-1.

13-3 FLOW CASCADE

By far, the most common slave variable is flow. Consider the furnace temperature control system shown in Figure 13-4. When the furnace temperature deviates from

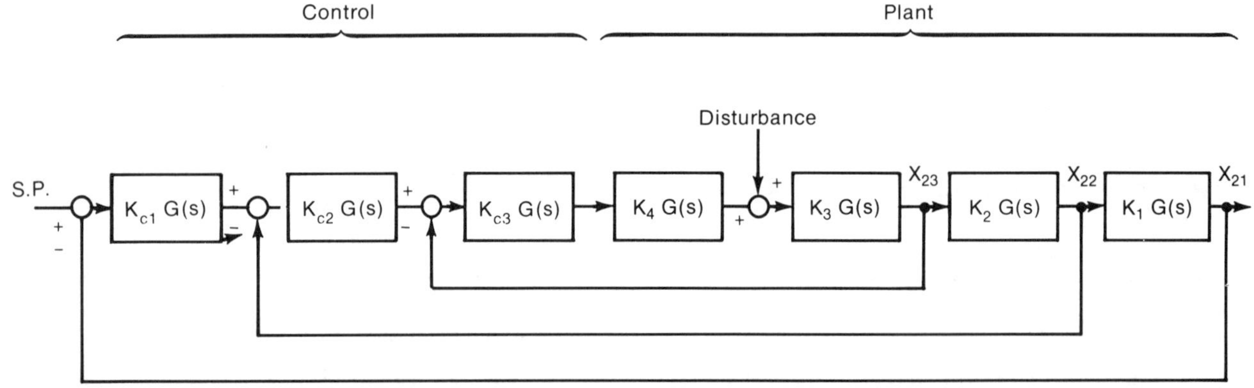

Figure 13-2. Chain-Like Plant Structure

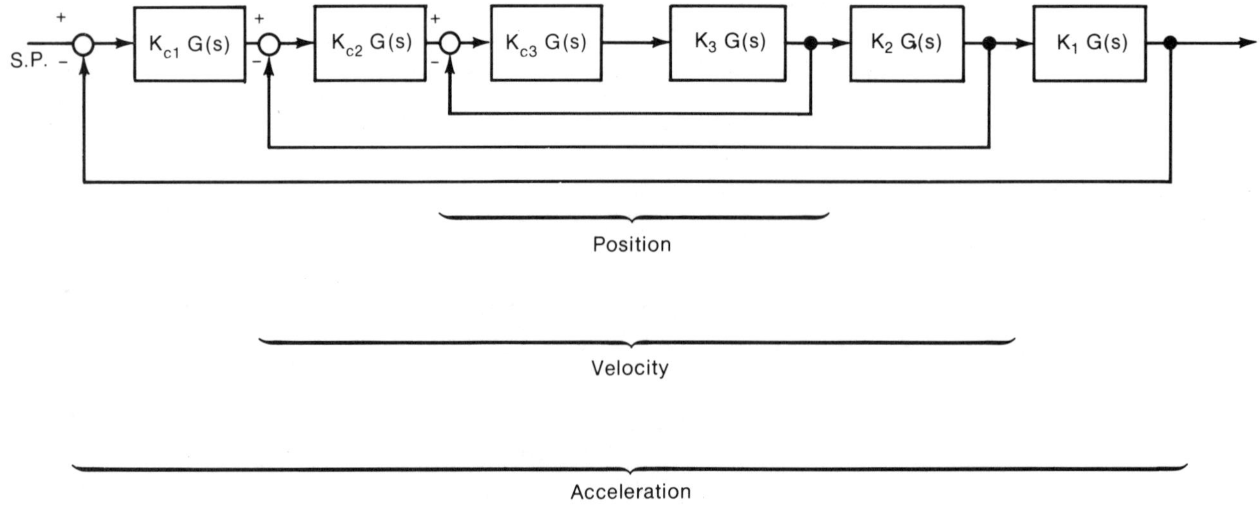

Figure 13-3. Position System Cascade Control

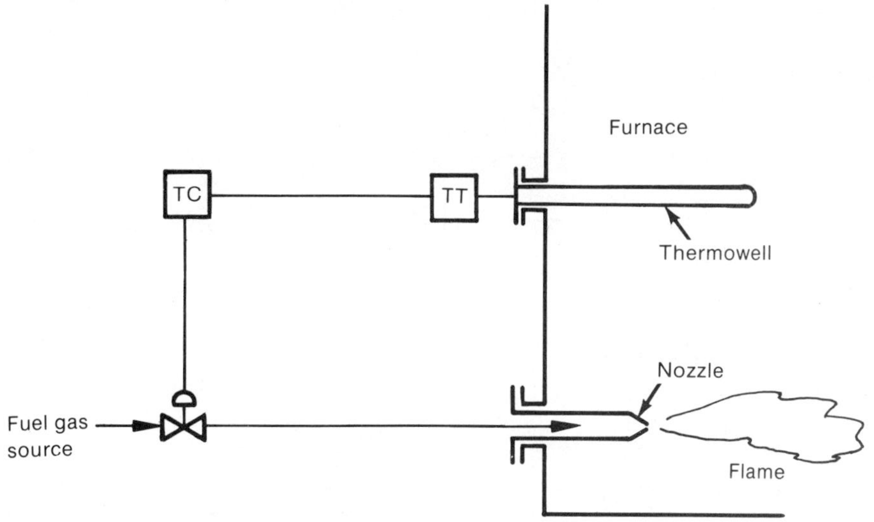

Figure 13-4. Temperature Controlled Furnace, Single Loop

the controller set point, the temperature controller (TC) will adjust the fuel valve until the measured temperature matches the set point. The system will be fully adequate as long as fuel characteristics (composition, pressure) remain constant. But wide deviations in fuel pressure will cause rapid variations in fuel flow, even though the valve stem does not move. The furnace temperature will change, albeit much more slowly. When the news that the temperature has changed reaches the controller, it repositions the valve. However, the temperature changes slowly; time elapses before the controller knows that any correction is needed; additional time elapses before it knows how much; and still more time elapses before it can correct the valve position. In the meantime, the pressure transient could have disappeared. Thus, flow changes very quickly and temperature very slowy. Temperature control will be poor because the valve correction is not made until the temperature has changed. Continuing changes in pressure will not only result in continuing valve correction, but will make temperature control quite unsatisfactory.

Control is poor because the temperature controller regulates the valve position, which is only one of several variables that sets fuel flow. But the temperature is a function of flow, not stem position, so improved control would be expected if the temperature control was to set flow rather than valve position. Flow can be controlled by a flow control system, as shown in Figure 13-5. Then, if temperature is used to set the set point of the flow control system, significantly improved temperature control will result. This arrangement is shown as the temperature cascade in Figure 13-5.

Figure 13-6 is a typical signal flow diagram for all cascade control systems. It illustrates the following points pertinent to all cascade control systems:

(1) Two elements, the secondary controller and the secondary measuring device, have been added to the simple control loop.

(2) The secondary controller has the special feature that its set point can be adjusted by a transmitted signal.

(3) The process has been partitioned into separate parts — the fuel flow process and the temperature process.

(4) The two control loops are separate entities.

The slave flow control system meets two of the major criteria for cascade control performance, namely, that the slave control loop should have fast response and the major disturbance should enter the slave control loop. Figure 13-7 shows D_1 as the disturbance that can be best attenuated by the cascade system.

13-4 BENEFITS OF CASCADE CONTROL

The improvement in control attributable to the cascade control will vary with the application, but will generally include the following:

(1) Decrease in deviation of the primary variable

(2) Shorter settling time of the primary variable

(3) Facilities for additional protective limiters and overrides

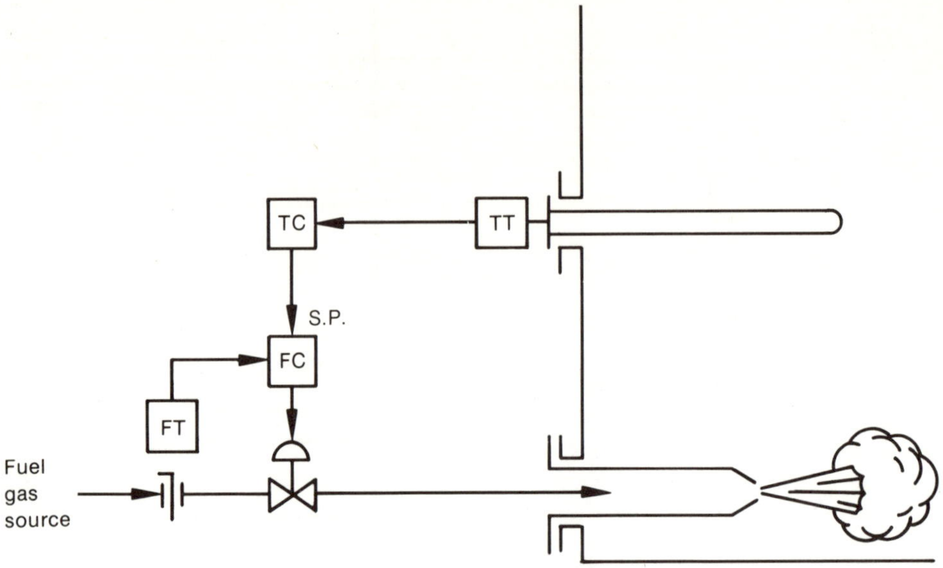

Figure 13-5. Temperature Controlled Furnace, Cascade Control

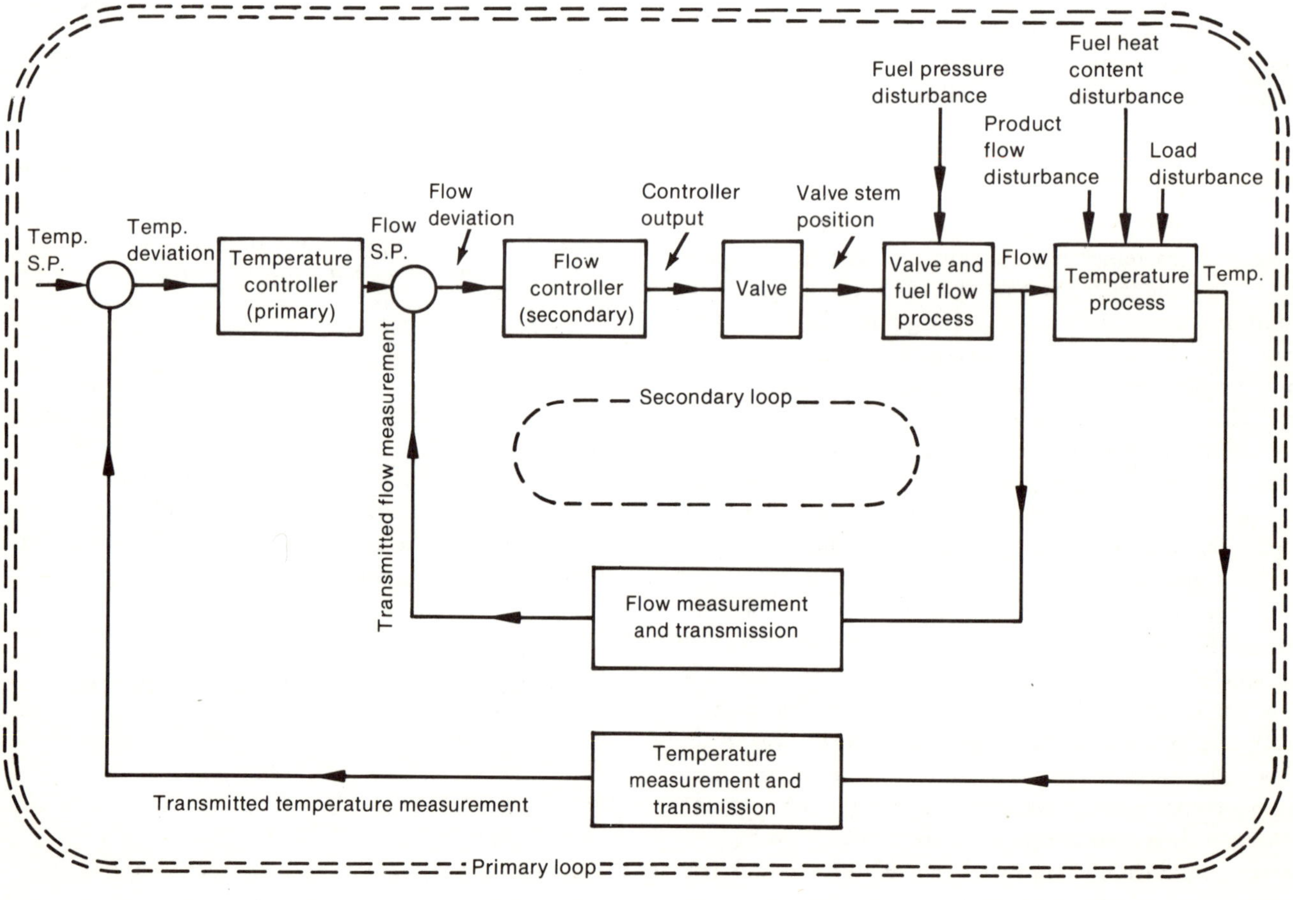

Figure 13-6. Block Diagram of Cascade Control of Temperature with Secondary Control of Fuel Flow to a Furnace[70]

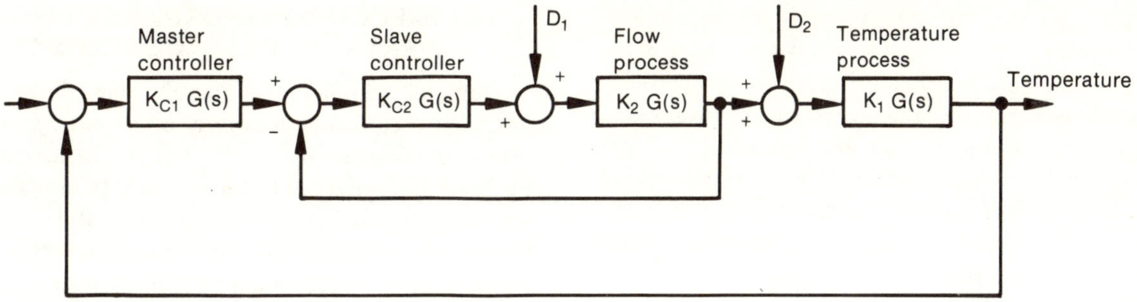

Figure 13-7. Temperature Cascade Signal Flow Diagram

13-5 CASCADE CONTROLLER TUNING

Implementation of cascade control involves the secondary feedback around individual processes, with set points of the inner or slave loops determined by the outer or master units. Standard analog controllers can be used if the slave controller can accept set point adjustment by a remote transmitted signal. Specially packaged dual analog controllers are also available for cascade applications. Cascade control is readily implemented digitally, and most packaged systems include this capability.

The greatest obstacle to the use of cascade control is the tuning of the controllers. Should PID controllers be selected, six adjustment knobs must be dealt with at start-up. Recursive techniques are used in tuning, wherein the standard procedure is to tune the slave controller with the master controller in the manual mode. The slave is then placed in the automatic mode, and the master is tuned. Techniques for simplifying the cascade signal flow diagram are shown in Section 6. The same recursive technique is used in applying frequency response to calculate controller parameter adjustments. The slave loop elements are combined on a Bode plot, and the controller parameters are calculated. The parameters are substituted into the simplified signal flow diagram, and a second Bode plot is made combining the elements in the master control loop. The master controller adjustments can then be determined.

Derivative action is seldom applied in the slave loop.(70) Actually, reset action is sometimes not required, either, since the reset in the primary controller will drive the slave controller to an output that brings the controlled variable to the set point. Section 9 showed that removal of reset action makes possible a faster loop response, so the absence of reset has the dual benefit of faster slave loop response and simplified tuning. The master controller requires both proportional and reset control modes with derivative added using the same criteria as for a simple control loop.

The limitation on the omission of reset action from the slave controller is whether or not stability will permit the proportional band to be set at 100 percent or less. Should it become necessary to widen the proportional band past 100 percent, then the controller will not fully stroke the control valve, and reset action is required. This question is especially pertinent with regard to slave flow controllers, where Section 10 described the many conditions that required wide proportional bands.

Should reset action be required in the slave controller, Shinskey(74) has suggested the circuit in Figure 13-8 as a means of avoiding saturation in the slave controller when the master controller drives it to the end of its span.

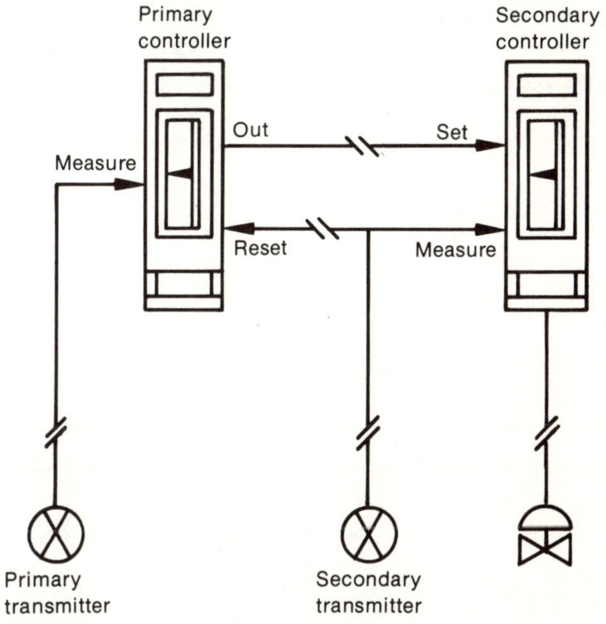

Figure 13-8. Avoiding Primary Controller Windup in Cascade Systems with External Reset Connections(74)

13-6 CASCADE OF pH CONTROL ON pH CONTROL

Occasionally a cascade control system will prove profitable in a completely unexpected application. Consider the chemical neutralization of acidic waste water in a ditch leaving the plant site.[75] Lime slurry is used for neutralization for reasons of economy. The slurry has a neutralization time constant made up of the time for the lime particles to dissolve (mass transfer) and the chemical reaction time of the acid-lime combination. The neutralization time constant can vary from 30 seconds to 30 minutes, depending on the composition of the lime. Suppose the neutralization time constant is 2 minutes and ditch flow velocity is 1 ft/second. Reaction is essentially complete after 3 time constants, or 6 minutes. The final pH measurement must therefore be some 350 feet downstream in order to sense the final pH for control.

Thus, the system dynamics consist entirely of 6 minutes of dead time. The controller must be tuned to respond slowly in order to stabilize, and at least "6 minutes worth" of every disturbance will get by the control system because the disturbance cannot be sensed until that much time has elapsed. Simple single-loop control system performance will be poor.

A cascade control system with a slave pH control system as shown in Figure 13-9 will improve performance. The slave pH sensor will be 30 to 50 feet downstream from the slurry injection and will respond much more quickly to disturbances. The pH at that point will not as yet have reached its final value, but the master pH control loop will set the slave set point to whatever it must be in order to make the final value conform to set point.

13-7 EXPECTED PERFORMANCE IMPROVEMENT WITH CASCADE CONTROL

Computer simulation of a multi-time constant system has been used to quantitatively assess the improvement provided by cascade control.[73] The ITAE (integrated time times absolute error) criterion was used to provide a quantitative comparison between single-loop and cascade control. The ITAE is composed of the absolute error between measurement and set point integrated over time. Every effort was made to optimally tune all controllers.

The theoretical improvement provided by well applied and well tuned cascade control over single-loop control is shown in Figure 13-10. Improvement under conditions where the disturbance enters the slave (secondary) loop is striking, thus verifying the widespread use of flow as the slave variable.

13-8 THE CONCEPT OF FEEDFORWARD

Feedback control is the most common type of control, finds widespread application, and has been extensively discussed in the foregoing sections. The basic principle of feedback control is shown in Figure 13-11.

In feedback control, the measurement is compared to the value of the set point; the difference or error between the two is fed into the controller modes; the controller output then acts on one of the process system inputs to

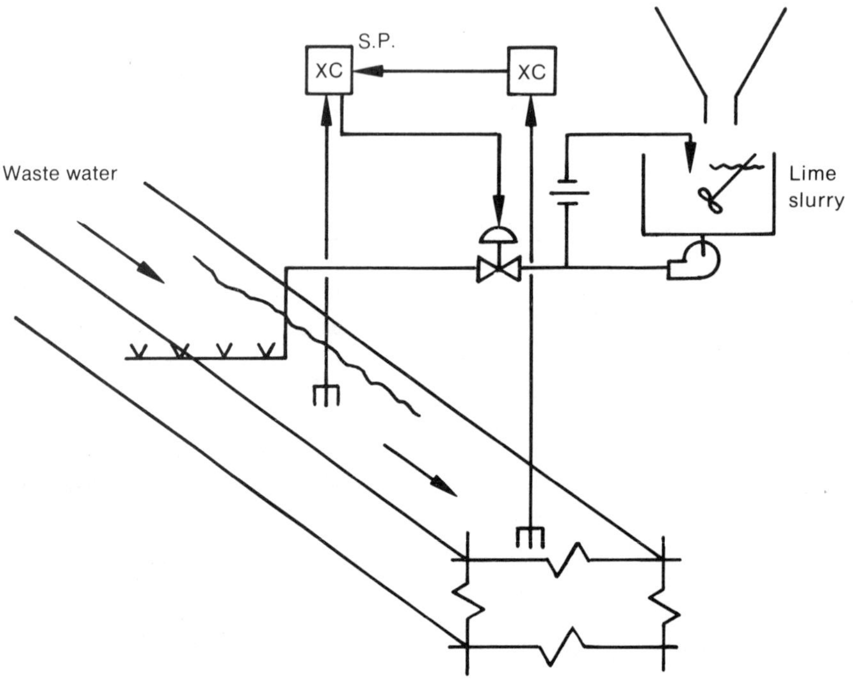

Figure 13-9. Cascade pH Control

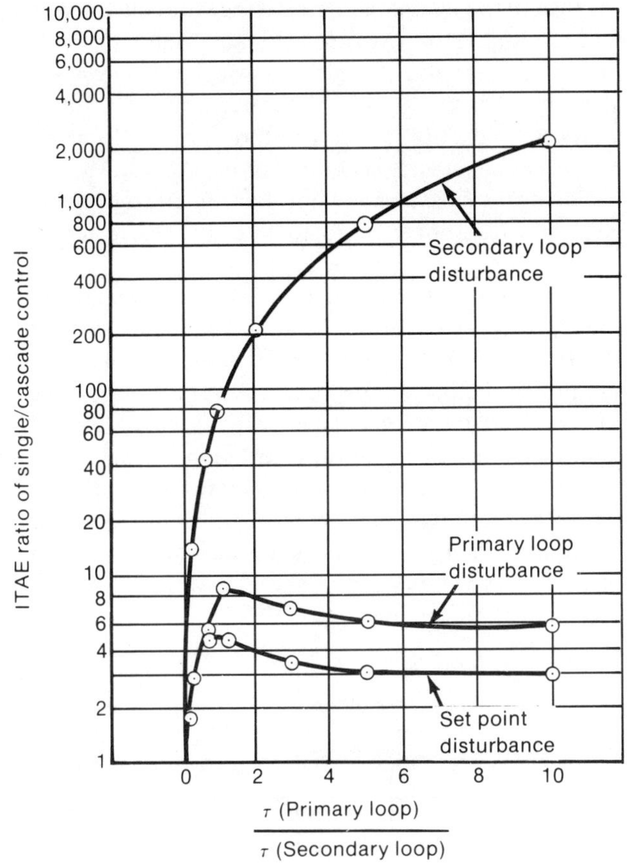

Figure 13-10. Relative Improvement of Cascade Control over Single-Loop Control[73]

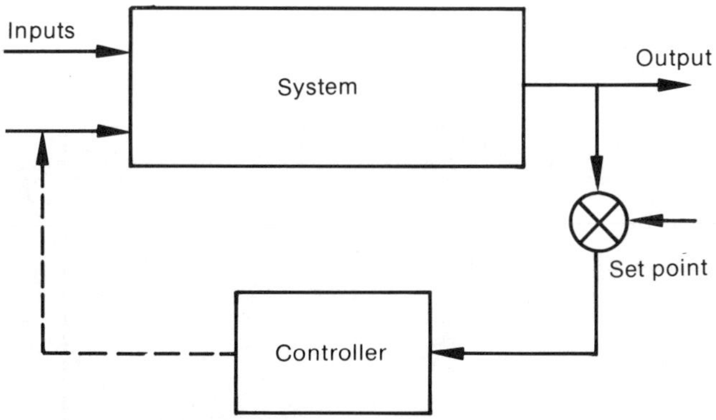

Figure 13-11. Feedback Control[76]

bring the measurement to the value of set point. Inherent in feedback system operation is the necessity for the deviation from set point to occur before control action takes place.

Feedforward control, shown in Figure 13-12, can be used in conjunction with feedback control or can be used alone.

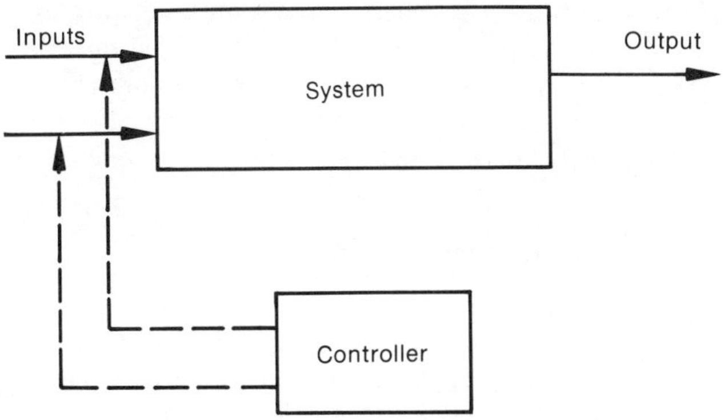

Figure 13-12. Feedforward Control[76]

In feedforward, disturbances are identified and measured before they affect the process system and the controlled variable. Corrective action is taken with the intention of cancelling the effect of the disturbance on the output. Since no deviation from set point is required, feedforward control can be theoretically perfect, depending only on the ability to measure the disturbance. Additional advantages of feedforward control are shown in Table 13-1.

Table 13-1
Advantages of Feedforward Control

Cannot be unstable.
Contributes nothing to feedback loop dynamics.
Does not wait for an error.
Reduces recovery time.
Improves product quality control of from 4:1 to 120:1 (per the literature).

13-9 APPLICATIONS OF FEEDFORWARD

Feedback has become well defined and well understood as a result of analytical developments during World War II. The feedforward concept burst on the scene in the mid 1960s, as a glance at the dates on the references on feedforward will indicate. But feedforward is not new, with the first patent in the area being issued in 1920.[78] So why did feedforward take some 45 years to reach the status of a commonly applied technique? There are two reasons. First, feedback can be applied with little *a priori* knowledge of process dynamics.[79] Then, feedforward came into its own when it was determined that the limits of what can be done by feedback alone were being approached.[80]

Another glance at the reference section will indicate another generality: The predominant applications of feedforward in the processing industries has been in the areas of distillation, reactors, and neutralization. While these unit processes appear to be widely diverse, they have one characteristic in common: they all have well defined disturbances — the feed streams. This discovery emphasizes a basic requirement of successful feedforward control — well defined, dominant disturbances that are measurable.

13-10 BASIC CONCEPTS

Basically, feedforward is a very straightforward concept. This is illustrated by the block diagram of Figure 13-13, which shows both feedforward and feedback control.

It will be remembered that in a block diagram the lines represent quantitative information, while the blocks represent dynamic operators that change the input information into the output information in magnitude and time. Thus the disturbance (μ) acts through its gain (N_μ) to change the temperature T, where temperature is arbitrarily chosen as a variable for illustration. The change in temperature is fed back through the transmitter to change the transmitter's pneumatic (or electronic) output. The controller compares the transmitted pneumatic signal to the set point and acts on the resulting error (e) to send another pneumatic signal to the valve. The valve changes its throughput in response to the new pneumatic signal, and the new flow acts through the process to change the temperature. The feedforward, on the other hand, consists of measuring the disturbance, transducing the measurement into a pneumatic signal, and applying the transducer output to the control valve in such a way as to reduce the change in temperature.

The brief analysis below Figure 13-13 shows that, theoretically, perfect control can be attained by the choice of the proper function in the feedforward transducer.[81] Quite obviously, perfect feedback control is not possible since the variable must change in order to initiate control action. Dynamically, the feedforward loop senses the disturbance and takes corrective action before the disturbance reaches the process, while the process has already responded dynamically when the feedback measurement senses a change. This factor alone makes feedforward very attractive in distillation column control.[82,83,84]

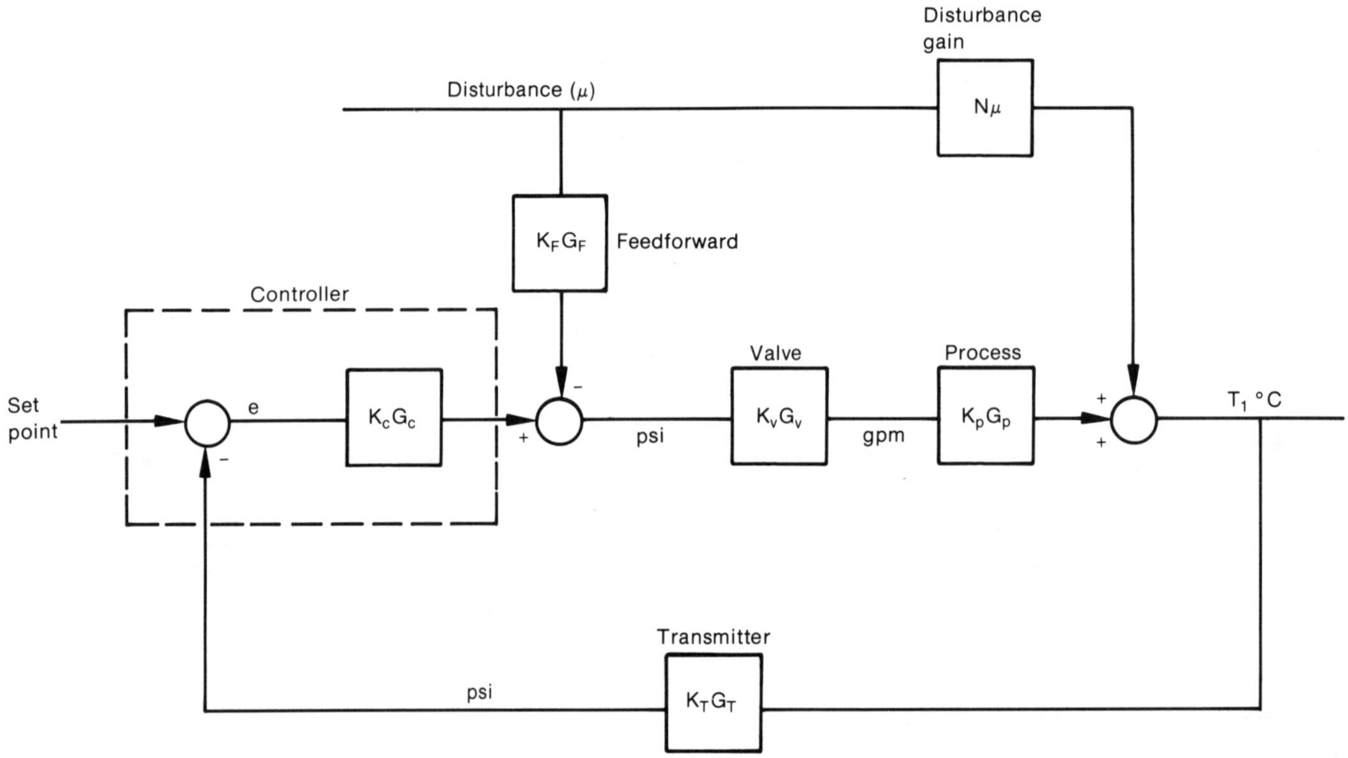

$$\Delta T_1 = [N\mu - K_p G_p K_v G_v K_F G_F]\ \Delta\mu$$

But, $\Delta T_1 = 0$

$$\text{Then, } K_F G_F = \frac{N\mu}{K_p G_p K_v G_v}$$

Figure 13-13. Block Diagram Used to Determine Feedforward Function

But the development of the "perfect" feedforward shown in Figure 13-13 is seldom attempted for several reasons. First, it is not economically feasible to develop the exact feedforward compensation function, and, if determined, to develop the hardware to implement it;[81,85] the feedforward function often requires high-order differentiation or dead time compensation (prediction of the future) and is not physically realizable;[86,87,80] physical limitations may constrain the manipulative inputs, so that they cannot be adjusted to the values required by the feedforward compensation;[88] and, the feedforward function is limited to compensation of the disturbance for which it was designed, and variations in unknown disturbances will still require feedback control for regulation.[85,86,89] The practical objective of feedforward is not perfect regulation, but reduction in the magnitude of deviations in the controlled variable to the extent that, with feedback trimming, the desired control objectives can be obtained.[81] This leads to more precise control, greater operating safety and greater throughputs.[90]

With the foregoing objective, much of the literature on feedforward control discusses the degree of improvement in regulation with various approximations to the "perfect" feedforward function. Generally, major improvements in regulation are achieved with the simplest of feedforward functions, while each additional increment of complexity of the function yields a decreasing increment of improvement in control. The literature discusses feedforward functions of pure gain,[79,84] first-order lags,[85] second-order lags,[83] general linear models,[80,81] and nonlinear models for batch reactors.[87] These analyses claim reductions in deviation of the controlled variable of 4:1,[82] 10:1,[81] and 120:1.[80] The Balchen's Q factor[85] has been developed as a quantitative measure of the degree of simplification.

13-11 THREE-ELEMENT FEEDWATER CONTROL

One of the earliest applications of feedforward control was the three-element boiler drum level control,[79] even though the feedforward feature was merely incidental in the original design.

The three-element regulator is shown in Figure 13-14, where the three elements are steam flow, make-up water flow, and drum level. The system was developed as a result of the failure of the control loop that regulated level by manipulating the feedwater control valve. The difficulty arose because, upon an increase in steam demand, drum pressure dropped, causing increased vaporization. The surge of vaporization caused a large number of vapor bubbles to appear in the liquid, resulting in a momentary apparent increase in level (called "swell") and calling for the level control to close the valve, whereupon the increased load quite obviously demanded that the valve open. Upon recognition of the problem, it was realized that level could be more readily regulated by matching input water flow to output steam flow, or a ratio of 1:0 in Figure 13-14. However, a secondary problem was inherent in the new concept. Since the mass balance on the drum is an integrating process, any inaccuracies in the primary flow elements or the associated instrumentation would be integrated over long periods of time to gradually fill or empty the drum. This problem was overcome by cascading a level control loop on the ratio and tuning the level controller to a very slow response to ignore the "swell." The control system worked extremely well and can reportedly regulate level to within ±1 inch for ±50 percent load changes.

It is now recognized that a part of this remarkable improvement is due to the feedforward feature of the configuration. The feedforward is somewhat subtle, since it

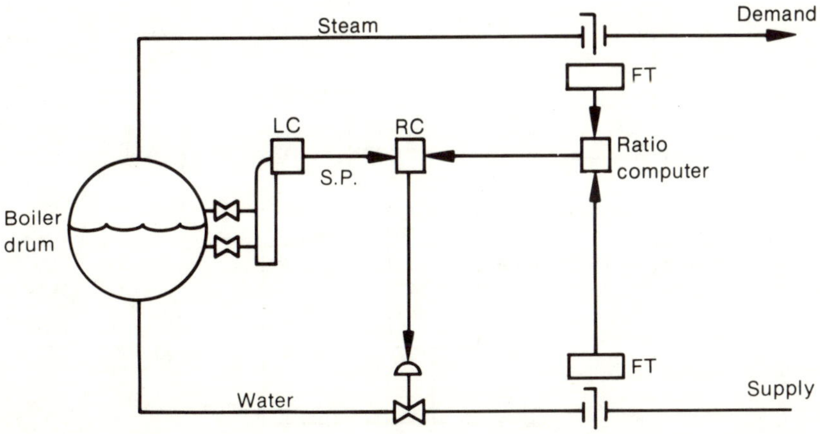

Figure 13-14. Three-Element Boiler Feedwater Control

has no specially designed feedforward function but is inherent in the mass balance equation. However, it is evident that the disturbance to the system is steam demand, which is measured and fed forward into the system. Without this measurement, increased demand would cause increased flow, decreasing drum pressure, increasing water vaporization, and finally decreasing level to actuate the control system.

13-12 DISTILLATION COLUMN

Figure 13-15 shows an application of feedforward control to a large distillation column. References to distillation column control listed in Section 13-10 agree that the objective of feedforward is, in general, to reduce changes in the equilibrium relationships represented by the McCabe Thiel diagram. This is accomplished by maintaining key ratios via the feedforward, namely, distillate flow to feed flow, bottoms flow to feed flow, and vapor boil-up to feed flow. Figure 13-15 shows the control of the latter of these three ratios.

The feedforward system in Figure 13-15 has not been applied to improve controllability *per se*, but for very obvious economic reasons. The column shown is a very large column and requires large quantities of utilities (steam, water). A natural operating tendency is to operate

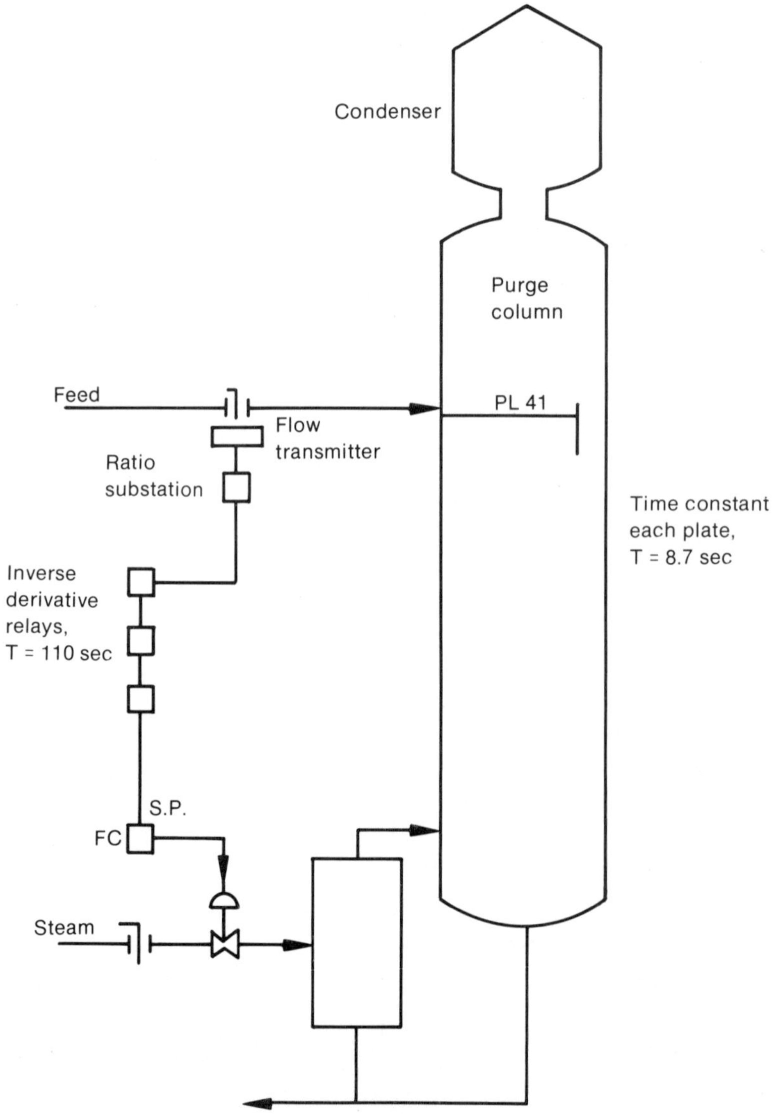

Figure 13-15. Feedforward Control Applied to a Distillation Column

at a steam rate higher than necessary to maintain a "good separation." This tendency is especially strong when column throughput is being reduced, wherein steam flow rate is permitted to remain the same. The feedforward system shown automatically reduces steam rate as feed rate is reduced.

Since the column is so large, dynamics were required in the feedforward loop to prevent a "wrong way" response due to a decrease in steam rate before the reduced feed reached the bottom of the column.[88] An extremely simple criterion was used to determine the time constants of the three delay elements in the feedforward loop: the coefficient of the first derivative in the response equation of the three delay elements equal the coefficient of the first derivative in the response equation for the 41 plates.

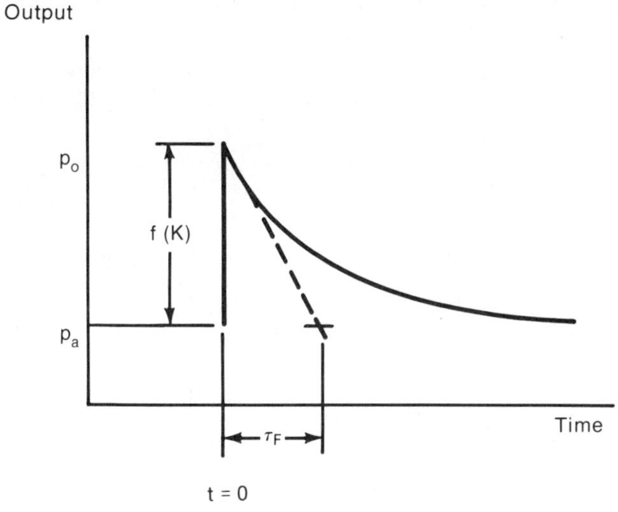

Figure 13-17. Impulse Function Response

13-13 IMPULSE FEEDFORWARD

The need to develop a feedforward function ($K_F G_F(s)$ in Figure 13-12) can be eliminated by the use of impulse feedforward.[77] The function is replaced by an impulse, which has the transfer function shown in Figure 13-16.

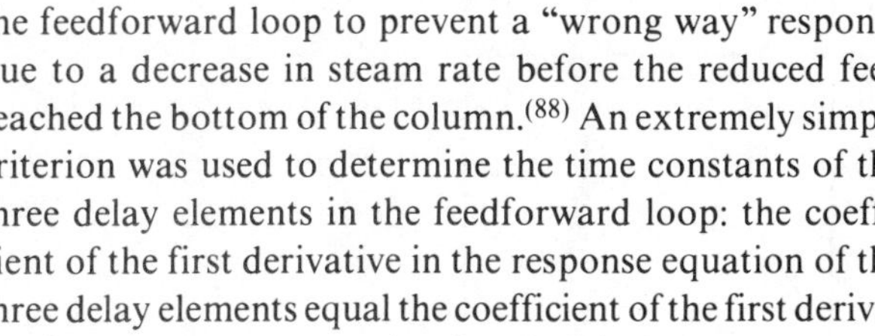

Figure 13-16. Impulse Function

For a step change in input p_c, the impulse relay will provide an output as shown in Figure 13-17.

A step change in input p_c produces an immediate change in the output. The magnitude of the output change is adjusted with the gain K. The output then decays exponentially back to the input p_a. The adjustable time constant τ_F sets the rate of decay of the output signal. Thus an impulse relay is characterized by adjustable magnitude and decay rates, and returns to the original input value after some six time constants of time have elapsed.

The return to the original value is the feature that eliminates the need for a feedforward function. When used in a feedforward application, the relay contributes a steady-state bias at all times except when its input is changed. Thus the feedback system operates as if the feedforward wasn't there except when a disturbance upsets the relay input. Consider the application of the impulse feedforward to the steam header pressure control system shown on Figure 13-18.

A number of users of superheated steam draw on a manifold connected to a common header. The users open and close the demand valves at random. Flow is regulated through the user jets by maintaining critical pressure drop across the jets. Thus pressure must be controlled very closely, and undesirable flow variations occur due to the pressure variations that result from the opening and closing of the demand valves.

The impulse feedforward system shown on Figure 13-18 is actuated by switches on the user demand valves. As the user demand valve is opened, it generates a positive step and a positive pulse, as shown on Figure 13-17. Closing the valve generates a negative pulse. Figure 13-19 shows the response of the feedback pressure control system as one of the demand valves is opened.

The impulse response in Figure 13-17 and the pressure controller response in Figure 13-19 depend on the tuning of the respective instruments. Given that they can be tuned to similar first-order responses, notice that adding them together through the summer results in a step change being transmitted to the control valve in response to a step change in demand. Thus pressure variations are minimized, and pressure control is improved.

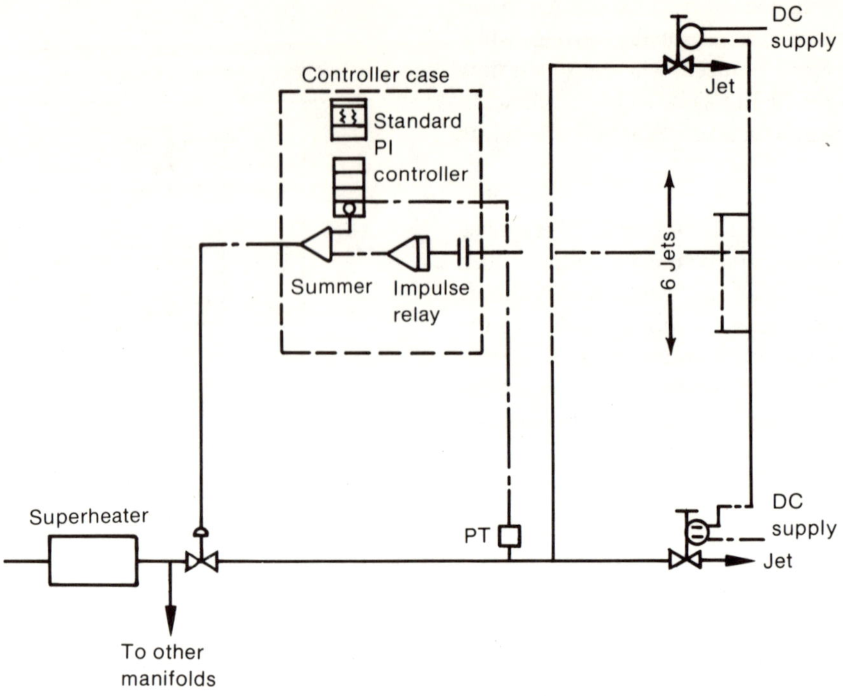

Figure 13-18. Pressure Control with Impulse Feedforward

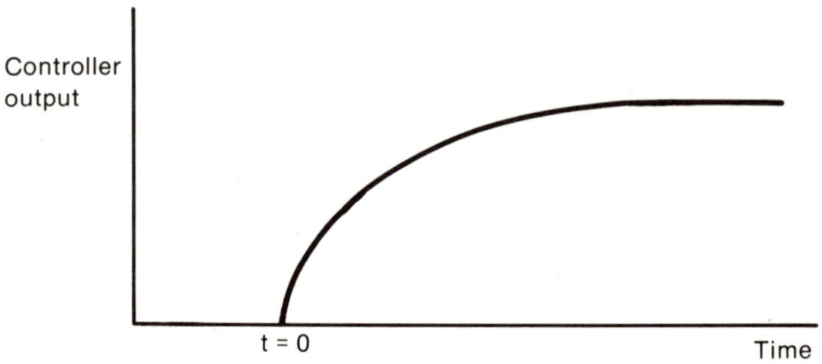

Figure 13-19. Pressure Controller Response

Problems

1. Consider a process represented by the block diagram below. What portion of the control loop would be enclosed by a secondary cascade loop?

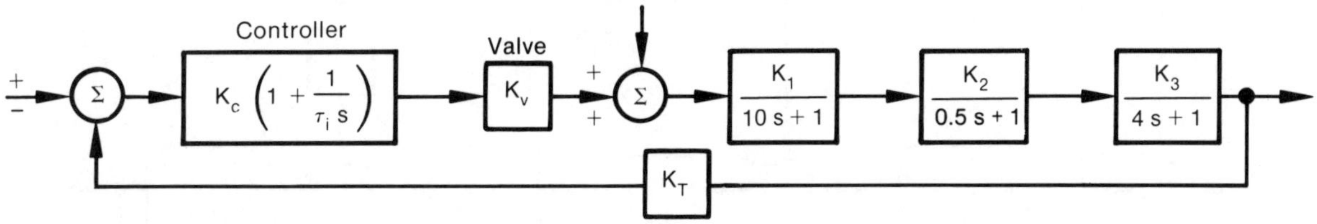

2. Why doesn't the offset due to the use of a proportional only secondary cascade controller become an undesirable feature? What is the advantage of proportional only control?

3. With regard to the lime neutralization system shown in Figure 13-19, suppose that flow velocity in the ditch is 1 ft/second, and that the lime reaction time constant is 1.5 minutes. What distance downstream from the lime slurry injection point should the primary pH measuring sensor be?

4. Determine the feedforward function ($KG(s)$) for the following block diagram.

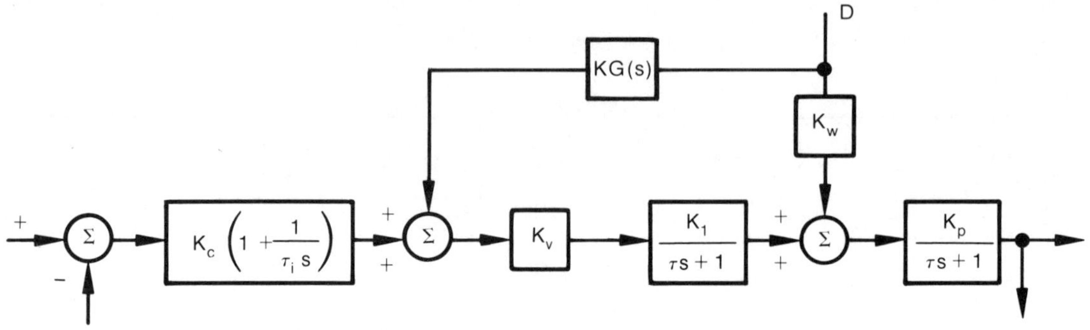

SECTION **14**

STATE SPACE TECHNIQUES

14-1 INTRODUCTION

Classical control theory has developed over the last century using the frequency and complex domains as computational aids for the design of control systems. The control system is considered to be a single loop made up of process, sensor, controller, and manipulated device. The theory is so highly developed that it lends itself to rather simple calculations for analysis and the choice of control components. The transfer function concept has the advantage of being a direct analog to the physical equipment, and the results are related directly to the operating process. Classical techniques can be applied anywhere with a few forms and graphs and provide substantial insight into the workings of the system.

Modern control theory, on the other hand, considers the system to be an entire process, unit operation, or an entire plant. Many control loops are included. The theory addresses the objectives by having all the loops operate together in some optimal fashion to minimize energy demand, maximize production, and meet product quality specifications. Modern theory considers the state of the system at any given time. The state is described as the observed value of all the variables that describe the system behavior. Dynamic behavior is described by the variables and their derivatives. The state of a system will usually be described by a substantial number of variables and the equations relating them to the other variables. The n-dimensional space defined by the state variables is called a state space.

The computational complexity of the repeated simultaneous solution of a large number of equations requires mathematical techniques to handle the equations and computational aid to solve them. The basic concept of state space formulation entails computer analysis of the system in the time domain.[91,92,93]

14-2 CONTROL STRATEGY

A classical challenge in the chemical plant is the control of a distillation column, as shown in Figure 14-1.

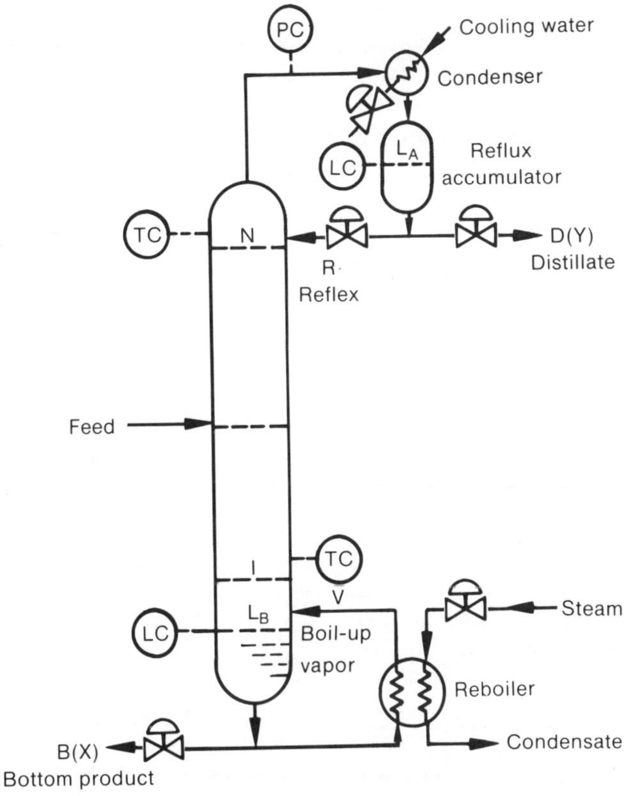

Figure 14-1. Representative Distillation Column[94]

Figure 14-1 shows five variables to be controlled and five control valves. The question is which variable should be connected to which valve. Many years ago, the tabulating technique shown in Figure 14-2 was found to provide an orderly means of making the decision.[3]

Measurements	Available Control Valves: V_1	V_2	V_3	V_4	V_5
TC_{top}	A	B	D	B	D
PC_{top}	B	A	D	C	D
LC_{cond}	B	C	A	D	D
TC_{bottom}	C	D	D	A	C
LC_{bottom}	D	D	C	B	A

Figure 14-2. Control Strategy Matrix

The point where a given valve column crosses a measurement row is given an index of strength of the relationship between the manipulated variable and measured variable, A for strong, D for weak, and so on. Quite obviously, an arrangement with one A and all Ds on any one row is ideal. Then the rows are interchanged, as shown, to position the figure representing the strongest relationship on the diagonal. Then it can be said that good control will result from a strong diagonal. Note, however, some strong off-diagonal terms that will cause interaction among the control systems. For example, movement of valve V_4 will cause a significant response in TC_{top} and LC_{bottom} as well as in TC_{bottom}.[3]

Matrices are used to depict and manipulate large numbers of equations. Matrix algebra is used as the basis for the manipulations, and they are performed on a digital computer. The computer can handle large numbers of equations and removes any limitation on the number of equations necessary to model any reasonable system. However, for simplicity, 2×2 matrices will be used in the following discussion. Consider the steam boiler in the following example.

Example

Mika[91] describes a steam generator with the transfer function shown in Figure 14-3.

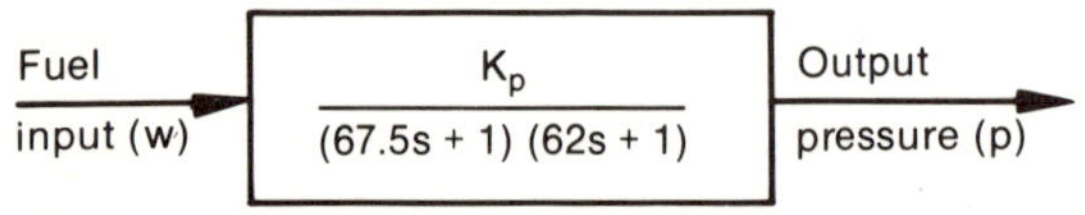

Figure 14-3. Signal Flow Diagram Steam Boiler

Multiplying out the transfer function:

$$4185 \frac{d^2p}{dt^2} + 129.5 \frac{dp}{dt} + p = K_p w \qquad (14\text{-}1)$$

But Equation 14-1 must be converted into two first-order differential equations:

$$\text{Let } P_1 = p \qquad (14\text{-}2)$$

$$\text{And } P_2 = \frac{dp}{dt} = \frac{dp_1}{dt} = \dot{P}_1 \qquad (14\text{-}3)$$

Then, substituting:

$$\dot{P}_2 = -\frac{1}{4185} P_1 - \frac{129.5}{4185} P_2 + \frac{K_p}{4185} w \qquad (14\text{-}4)$$

Equations 14-2, 14-3, and 14-4 can now be put into matrix notation:

$$\begin{bmatrix} \dot{P}_1 \\ \dot{P}_2 \end{bmatrix} = \begin{bmatrix} 0 & 1 \\ \frac{1}{-4185} & \frac{-129.5}{4185} \end{bmatrix} \begin{bmatrix} P_1 \\ P_2 \end{bmatrix} + \begin{bmatrix} 0 \\ \frac{K_p}{4185} \end{bmatrix} w$$

Then let

$\dot{X}$ equal $\begin{bmatrix} \dot{P}_1 \\ \dot{P}_2 \end{bmatrix}$

A equal $\begin{bmatrix} 0 & 1 \\ \frac{1}{-4185} & \frac{-129.5}{4185} \end{bmatrix}$ which is called the system matrix

X equal $\begin{bmatrix} P_1 \\ P_2 \end{bmatrix}$ which is called the state vector

B equal $\begin{bmatrix} 0 \\ \frac{K_p}{4185} \end{bmatrix}$ which is called the gain matrix

u equal w, which is called the control vector

The system can then be described in terms of the normal state variable form

$$\dot{X} = AX + Bu \qquad (14\text{-}5)$$

The A matrix contains all the information required to determine the stability of the system. The state vector X is defined as one containing the state variables that completely define the state of the system at any time t. The steam pressure and rate of change of steam pressure are the state variables for the steam boiler.

A perplexing question in the use of state variables is the establishment of the number of variables needed to specify the system uniquely.[95] One state variable is required for each degree of freedom of the system. Various guidelines, such as the Gibbs phase rule,[22] are helpful in determining the degrees of freedom. A more convenient way to establish the state variables in a system is to associate one with each integration in the block diagram of the system. Conversely, if more proportional-reset controllers are applied to a system than there are degrees of freedom, the "excess" controllers will drive their control valves either fully open or fully closed. Practically, the number of manipulated variables available is accepted as the degrees of freedom. It is by this reasoning that the degrees of freedom in the distillation column in Figure 14-1 are five.

Matrices such as the state space matrix, Equation 14-5, can be used to determine an optimal controller, where system response is specified and a controller is synthesized to meet that criterion; to determine adaptive controller settings, where controller tuning changes in response to system changes; and to determine variable pairing, as was done quantitatively in Figure 14-2.

14-3 OPTIMAL CONTROLLER

One of the promising attributes of modern control theory is that it provides an explicit method to design an optimal controller even with the most complex multiple input-multiple output system. Controller synthesis using the methods in earlier sections requires a guess as to the probable form of the controller, which is then followed by an analysis as to the appropriateness of the guess. If the system performance is not satisfactory, a second guess for the controller is made and the procedure is iterated until it converges on the desired form of the controller. Convergence of this procedure is not guaranteed, and there is no assurance that the final choice of a controller is the best choice. If, on the other hand, a mathematical model of the system is available in terms of all its inputs and outputs and their interactions, then a controller can be designed with state space techniques, which optimizes any chosen index of performance for the system.

A second vector Y is defined for purposes of controller design as:

$$Y = CX \tag{14-6}$$

Equations 14-5 and 14-6 are called the state model. The term Y is some desired output vector, and C is a matrix.

Then define:

e_1 = actual value of P_1 minus desired value of P_1

e_2 = actual value of P_2 minus desired value of P_2

f = control effort at actual values of P_1 and P_2

f_0 = controller effort at desired values of P_1 and P_2

$u = f - f_0$

A performance criterion, J, must be chosen. Let the criterion be that the system be controlled with the least squared error defined by

$$J = \frac{1}{2} \int_o^T \left[q_1 e_1^2 + q_2 e_2^2 + u^2 \right] dt$$

Finally, the form of an optimal controller has been shown to be[91]

$$u = -\begin{bmatrix} 0 & \dfrac{K_p}{4185} \end{bmatrix} \begin{bmatrix} K_{11} & K_{12} \\ K_{21} & K_{22} \end{bmatrix} \begin{bmatrix} e_1 \\ e_2 \end{bmatrix}$$

The matrix K can be readily calculated with a digital computer. The weighing factors q_1 and q_2 are adjusted during the calculation of the K matrix so that the constraints on system response such as overshoot are accommodated. Since the terms of the K matrix are simple gain settings, the controller will usually not be of the familiar PI or PID type and might even be nonlinear. Feedforward is automatically introduced as needed by the solution for the terms in the K matrix. A block diagram of the optimal control system is shown in Figure 14-4.

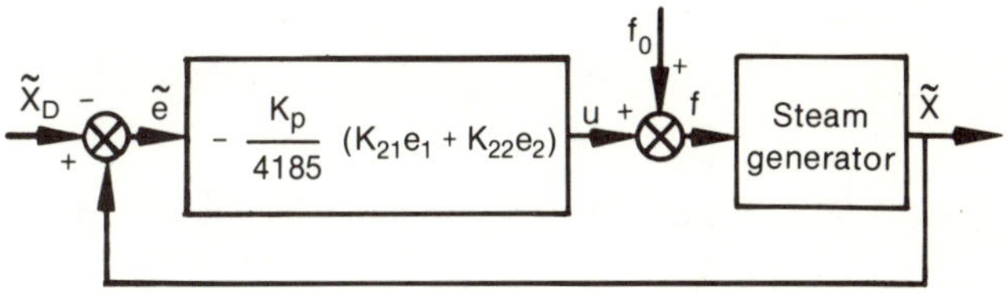

Figure 14-4. Optimal Controller[91]

In this optimal controller, feedforward was introduced automatically by solution of the K matrix.

14-4 ADAPTIVE CONTROL

Processes and unit operations change with time due to catalyst aging, heat transfer surface fouling, product transitions, throughput changes, chemical buffering, and a host of other phenomena. The changing process parameters result in poor control performance even though the controller was originally well tuned. Poorly tuned controllers invoke an economic penalty in terms of process failure to meet product specifications or failure to maintain operating limits. Consider the record of distillation column overhead purity shown in Figure 14-5.

A control strategy that requires purity regulation of both overhead and bottoms products results in a change in purity gain as purity changes. Thus controller gain is made very low to ensure stability as high purity is approached. But low controller gain permits wide deviations of purity, and demands that purity set point be set well above the purity specification to ensure that transient variations do not produce unsaleable product. The high set point incurs

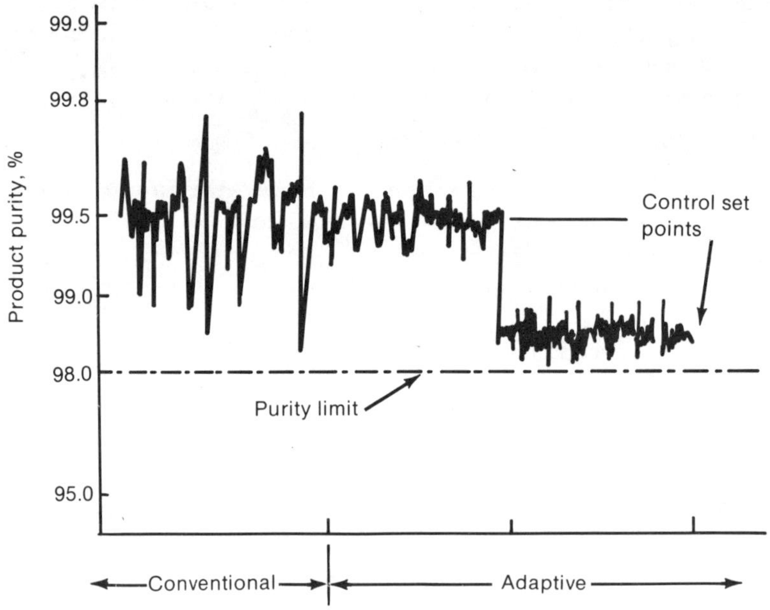

Figure 14-5. Distillation Column Adaptive Control Advantages

costs of give-away of the overhead product and additional steam to distill to higher purity. The adaptive controller readjusts the purity controller parameters to keep it well tuned with changing purity gain. Less deviation in purity results from better tuning, as shown in Figure 14-5. Set point can then be set closer to the purity limit, reducing production costs by reducing give-away and reducing energy consumption.

Safety or environmental considerations often require that a process operate within strictly enforced limits. The Environmental Protection Agency, for example, requires that pH be between 6 and 9 pH units before wastewater can be discharged from a plant site. A system such as that shown in Figure 13-8 is installed to control the pH. The process gain is the slope of the titration curve, as shown in Figure 14-6. The curves in Figure 14-6 show a tremendous change in gain as different chemical mixes pass the control system. The control proportional band should be changed to accommodate each change in process gain. Failing this, the pH control system performance will be poor, and pH will vary outside the prescribed limits.[75] Large, costly averaging basins must be installed to attenuate the transient variations in pH. An adaptive controller would automatically retune the controller as needed, maintain system performance as required, and eliminate the cost of the averaging basins.

The adaptive controllers described in the foregoing examples retune the controllers in a single control loop and are called self-tuning regulators. Several self-tuning techniques are used, one of which is shown in the continuous adaption scheme shown in Figure 14-7.

The heart of the self-tuning feature is the identification block. It accumulates historical information in the form of repeated sensor readings on process operation. The information is used to determine the parameters of a simple process model, such as shown in Figure 8-1. Parameters of the process model are derived by least squares estimation. This means that the sum of the squares of the deviations of the process history data points from the line representing the model are minimized.

Policy decisions for the identification block in the areas of the length of time to retain historical data and the discrimination of valid data from noise are difficult. If historical data are not kept long enough, the values of the parameters will not be the best estimates; if kept too long, the values will not be responsive to new conditions. Data validity is verified in Figure 14-7 by the sensitivity analysis block, which calculates the worst case effect of controlled variable noise on the identification. Parameters are average with past values weighted according to sensitivity.[96]

The process block includes control valve, process, and sensor. The inputs to the identification block are the controller input and output, both signals that are readily available.

Upon identification of the state of the process in terms of the simplified process transfer function, controller tun-

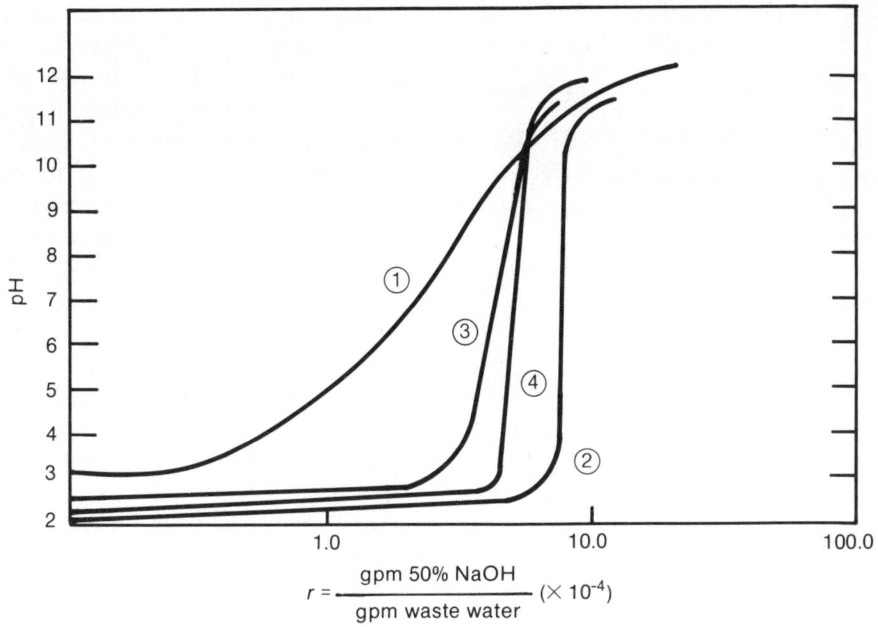

Experimental titration curves from samples taken over a 48-hour period at the same facility. Gains at pH = 7 are as follows:

Curve	Reagent	Gain
1	Caustic	21,700 ph/*r*
2	Caustic	126,750 ph/*r*
3	Caustic	57,870 ph/*r*
4	Caustic	380,230 ph/*r*

Figure 14-6. Experimental Titration Curves[75]

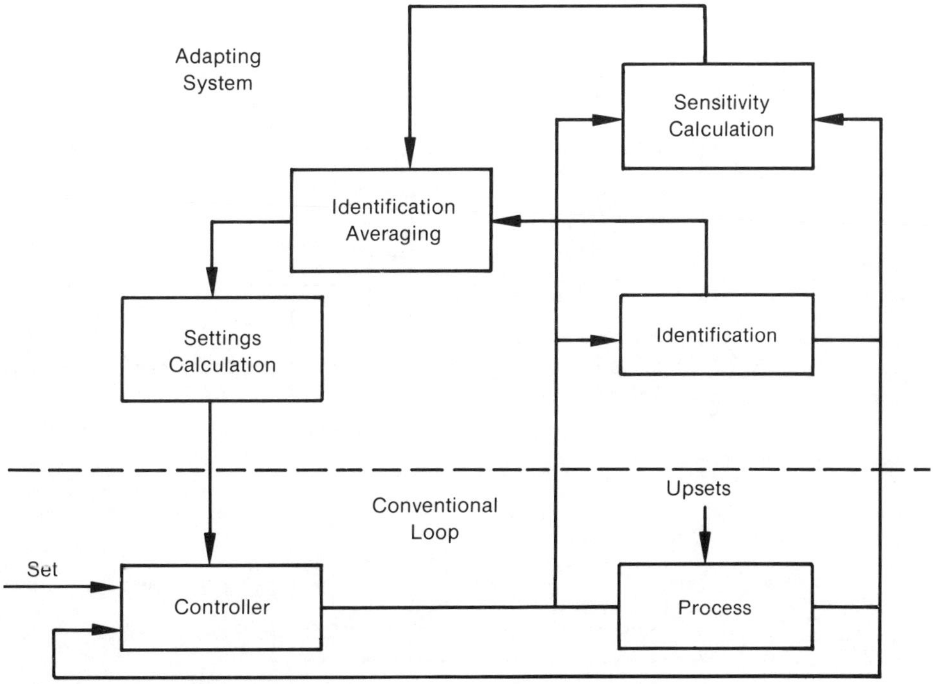

Figure 14-7. Adaptive Control System[96]

ing parameters are calculated by much the same techniques as described in Section 7. The calculated settings are then automatically applied to the controller.

14-5 PAIRING OF VARIABLES FOR CONTROL

Two independent control loops controlling two state variables for a system are shown in Figure 14-8(a). The matrix block diagram for the two loops is shown as Figure 14-8(b). The same rules for matrix and vector notation as previously discussed apply to the block diagram. Realistically, each of the two state variables will probably be sensitive to both of the manipulated variables as shown in Figure 14-9(b).

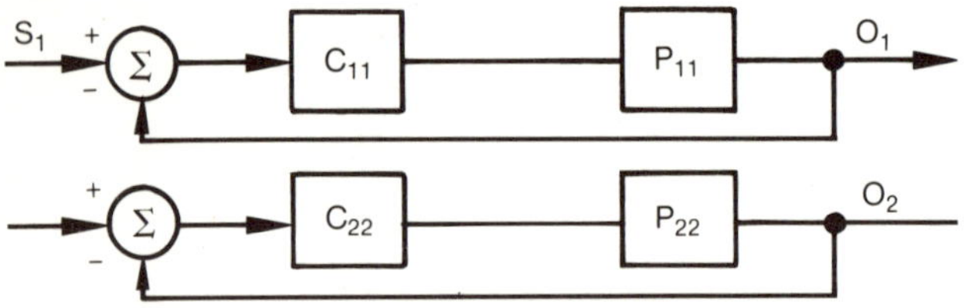

Figure 14-8(a). Two Independent Control Loops Acting on Same Process(104)

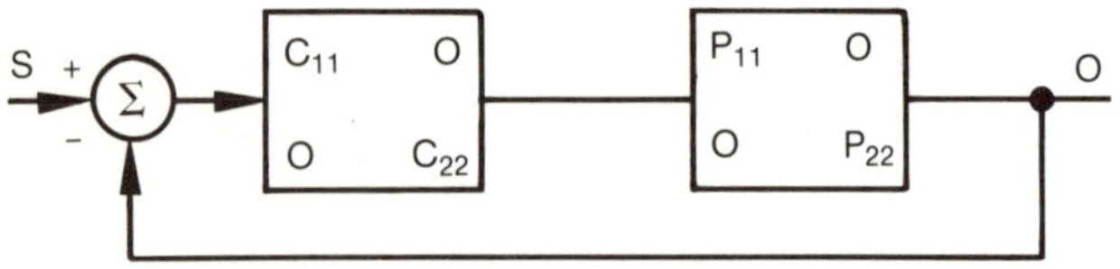

Figure 14-8(b). Matrix Notation for Control Loops of Figure 14-8(a)

Figure 14-9(b) shows the matrix block diagram for this interacting system. The five control loops on the distillation column in Figure 14-1 could be drawn in this same fashion, except that the process matrix would have a rank of five. Figure 14-2 shows the desirability of having a matrix with high gains on the diagonal and off-diagonal terms with low gains. Ideally, referring to Figure 14-8(b), the off-diagonal terms should be zero, eliminating all interaction. Seldom is this realizable, and controlled variables must be paired with manipulated variables in such a way as to result in a matrix with a strong diagonal.

E. H. Bristol's[97] relative gain matrix is one of the most useful measures of interaction. It normalizes the process gain matrix in such a way as to make diagonalization more convenient. Remember that the process gains are partial derivatives. The relative gain matrix specifies that each term in the matrix be of the following form:

$$\lambda ij = \frac{\left(\dfrac{\partial c_i}{\partial mj}\right)\Bigg|_m}{\left(\dfrac{\partial c_i}{\partial mj}\right)\Bigg|_c}$$

where: λij = relative gain at row i, column j

c = controlled variable

m = manipulated variable

The numerator is the process gain of the controlled variable with all other loops open or with all other manipulated variables constant. The denominator is the process

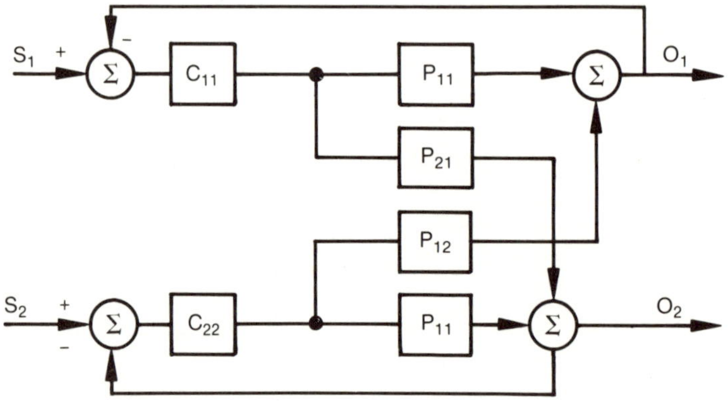

Figure 14-9(a). Conventional Control of an Interesting Process(104)

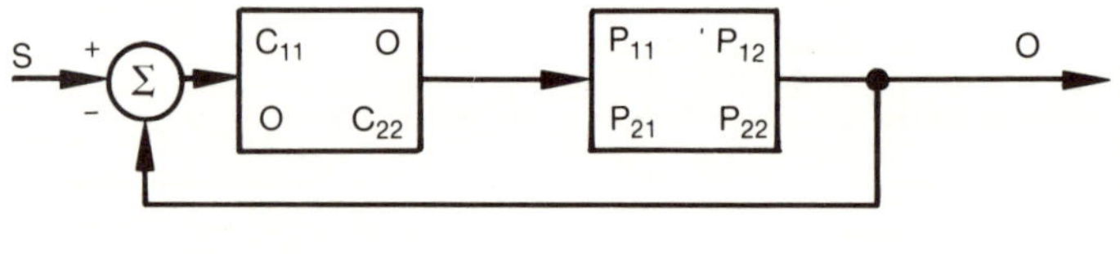

Figure 14-9(b). Matrix Notation for Control System of Figure 14-9(a)

gain with all other loops closed and controlled perfectly, or with all other controlled variables constant. To hold all of the other controlled variables constant, some or all of the remaining variables must change. The changes in other manipulated variables result in a change in the steady-state value of the controlled variable.

Since m is held constant in calculating the numerator, the numerator represents an open-loop situation for the second loop. Similarly, because c is constant in the calculation of the denominator, the denominator represents the special closed-loop situation in which perfect control is achieved in the second loop. Thus λ represents a dimensionless ratio of open- and closed-loop sensitivity and, as such, is a comparison of how a loop would respond as a simple, isolated loop to how it would respond as part of an interacting, multi-variable process.[94,98,99]

The number of partial derivatives that have to be evaluated for an $n \times n$ relative gain matrix is $2n^2$, or, for the distillation column of Figure 14-1, a total of 50 partial derivatives. The relative gain matrix can be computed from the process matrix by matrix algebra techniques. However, this is a major computational task, and the dimensions of the system are almost always reduced by pairing some controlled and manipulated variables by inspection or experience. Shinskey[100] and McAvoy[99] have used the system shown in Figure 14-10, which results in a 2×2 relative gain matrix.

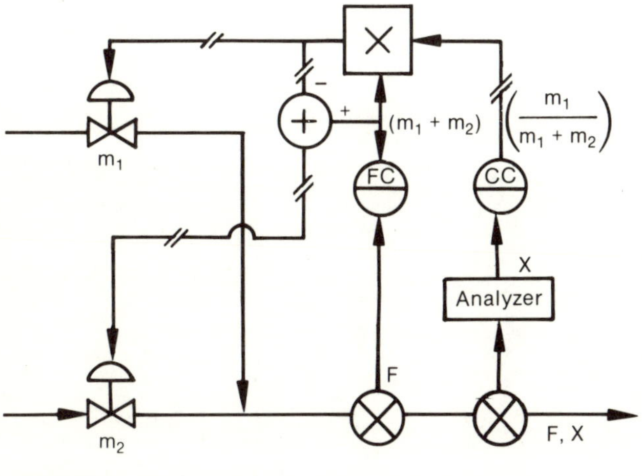

Figure 14-10. Completely Decoupled Control Scheme[99]

Two mass balances can be written. First, the total mass balance:

$$F = m_1 + m_2$$

For the component X,

$$X = \frac{m_1}{m_1 + m_2}$$

Since each row and each column in the normalized matrix adds up to 1, only one element of the 2×2 matrix must be evaluated. Consider:

$$\left.\frac{\partial x}{\partial m_1}\right|_{m_2} = \frac{m_2}{(m_1 + m_2)^2} \qquad (14\text{-}7)$$

$$\left.\frac{\partial x}{\partial m_1}\right|_{F} = \frac{1}{m_1 + m_2} \qquad (14\text{-}8)$$

One element of the relative gain matrix can be determined by dividing Equation 14-7 by Equation 14-8:

$$\lambda_{x,m_1} = \frac{\left.\dfrac{\partial x}{\varphi m_1}\right|_{m_2}}{\dfrac{\partial x}{\partial m_1} \quad F} = \frac{\dfrac{m_2}{(m_1 + m_2)^2}}{\dfrac{1}{m_1 + m_2}} = 1 - x$$

And the matrix becomes

	m_1	m_2
X	λ_{x,m_1}	λ_{x,m_2}
F	λ_{F,m_1}	λ_{F,m_2}

The resulting gain matrix is

	m_1	m_2
X	$1-X$	X
F	X	$1-X$

The largest positive element in the matrix dictates the pairing of controlled and manipulated variables. Suppose that $X = 0.2$:

	m_1	m_2
X	0.8	0.2
F	0.2	0.8

For this condition, the analyzer control system would be connected to m_1 and the flow control system to m_2. But suppose that composition $X = 0.8$:

	m_1	m_2
X	0.2	0.8
F	0.8	0.2

Now the relationship between X and m_2 is strongest, and that pairing would be made. Finally, if $X = 0.5$, severe interaction would be expected, and a decoupling network for this process is shown in Figure 14-10.

APPENDIX **A**

ELECTRIC CIRCUIT ANALYSIS BY CLASSICAL METHOD

The transient response of an electrical circuit can be determined in the time domain by the formal solution of the differential equations describing the circuit. Consider Figure A-1.

A common group of control problems involves the correlation of potential as a function of time (e.g., level, pressure, temperature). The circuit in Figure A-1 shows potential in the element of storage — the capacitor.

The equations below show how the circuit parameters — E_B, V_b, R and C — combine to describe the "transient response" of this system after a step change. It will be seen the R and C values establish the "time constant".

Assume the switch has been in the position shown. There will be no current flowing, and

$$V_c = V_b \text{ and } V_R = 0$$

This describes the initial conditions, for $t = < 0$. If the switch is now thrown to connect E_B across the capacitor and resistor, a current will flow until the capacitor is fully charged for this new condition, and then

$$V_c = E_B \text{ and } V_R = 0$$

This describes the ultimate, or steady-state, condition, at t = ∞.

I and V_c must change until this steady-state condition is reached. This transient response can only be described mathematically by differential equations. So:

$$\text{for } \infty > t > 0 \quad V_R = IR = E_B - V_c$$

and

$$I = \frac{E_B - V_c}{R}$$

which is easy to write down, but doesn't help much since the storage element, C, is not yet represented.

The basic concept of a capacitor is that of removing electrons from one plate and piling up these electrons on an adjacent plant, thereby establishing difference of potential between displaced electrons, q, and potential, e, or

$$C = \frac{q}{e}$$

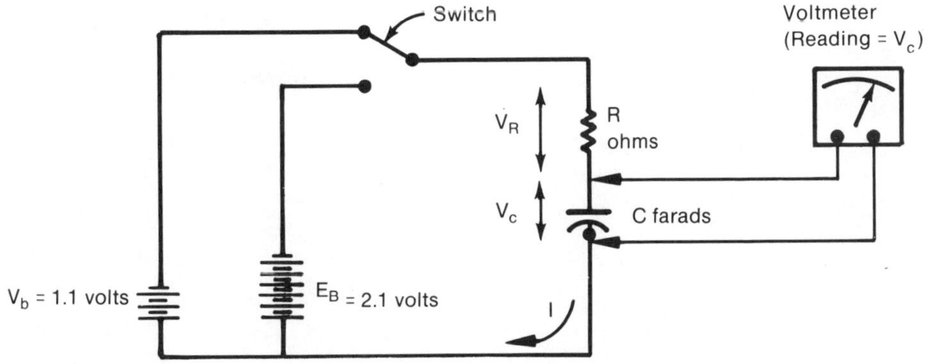

Figure A-1. RC Electric Circuit

The usual units of electrical work are volts and amperes. Amperes carry the sense of coulombs flowing per second, i.e., $\frac{dq}{dt}$. Then, in more usual terms

$$C = \frac{q}{e} = \frac{dq/dt}{de/dt}$$

or,

$$C = \frac{I}{dV_c/dt} \text{ as farads} = \frac{\text{amperes}}{\text{volts/sec}}$$

Then, from the previous $I = \frac{E_B - V_c}{R}$

and the above $I = C\frac{dV_c}{dt}$

$$\frac{E_B - V_c}{R} = C\frac{dV_c}{dt}$$

Sorting out like terms:

$$\frac{1}{RC}dt = \frac{dV_c}{E_B - V_c}$$

This differential equation must be solved (i.e., integrated) to express the relationship in a form suited to engineering design. Then,

$$\frac{1}{RC}\int_0^t dt = \int_{V_b}^{V_c} \frac{dV_c}{E_B - V_c}$$

which for this problem can be evaluated, as shown, between specific limits: V_B to V_c, for V_c, while t goes from 0 to t. This can be integrated from tables of integrals, or if some of the simpler integrals are remembered, let $u = (E_B - V_c)$.

Then du $= -dV_c$ and $\int_{V_B}^{V_c} \frac{dV_c}{E_B - V_c}$ is represented by $\int \frac{-du}{u}$. But $-\int \frac{du}{u}$ gives $-\ln u$, so, $\int_{V_B}^{V_c} \frac{du}{u} =$

$$\int_{V_B}^{V_c} \frac{dV_c}{E_B - V_c} = \frac{1}{RC}\int_0^t dt \text{ gives}$$

$$-\ln(E_B - V_c) - [-\ln(E_B - V_B)] = \frac{1}{RC}t - \frac{1}{RC}(0)$$

$$\text{and} \quad \ln\frac{(E_B - V_c)}{E_B - V_B} = -\frac{1}{RC}t$$

or, since by definition

$\ln N = x$, where x is the exponent in

$N = e^x$

$$\text{Then,} \quad \frac{E_B - V_c}{E_B - V_B} = e^{\frac{-t}{RC}}$$

$$\text{or} \quad E_B - V_c = (E_B - V_B)\, e^{\frac{-t}{RC}}$$

$$\text{i.e., unbalance} = \text{upset } x \frac{1}{e^{\frac{t}{RC}}}$$

Note that the only element reflecting time is the product RC, and that as RC becomes larger the unbalance changes (i.e., the system responds) more slowly. This latter is most readily evident by differentiating the equation above:

$$(E_B - V_c) = (E_B - V_B)\, e^{\frac{-t}{RC}}$$

$$-\frac{dV_c}{dt} = [(E_B - V_B)\, e^{\frac{-t}{RC}}]\,[-\frac{1}{RC}]$$

i.e., at any selected value of t, unbalance is changing at the rate V_c is changing, so:

$$\text{Rate of response} = \frac{1}{RC} \text{ K}$$

APPENDIX **B**

DYNAMICS OF A TEMPERATURE SENSOR

Following Lefkowitz,[25] let a bare thermocouple be used to measure the temperature T of a process. Under equilibrium or steady-state conditions, the temperature of the thermocouple element T_m is essentially the same as that of the process. The signal transmitted to the recorder or controller is a correct measure of the process temperature. If the process temperature T was to change suddenly, a transfer of energy (heat flow) between process and thermal element is required before T_m can respond. The heat flow, Q, is proportional to the temperature gradient, $T - T_m$, the heat transfer coefficient U, and the heat transfer area A, as follows:

$$Q = UA\,(T - T_m) \qquad \text{(B-1)}$$

The rate of change of the thermocouple temperature is proportional to the heat flow:

$$\frac{dT_m}{dt} = \frac{1}{C}\,Q \qquad \text{(B-2)}$$

where C is the thermal capacity of the thermocouple.

Combining,

$$\tau_1 \frac{dT_m}{dt} + T_m = T \qquad \text{(B-3)}$$

$$\text{where } \tau_1 = \frac{C}{UA}$$

Let T change suddenly from some initial value T_i to some final value T_f. Then Equation B-3 can be solved by conventional differential equation methods to yield the response of the thermocouple as a function of time t, measured from the time of the step change in T:

$$T_m = T_f - (T_f - T_i)\,e^{-t/\tau_1} \qquad \text{(B-4)}$$

This equation is plotted on Figure B-2 where the response T_m to a step change in I is shown as curves A and C. The difference between the two curves is the dynamic error of the temperature sensor under the input conditions stipulated. The magnitude of error depends both on the nature of the disturbance and the value of the time constant τ_1. Curve C of Figure B-2 shows the response of the thermal element with a time constant of twice the value as that of curve B.

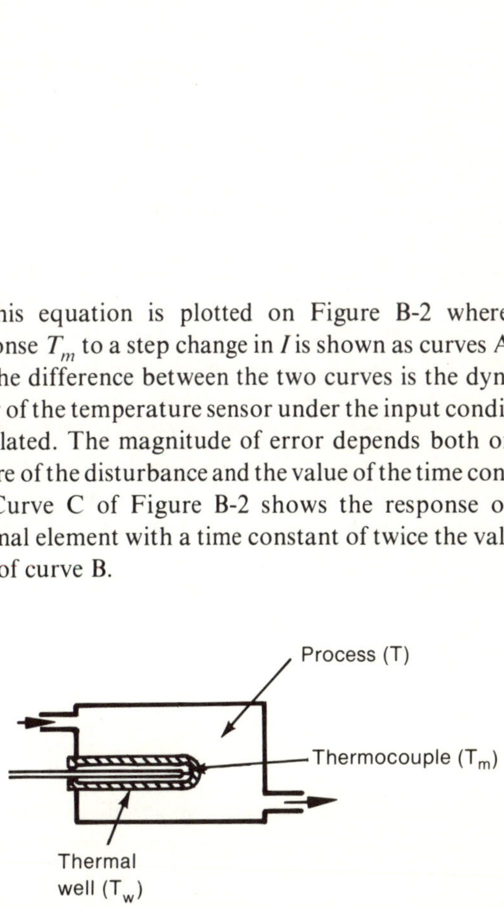

Figure B-1.

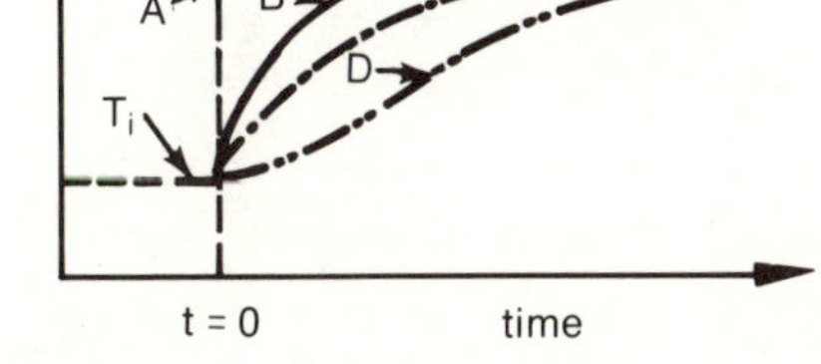

Figure B-2.

If the thermal sensor is inserted in a thermowell, as it almost always is in an industrial plant, then the response of the well must also be taken into account. From Figure B-1, the following derivation is analogous to that of Equation B-3:

$$Q_1 = U_w A_w (T - T_w) \qquad \text{(B-5)}$$

$$\frac{dT_w}{dt} = \frac{1}{C_w} Q_1 \qquad \text{(B-6)}$$

$$Q_2 = K(T_w - T_m) \qquad \text{(B-7)}$$

$$\frac{dT_m}{dt} = \frac{1}{C} Q_2 \qquad \text{(B-8)}$$

where subscript w refers to thermowell.

Combining Equations B-5 through B-8 and eliminating the heat flow terms Q_1 and Q_2:

$$\tau_w \frac{dT_w}{dt} + T_w = T \qquad \text{(B-9)}$$

$$\tau \frac{dT_m}{dt} + T_m = T_w \qquad \text{(B-10)}$$

where $\tau_w = \dfrac{C_w}{U_w A_w}$

$\tau = \dfrac{C}{K}$

Then, substituting for T_w,

$$\tau \tau_w \frac{d^2 T_m}{dt^2} + (\tau + \tau_w) \frac{dT_m}{dt} + T_m = T \qquad \text{(B-11)}$$

Taking advantage of the linearity of the equations and assigning the initial value T_i the value of zero, Equation B-11 is easily transformed into the Laplace or complex domain

$$T_m = \frac{T_f - T_o}{\tau \tau_w s^2 + (\tau + \tau_w) \quad s + 1} \qquad \text{(B-12)}$$

This can be factored:

$$T_m = \frac{T_f - T_o}{s(\tau_w s + 1)(\tau s + 1)} \qquad \text{(B-13)}$$

where $\dfrac{T_f - T_o}{s}$ is the Laplace transform of a step disturbance.

Equation B-13 can be expanded by partial fraction techniques, resulting in:

$$T_m = \frac{T_f - T_o}{s} - \frac{T_f - T_o}{\tau_w - \tau} \left[\frac{\tau_w}{s + 1/\tau_w} - \frac{\tau}{s + 1/\tau} \right] \qquad \text{(B-14)}$$

This can then be converted to the time domain by looking up the inverse transform in a table of transforms.[(23)]

$$T_m = (T_f - T_o) \left[1 - \frac{1}{\tau_w - \tau} (\tau_w e^{-t/\tau_w} - \tau e^{-t/\tau}) \right] \qquad \text{(B-15)}$$

This solution is shown as curve D in Figure B-2.

Equation B-11 can also be written in the form of a transfer function:

$$T_m = \frac{T}{(\tau_w s + 1)(\tau s + 1)} \qquad \text{(B-16)}$$

and in block diagram form as shown in Figure B-3.

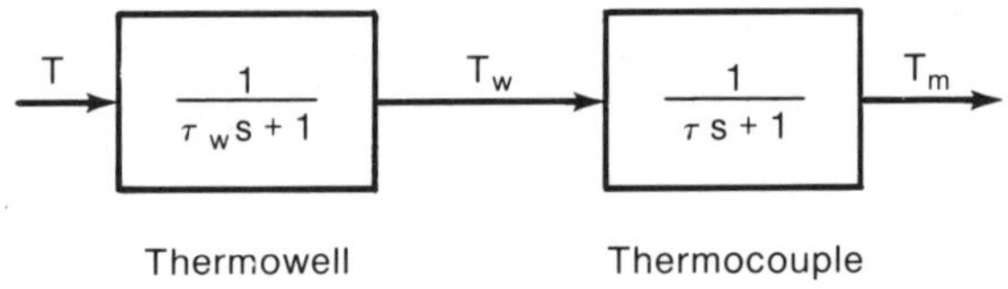

Figure B-3.

Usually, it is not found to be necessary to apply the inverse transformation since the nature of the dynamic response can be deduced from the form of the transfer function and the relative values of the time constants. In fact, criteria for control stability and dynamic stability can be quickly determined directly from the transfer function without reference to the time domain.

The time constants (τ) can be determined experimentally by transient or frequency response techniques, or analytically by the application of the defining equations. Ranges of expected time constants can be determined from Figure B-4.

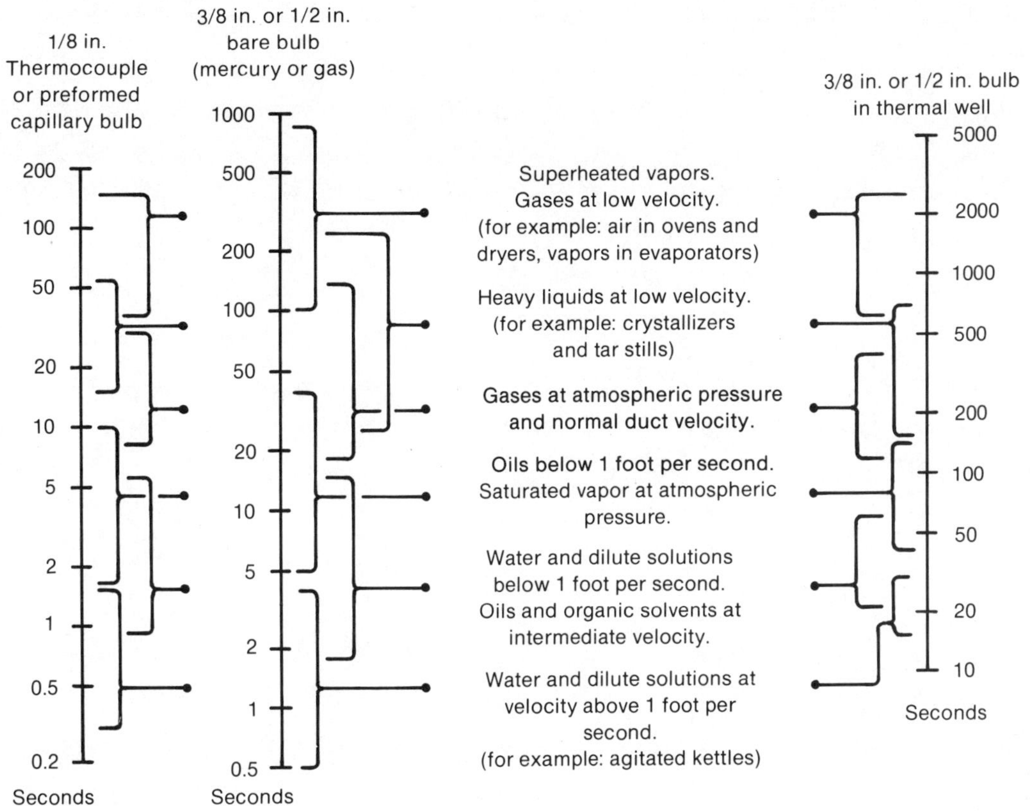

Figure B-4. Range of Temperature Measurement Lags
(Courtesy Taylor Instruments)

APPENDIX **C**

RESPONSE OF MULTI-CAPACITY SYSTEMS

The effects of multiple "time constants" may combine in several ways. Thus, two capacities might combine as a non-interacting system as in Figure C-1(a), as an interacting series as in Figure C-1(b), or as a single-capacity system with side capacity, as in Figure C-1(c).

The level (potential) of major interest will depend upon the particular installation. Also, the source of disturbance will vary. Thus, in the case shown in Figure C-1(b), control might be desired on level h_{10} or h_{11} (h_{11} is an unlikely case). Disturbances might arise from changes in q_i, in $R11$ or in $R12$.

It is common to say, "Let's follow what happens when there is a sudden change in (some variable)." Therefore, compare the relationship between step upsets of q_i and the

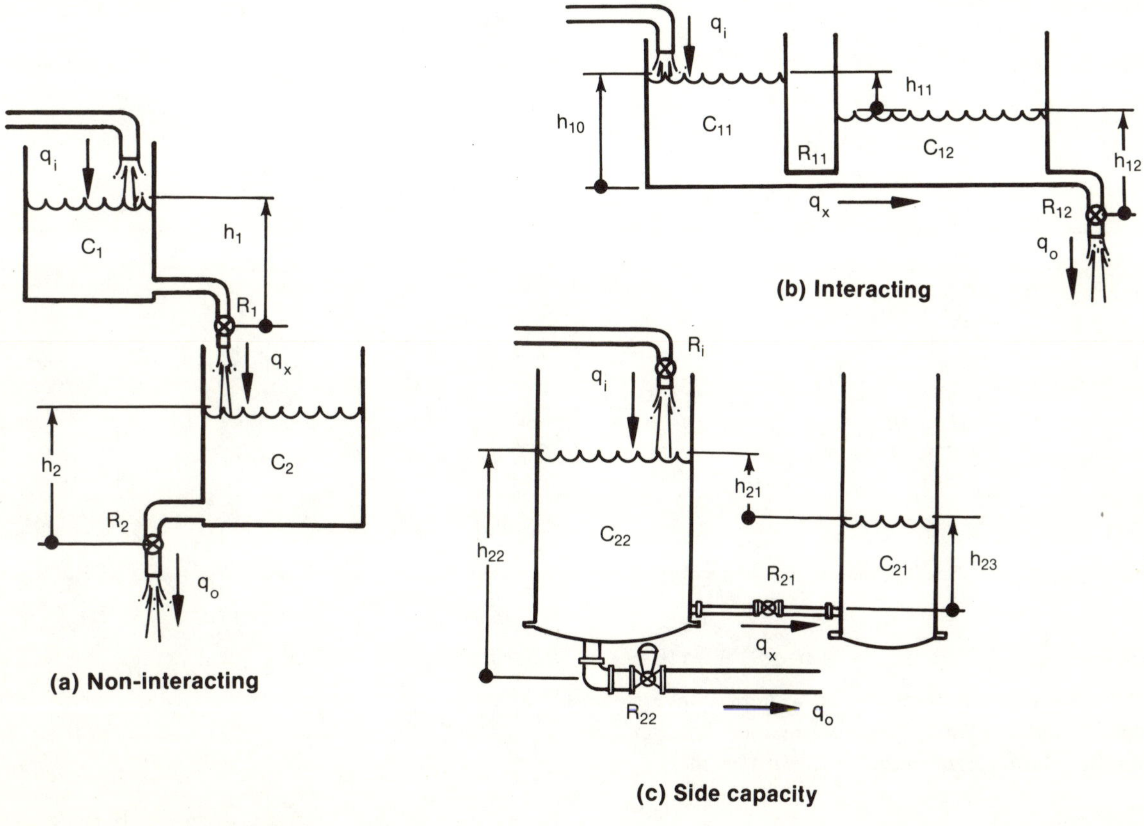

Figure C-1. Two-Capacity Systems

comparable level h_2, h_{12} and h_{22} for the different situations. Assume all other constants (capital letters on the diagram) remain unchanged during the analysis, and that all Rs are linear (i.e., $R = q/h$).

Case C-1 (a)

The describing relationships are:

$$\text{Storage 1} = q_i - q_x = C_1 \frac{dh_1}{dt} \quad \text{(C-1)}$$

$$\text{Storage 2} = q_x - q_o = C_2 \frac{dh_2}{dt} \text{ or}$$

$$q_x = C_2 \frac{dh_2}{dt} + q_o \quad \text{(C-2)}$$

$$q_x = \frac{h_1}{R_1} \text{ and } \frac{dh_1}{dt} = R_1 \frac{dq_x}{dt} \quad \text{(C-3)}$$

$$q_o = \frac{h_2}{R_2} \text{ and } \frac{dh_2}{dt} = R_2 \frac{dq_o}{dt} \quad \text{(C-4)}$$

A relationship involving only q_i and h_2 is desired. So select from these observations as necessary to form one equation in q_i and h_2:

$$q_i - q_x = C_1 \frac{dh_1}{dt} = q_i - C_2 \frac{dh_2}{dt} - \frac{h_2}{R_2} = C_1 R_1 \frac{dq_x}{dt} \quad \text{(C-5)}$$

But the same q_x from Equation C-2 gives:

$$\frac{dq_x}{dt} = C_2 \frac{d^2h_2}{dt^2} + \frac{dq_o}{dt} = C_2 \frac{d^2h_2}{dt^2} + \frac{1}{R_2} \frac{dh_2}{dt} \quad \text{(C-6)}$$

Then from Equation C-5,

$$q_i - C_2 \frac{dh_2}{dt} - \frac{h_2}{R_2} =$$

$$C_1 R_1 C_2 \frac{d^2h_2}{dt_2} + \frac{C_1 R_1}{R_2} \frac{dh_2}{dt} \quad \text{(C-7)}$$

and rearranging:

$$R_1 C_1 R_2 C_2 \frac{d^2h_2}{dt^2} + [C_1 R_1 + C_2 R_2] \frac{dh_2}{dt} =$$

$$q_i R_2 - h_2 \quad \text{(C-8)}$$

But $[q_i R_2]$ is the ultimate level toward which the investigated level h_2 is progressing. Let this be called h_F; then, paralleling the relation used in the electrical system:

instananeous error = $h_F - h_2 = E$ (C-9)
= ultimate potential – instantaneous potential

Equation C-8 can be made a little easier to write assuming

$$R_1 C_1 = \tau_1 \qquad u_2 = h_2 - q_i R_2 = h_2 - h_F$$

$$R_2 C_2 = \tau_2 \qquad du_2 = dh_2$$

Then

$$\tau_1 \tau_2 \frac{d^2u_2}{dt^2} + [\tau_1 + \tau_2] \frac{du_2}{dt} + u_2 = 0 \quad \text{(C-10)}$$

Case C-1(b)

$$\text{Storage 11} = q_i - q_x = C_{11} \frac{dh_{10}}{dt} \quad \text{(C-11)}$$

$$\text{Storage 12} = q_x - q_o = C_{12} \frac{dh_{12}}{dt} \text{ or}$$

$$q_x = C_{12} \frac{dh_{12}}{dt} + q_o \quad \text{(C-12)}$$

$$q_x = \frac{h_0 - h_{12}}{R_{11}} = \frac{h_{11}}{R_{11}} \text{ and}$$

$$\frac{dh_{11}}{dt} = R_{11} \frac{dq_x}{dt} = \frac{dh_{10}}{dt} - \frac{dh_{12}}{dt} \quad \text{(C-13)}$$

$$q_o = \frac{h_{12}}{R_{12}} \text{ and } \frac{dh_{12}}{dt} = R_{12} \frac{dq_o}{dt} \quad \text{(C-14)}$$

$$q_i - q_x = C_{11} \frac{dh_{10}}{dt} = q_i - C_{12} \frac{dh_{12}}{dt} - \frac{h_{12}}{R_{12}} =$$

$$C_{11} R_{11} \frac{dq_x}{dt} + C_{11} \frac{dh_{12}}{dt} \quad \text{(C-15)}$$

$$\frac{dq_x}{dt} = C_{12} \frac{d^2h_{12}}{dt^2} + \frac{dq_o}{dt} =$$

$$C_{12} \frac{d^2h_{12}}{dt^2} + \frac{1}{R_{12}} \frac{dh_{12}}{dt} \quad \text{(C-16)}$$

$$q_i + C_{12} \frac{dh_{12}}{dt} - \frac{h_{12}}{dt} = C_{11} R_{11} C_{12} \frac{d^2h_{12}}{dt^2} +$$

$$\frac{C_{11} R_{11}}{R_{12}} \frac{dh_{12}}{dt} + C_{11} \frac{dh_{12}}{dt} \quad \text{(C-17)}$$

$$C_{11} R_{11} C_{12} R_{12} \frac{d^2h_{12}}{dt^2} +$$

$$[C_{11} R_{11} + C_{11} R_{12} + C_{12} R_{12}] \frac{dh_{12}}{dt} = q_i R_{12} - h_{12} \quad \text{(C-18)}$$

$$E = h_F - h_{12} \qquad u_{12} = h_{12} - q_i R_{12} \quad \text{(C-19)}$$

$$\tau_{11} \tau_{12} \frac{d^2u_{12}}{dt^2} +$$

$$[\tau_{11} + \tau_{12} + C_{11} R_{12}] \frac{du_{12}}{dt} + u_{12} = 0 \quad \text{(C-20)}$$

Case C-1(c)

$$\text{Storage 21} = q_x = C_{21} \frac{dh_{23}}{dt} \qquad \text{(C-21)}$$

$$\text{Storagc 22} = q_i - q_x - q_o = C_{22} \frac{dh_{22}}{dt} \; ;$$

$$q_x = q_i - q_o - C_{22} \frac{dh_{22}}{dt} \qquad \text{(C-22)}$$

$$q_x = \frac{h_{22} - h_{23}}{R_{21}} = \frac{h_{21}}{R_{21}} \quad \text{and}$$

$$\frac{dh_{21}}{dt} = R_{21} \frac{dq_x}{dt} = \frac{dh_{22}}{dt} - \frac{dh_{23}}{dt} \qquad \text{(C-23)}$$

$$q_o = \frac{h_{22}}{R_{22}} \quad \text{and} \quad \frac{dh_{22}}{dt} = R_{22} \frac{dq_o}{dt} \qquad \text{(C-24)}$$

$$q_x = C_{21} \frac{dh_{23}}{dt} = q_i - q_o - C_{22} \frac{dh_{22}}{dt} =$$

$$C_{21} \frac{dh_{22}}{dt} - C_{21} R_{21} \frac{dq_x}{dt} =$$

$$q_i - \frac{h_{22}}{R_{22}} - C_{22} \frac{dh_{22}}{dt} \qquad \text{(C-25)}$$

$$\frac{dq_x}{dt} = \frac{dq_i}{dt} - \frac{1}{R_{22}} \frac{dh_{22}}{dt} - C_{22} \frac{d^2 h_{22}}{dt^2} \qquad \text{(C-26)}$$

$$C_{21} \frac{dh_{22}}{dt} - C_{21} R_{21} \frac{dq_i}{dt} - \frac{C_{21} R_{21}}{R_{22}} \frac{dh_{22}}{dt} -$$

$$C_{21} R_{21} C_{22} \frac{d^2 h_{22}}{dt^2} = q_i - \frac{h_{22}}{R_{22}} - C_{22} \frac{dh_{22}}{dt} \qquad \text{(C-27)}$$

$$C_{21} R_{21} C_{22} R_{22} \frac{d^2 h_{22}}{dt^2} +$$

$$[\, C_{21} R_{21} - C_{21} R_{21} + C_{22} R_{22} \,] \frac{dh_{22}}{dt} +$$

$$h_{22} = R_{22}\, q_i + C_{21} R_{21} R_{22} \frac{dq_i}{dt} \qquad \text{(C-28)}$$

$$E = R_{22}\, q_i - h_{22} = h_F - h_{22}$$

$$u_{22} = h_{22} - R_{22}\, q_i \qquad \text{(C-29)}$$

$$\tau_{21} \tau_{22} \frac{d^2 u_{22}}{dt^2} + [\, \tau_{22} - \tau_{21} + C_{21} R_{22}] \frac{du_{22}}{dt} +$$

$$u_{22} = \tau_{21} R_{22} \frac{dq_i}{dt} \qquad \text{(C-30)}$$

Equations C-10, C-20, and C-30 could be transformed to the Laplace notation and used as a design tool to choose tank parameters to provide the desired dynamic response.

APPENDIX **D**

CHARACTERIZATION OF A LIQUID LEVEL PROCESS CONTROL SYSTEM

While the concepts of heat and mass balances used to characterize an industrial process have general application, the liquid level process previously described will be used as an explicit example to illustrate the applied techniques. Consider the material balance control of the liquid in an open tank as shown in Figure D-1.

Following Valstar,[101] a signal flow or block diagram of the control system can be developed from qualitative concepts. Consider Figure D-2. The pneumatic controller is actuated by an error signal that represents the difference between the set point and the level measurement. Controller output is transmitted via tubing to the control valve. The valve modulates flow in the piping from the tank. The tank is acted on by the inflow and the disturbance — the outflow. The level in the tank is measured in the float chamber. The level transmitter senses the level in the chamber and transmits it to the indicator and the controller via pneumatic tubing. Thus it becomes apparent that the lines connecting the blocks on the signal flow diagram represent the transmission of information.

Figure D-3 emphasizes that information is being transmitted. The indicator has been omitted since it is not a part of the control loop. The pneumatic tubing and the piping can be neglected unless they are greater than 200 to 300 feet in length. Directional arrows and sources of energy have been added for clarity in Figure D-3.

The spans of information being transmitted are shown on Figure D-4 to introduce the quantitative nature of the signal flow diagram. The span of each variable shown permits determination of the static gain of each block. For example, for the control valve:

$$K_v = \frac{(100\text{-}0)\ \text{gpm}}{(15\text{-}3)\ \text{psi}} = 8.35\ \frac{\text{gpm}}{\text{psi}}$$

The variables denoted on Figure D-4 are those shown on the foregoing physical drawing of the level process. Thus all static gains are determined except that of the controller, which has an adjustable gain that must be set to provide stable control. The dynamic response of each element in the signal flow diagram must be determined either by the mathematical formulation of its known physical and chemical properties or by physical testing such as the frequency response method previously demonstrated.

The dynamic response of the level tank shown is denoted by the mass balance on the liquid in the tank:

Accumulation = Inflow − Outflow

Quantitatively,

$$(60)\ (0.0043)\ A\frac{dh}{dt} = w_i - w_o$$

where 60 = conversion, seconds/minute

0.0043 = conversion, gallons/cu. in.

A = cross-sectional area of tank = 1960 sq. in.

h = height from lower float chamber tap, in.

t = time, seconds

w_i, w_o = flow in and out, gpm.

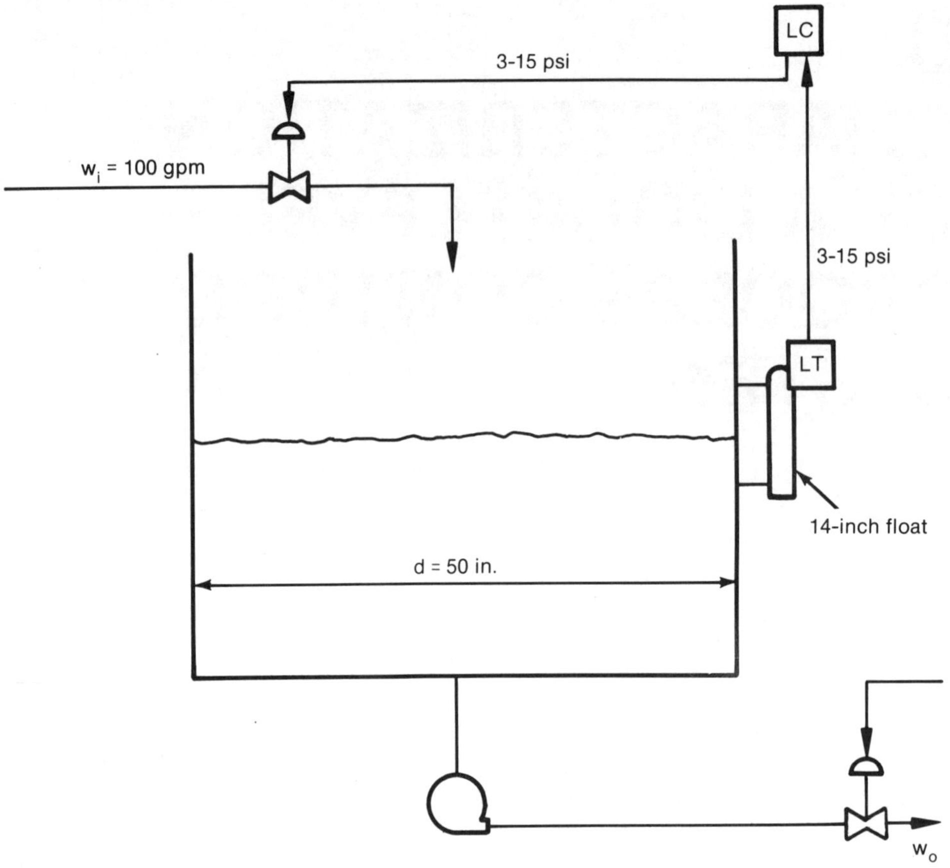

Figure D-1.

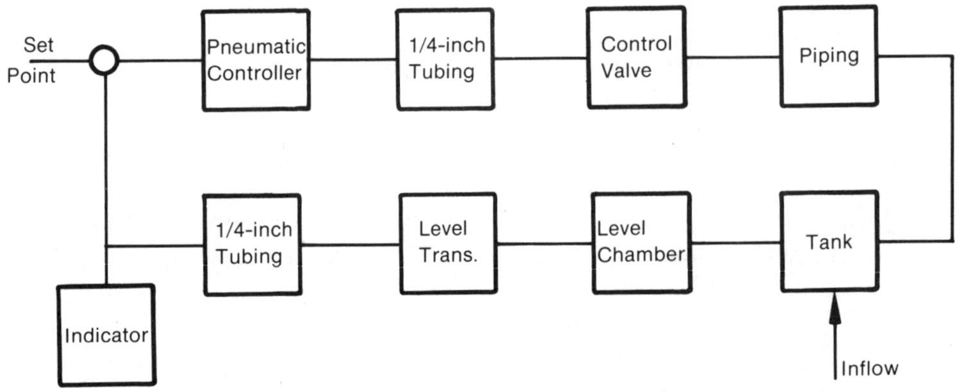

Figure D-2.

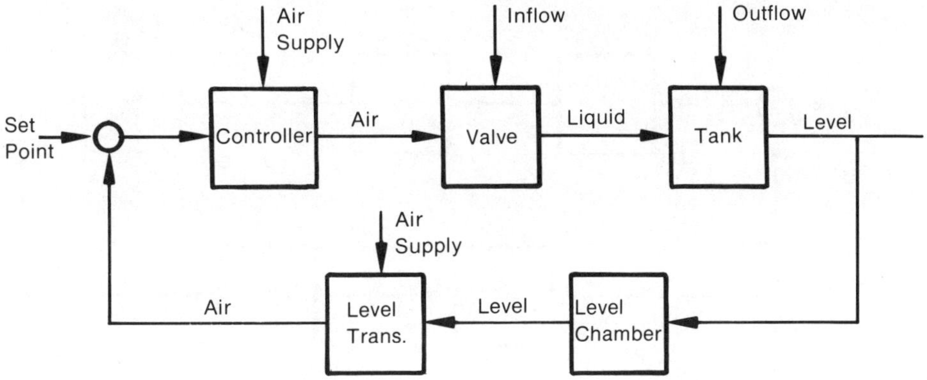

Figure D-3.

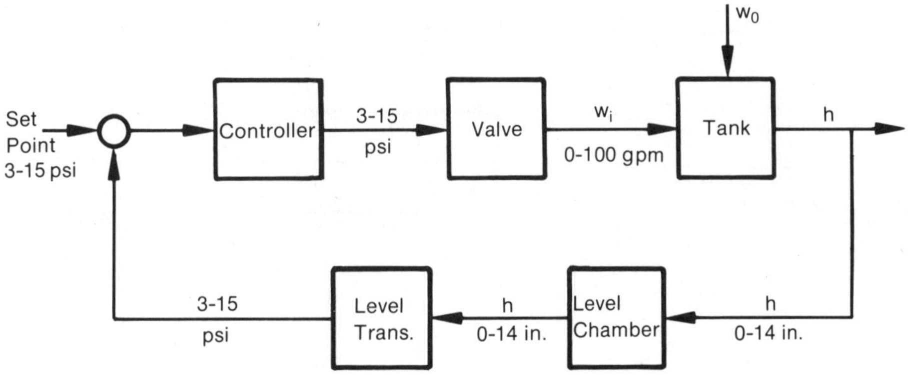

Figure D-4.

The relationship among the time domain, the frequency domain, and the complex domain has been defined as:

$$\frac{d}{dt} = s = j\omega$$

This relationship is used to reduce the differential equation to an algebraic equation. In the complex domain:

$$h = \frac{1}{510s}(w_i - w_o) \qquad \text{(D-1)}$$

Equation D-1 is shown in Figure D-5, where the disturbance inflow (w_i) is shown at a summing junction as indicated by Equation D-1. Letters have been arbitrarily assigned to other lines of information flow in Figure D-5 to identify them for further discussion.

The dynamics of the valve (G_v) and the level chamber (G_T) are to be found in the literature and are seldom developed analytically. In complex notation, the response of a 1- to 2-inch control valve will typically be:

$$w_i = \frac{K_v}{0.25s + 1}\, C = \frac{8.35\ \frac{\text{gpm}}{\text{psi}}}{0.25s + 1}\, C \qquad \text{(D-2)}$$

Per Schuder,[102] the combined level transmitter and float chamber have a transfer function:

$$b = \frac{K_T}{0.12s^2 + 0.035s + 1}\, h$$

$$= \frac{\frac{(15-3)\ \text{psi}}{(14-0)\ \text{inches}}}{0.12s^2 + 0.035s + 1}\, h$$

$$b = \frac{0.86\ \frac{\text{psi}}{\text{in.}}}{0.12s^2 + 0.035s + 1}\, h \qquad \text{(D-3)}$$

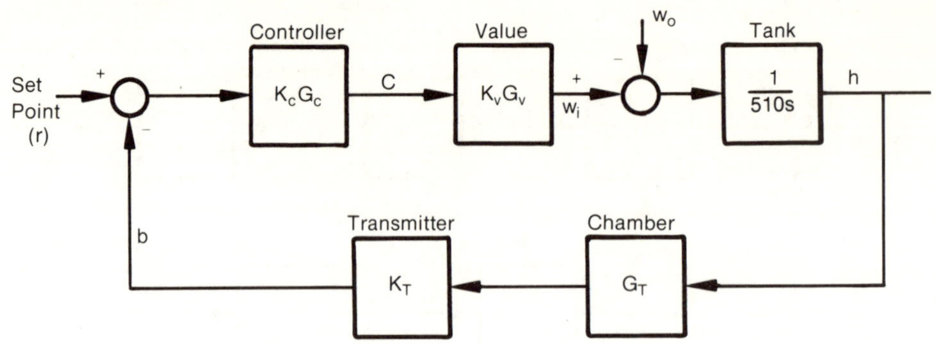

Figure D-5.

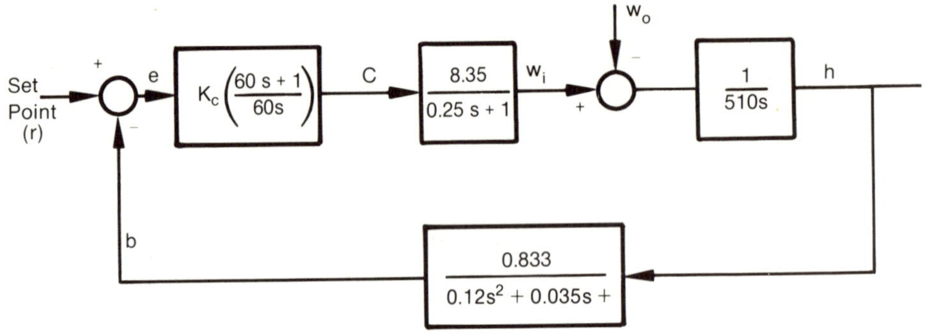

Figure D-6.

Equations D-2 and D-3 have been combined into Figure D-6, which is the signal flow diagram describing the level control system illustrated in the foregoing sketch.

Since the gains in the squares represent multiplications and the circles represent algebraic summing points, the block diagram can be combined into the following equation:

$$h = \frac{1}{510s} w_o + \frac{8.35\,K_cG_c}{510s\,(0.25s+1)} r - \frac{(8.35)\,(0.86)\,K_cG_c}{510s\,(0.25s+1)\,(0.12s^2+0.035s+1)} h$$

Since the set point (r) will normally be constant, this equation can be manipulated algebraically:

$$h = \frac{\frac{1}{510s} w_o}{1 + \frac{(8.35)\,(0.86)\,K_cG_c}{510s\,(0.25s+1)\,(0.12s^2+0.035s+1)}} \quad \text{(D-4)}$$

The demoninator of Equation D-4 is called the characteristic equation and dictates stability. Quite obviously, the equation is in trouble if the denominator becomes equal to zero, or

$$\frac{0.014\,K_cG_c}{s\,(0.25s+1)\,(0.12s^2+0.035s+1)} = -1 \quad \text{(D-5)}$$

The significance of Equation D-5 becomes apparent if it is written in the frequency domain with the reset time, τ_i, equal to 60 seconds:

$$\frac{0.00023\,K_c\,(1+j60\omega)}{j\omega^2\,(0.25j\omega+1)\,(-0.12\omega^2+0.035j\omega+1)} = -1 \quad \text{(D-6)}$$

Following Section 7, the real and imaginary parts of the transfer functions are of interest. It will be remembered that the real and imaginary parts can be plotted on two-dimensional axes:

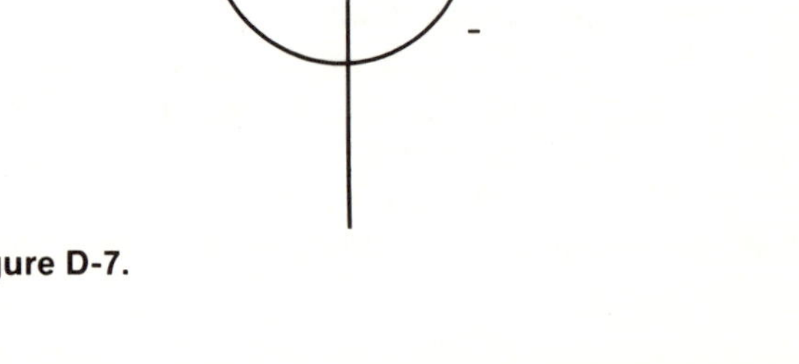

Figure D-7.

The number denoted by points a and b can be stated as $a + jb$ or as a magnitude (M) and a phase angle θ. The value -1 in Equation D-6 is thus described as a magnitude (M) of 1 and an angle of 180°. For the purpose of determining stability by frequency response, the 180° is minus as shown by the convention in the figure because the outputs of most physical processes lag the inputs.

Multiplying out Equation D-6:

$$\frac{0.00023K_c\,(1 + j60\omega)}{\omega^2\,[(0.12875\,\omega^2 - 1) + j\,(0.03\omega^3 - 0.285\,\omega)]} \qquad \text{(D-7)}$$

Resorting to the trickery of multiplying by the conjugate:

$$\left[\frac{0.00023K_c\,(1 + j60\omega)}{\omega^2\,(0.12875\,\omega^2 - 1) + j\,(0.03\omega^3 - 0.285\omega)}\right]\left[\frac{(0.12875\omega^2 - 1) - j\,(0.03\omega^3 - 0.285\omega)}{(0.12875\omega^2 - 1) - j\,(0.03\omega^3 - 0.285\omega)}\right]$$

Results in:

$$\frac{0.00023K_c\,(1.8\omega^4 - 16.9\omega^2 - 1)}{\omega^2\,[(0.12875\omega^2 - 1)^2 + (0.03\omega^2 - 0.285\omega)^2]} + j\,\frac{0.00023K_c\,(7.695\omega^3 - 59.7\omega)}{\omega^2\,[(0.12875\omega^2 - 1)^2 + (0.03\omega^3 - 0.285\omega)^2]} \qquad \text{(D-8)}$$

and

$$M = \frac{0.00023K_c\,[(1.8\omega^4 - 16.9\omega^2 - 1)^2 + (7.695\omega^3 - 59.7\omega)^2]^{1/2}}{\omega^2\,[(0.12875\omega^2 - 1)^2 + (0.03\omega^3 - 0.285\omega)^2]}$$

$$\theta = \text{Tan}^{-1}\,\frac{7.695\omega^3 - 59.7\omega}{1.8\omega^4 - 16.9\omega^2 - 1} \qquad \text{(D-9)}$$

where M and θ are defined in Figure D-7. Obviously, the magnitude (M) is proportional to controller gain (K_c), as shown in Figures D-9 and D-10.

Equation 9 is evaluated in attached Table D-1 and plotted in Figure D-9, which is known as a Nyquist diagram. This diagram illustrates the stability condition shown in Equation D-5, namely, that the function is unstable if it equals or is greater than -1. In terms of the Nyquist diagram, the criterion is stated graphically that the locus of the frequency plot shall not enclose the -1 point. Figure D-9 shows that controller gain can be raised (proportional band narrowed) to where the level control system will become unstable. From Table D-1, the point where the locus passes through -1 is at 2.85 cps, where controller gain would have to be

$$(0.00023K_c)\,(166.66) = 1$$

$$K_c = 26$$

or

$$\text{P.B.} = 3.8\%$$

Seldom would such an extreme value of proportional band be used, and any reasonable value of gain will be quite stable. The simplifications of the development shown in Section 11 are thus shown to have a limitation, whereupon that section showed the system to become more stable as controller gain was increased. As gain is increased to a very high value, minor time constants contribute to the phase lag, and the level control system will become unstable. Stability difficulties are almost entirely due to the reset in the controller, however, as discussed in Section 11.

Suppose, for the moment, that the control valve has an unusually slow response, as shown in Figure D-8, that derivative action had been added to the controller to compensate, and the reset action is set to a much faster value. The same analysis for magnitude and phase angle leads to the computer listing of Table D-2 and the Nyquist diagram, Figure D-10. Here it is apparent that the level control system will become unstable if controller

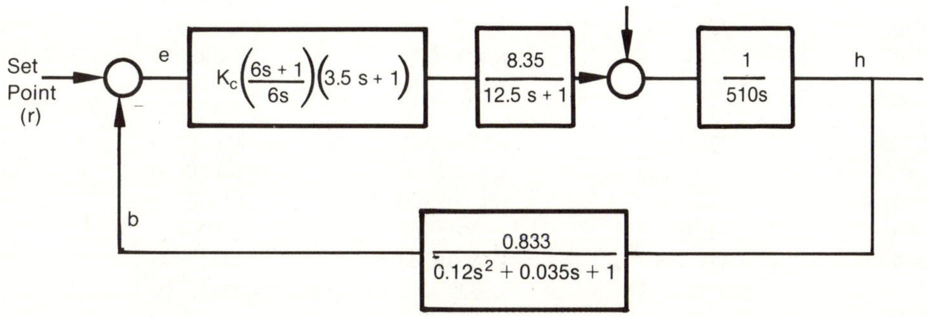

Figure D-8.

Table D-1
Computer Listing for Nyquist Plot

Imaginary part "b"	Real part "a"	Frequency, rad/sec	Dynamic magnitude ratio, M	Phase angle, θ
-2.9840E+00	-1.0424E+00	5.0000E-02	1.2649E+03	1.9661E-01
-8.9290E+00	-1.3809E+00	1.5000E-01	4.0316E+02	2.2569E-01
-1.4805E+01	-2.0536E+00	2.5000E-01	2.4181E+02	2.2818E-01
-2.0565E+01	-3.0518E+00	3.5000E-01	1.7346E+02	2.2667E-01
-2.6164E+01	-4.3626E+00	4.5000E-01	1.3584E+02	2.2382E-01
-3.1555E+01	-5.9687E+00	5.5000E-01	1.1215E+02	2.2036E-01
-3.6692E+01	-7.8485E+00	6.5000E-01	9.5958E+01	2.1657E-01
-4.1529E+01	-9.9761E+00	7.5000E-01	8.4287E+01	2.1259E-01
-4.6019E+01	-1.2321E+01	8.5000E-01	7.5556E+01	2.0847E-01
-5.0118E+01	-1.4849E+01	9.5000E-01	6.8856E+01	2.0426E-01
-5.3777E+01	-1.7522E+01	1.0500E+00	6.3631E+01	1.9997E-01
-5.6952E+01	-2.0295E+01	1.1500E+00	5.9523E+01	1.9562E-01
-5.9596E+01	-2.3121E+01	1.2500E+00	5.6295E+01	1.9120E-01
-6.1662E+01	-2.5949E+01	1.3500E+00	5.3788E+01	1.8670E-01
-6.3106E+01	-2.8723E+01	1.4500E+00	5.1897E+01	1.8211E-01
-6.3880E+01	-3.1381E+01	1.5500E+00	5.0556E+01	1.7742E-01
-6.3938E+01	-3.3859E+01	1.6500E+00	4.9731E+01	1.7258E-01
-6.3235E+01	-3.6089E+01	1.7500E+00	4.9416E+01	1.6755E-01
-6.1723E+01	-3.7996E+01	1.8500E+00	4.9637E+01	1.6226E-01
-5.9358E+01	-3.9502E+01	1.9500E+00	5.0456E+01	1.5662E-01
-5.6092E+01	-4.0527E+01	2.0500E+00	5.1984E+01	1.5050E-01
-5.1879E+01	-4.0982E+01	2.1500E+00	5.4407E+01	1.4366E-01
-4.6674E+01	-4.0779E+01	2.2500E+00	5.8032E+01	1.3578E-01
-4.0430E+01	-3.9820E+01	2.3500E+00	6.3375E+01	1.2627E-01
-3.3101E+01	-3.8008E+01	2.4500E+00	7.1345E+01	1.1409E-01
-2.4641E+01	-3.5239E+01	2.5500E+00	8.3614E+01	9.7171E-02
-1.5004E+01	-3.1404E+01	2.6500E+00	1.0329E+02	7.0972E-02
-4.1430E+00	-2.6391E+01	2.7500E+00	1.3461E+02	2.4795E-02
7.9876E+00	-2.0084E+01	2.8500E+00	1.6666E+02	-6.0276E-02
2.1434E+01	-1.2361E+01	2.9500E+00	1.4589E+02	-1.6683E-01
3.6242E+01	-3.0976E+00	3.0500E+00	9.9216E+01	-2.3655E-01
5.2495E+01	7.8361E+00	3.1500E+00	6.7998E+01	2.2652E-01
7.0130E+01	2.0574E+01	3.2500E+00	4.9327E+01	2.0469E-01
8.9302E+01	3.5254E+01	3.3500E+00	3.7538E+01	1.9025E-01
1.1002E+02	5.2020E+01	3.4500E+00	2.9608E+01	1.7980E-01
1.3233E+02	7.1017E+01	3.5500E+00	2.3989E+01	1.7170E-01
1.5628E+02	9.2398E+01	3.6500E+00	1.9842E+01	1.6510E-01
1.8192E+02	1.1632E+02	3.7500E+00	1.6682E+01	1.5954E-01
2.0928E+02	1.4293E+02	3.8500E+00	1.4212E+01	1.5471E-01
2.3843E+02	1.7241E+02	3.9500E+00	1.2240E+01	1.5043E-01
2.6940E+02	2.0493E+02	4.0500E+00	1.0640E+01	1.4657E-01
3.0223E+02	2.4064E+02	4.1500E+00	9.3215E+00	1.4305E-01
3.3699E+02	2.7974E+02	4.2500E+00	8.2224E+00	1.3980E-01
3.7370E+02	3.2240E+02	4.3500E+00	7.8961E+00	1.3678E-01
4.1243E+02	3.6880E+02	4.4500E+00	6.5084E+00	1.3395E-01
4.5321E+02	4.1915E+02	4.5500E+00	5.8330E+00	1.3128E-01
4.9609E+02	4.7363E+02	4.6500E+00	5.2499E+00	1.2875E-01
5.4111E+02	5.3243E+02	4.7500E+00	4.7432E+00	1.2635E-01
5.8833E+02	5.9578E+02	4.8500E+00	4.3003E+00	1.2406E-01
6.3779E+02	6.6386E+02	4.9500E+00	3.9112E+00	1.2187E-01
6.8954E+02	7.3690E+02	5.0500E+00	3.5677E+00	1.1978E-01

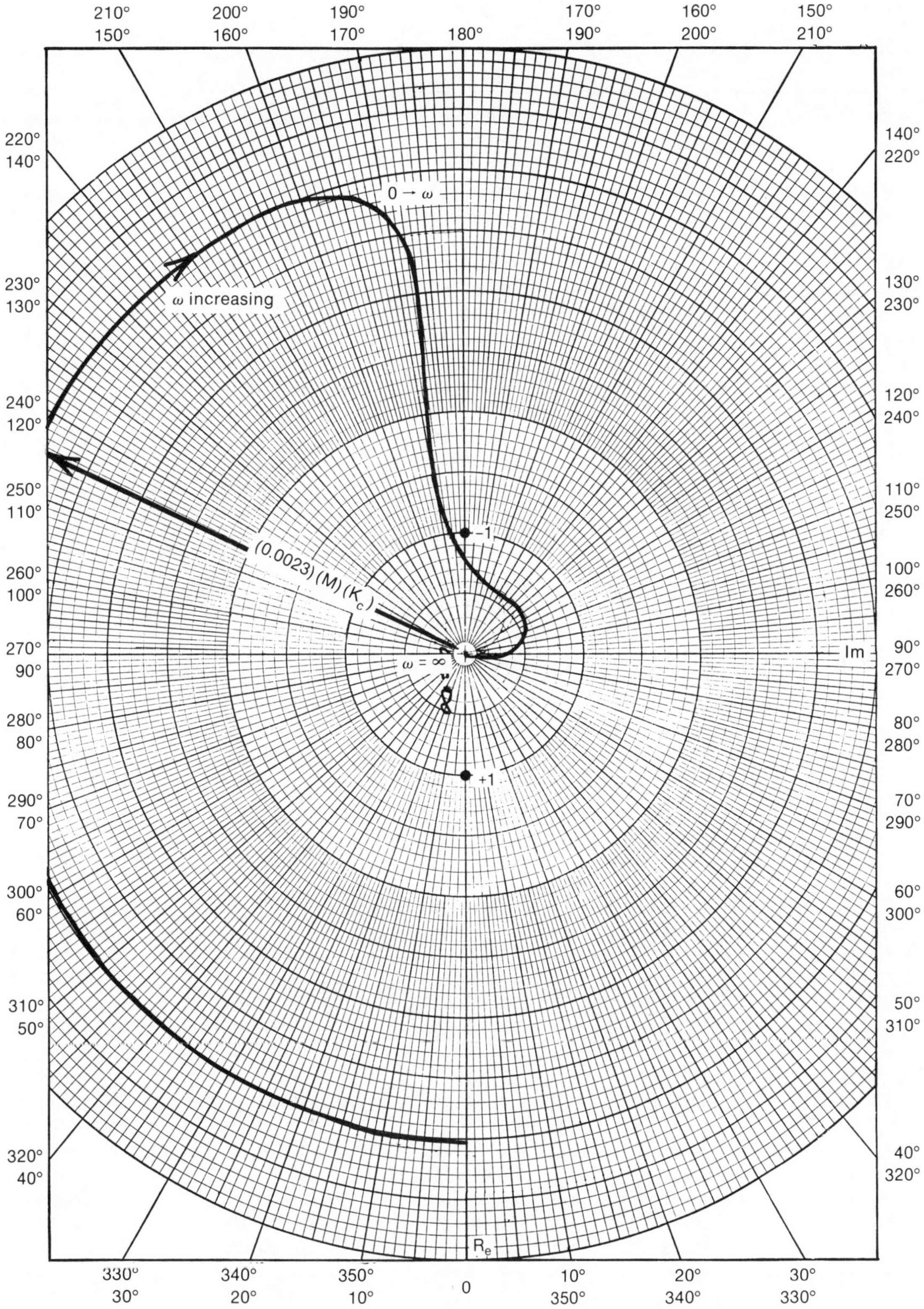

Figure D-9. Nyquist Diagram from Block Diagram of Figure D-6 and Table D-1

Table D-2
Computer Listing for Nyquist Plot with Controller Derivative Action

Imaginary part "b"	Real part "a"	Frequency, rad/sec	Dynamic magnitude ratio, M	Phase angle, θ
1.1767E-01	-1.2430E+00	5.0000E-02	3.5991E+02	-1.5030E-02
-4.2072E-01	-3.1857E+00	1.5000E-01	3.1937E+01	2.0908E-02
-3.2614E+00	-7.0650E+00	2.5000E-01	1.1800E+01	6.8867E-02
-9.8763E+00	-1.2869E+01	3.5000E-01	6.8083E+00	1.0423E-01
-2.1643E+01	-2.0578E+01	4.5000E-01	4.7714E+00	1.2908E-01
-3.9806E+01	-3.0170E+01	5.5000E-01	3.7031E+00	1.4685E-01
-6.5441E+01	-4.1612E+01	6.5000E-01	3.0553E+00	1.5994E-01
-9.9414E+01	-5.4868E+01	7.5000E-01	2.6251E+00	1.6982E-01
-1.4235E+02	-6.9896E+01	8.5000E-01	2.3221E+00	1.7745E-01
-1.9458E+02	-8.6646E+01	9.5000E-01	2.1002E+00	1.8342E-01
-2.5612E+02	-1.0506E+02	1.0500E+00	1.9338E+00	1.8814E-01
-3.2664E+02	-1.2509E+02	1.1500E+00	1.8075E+00	1.9189E-01
-4.0539E+02	-1.4665E+02	1.2500E+00	1.7117E+00	1.9486E-01
-4.9120E+02	-1.6968E+02	1.3500E+00	1.6404E+00	1.9717E-01
-5.8242E+02	-1.9409E+02	1.4500E+00	1.5896E+00	1.9890E-01
-6.7690E+02	-2.1980E+02	1.5500E+00	1.5570E+00	2.0013E-01
-7.7192E+02	-2.4673E+02	1.6500E+00	1.5414E+00	2.0087E-01
-8.6421E+02	-2.7476E+02	1.7500E+00	1.5428E+00	2.0111E-01
-9.4984E+02	-3.0380E+02	1.8500E+00	1.5621E+00	2.0083E-01
-1.0242E+03	-3.3373E+02	1.9500E+00	1.6016E+00	1.9997E-01
-1.0821E+03	-3.6445E+02	2.0500E+00	1.6654E+00	1.9840E-01
-1.1175E+03	-3.9582E+02	2.1500E+00	1.7603E+00	1.9592E-01
-1.1236E+03	-4.2772E+02	2.2500E+00	1.8972E+00	1.9221E-01
-1.0927E+03	-4.6002E+02	2.3500E+00	2.0948E+00	1.8668E-01
-1.0165E+03	-4.9257E+02	2.4500E+00	2.3861E+00	1.7828E-01
-8.8561E+02	-5.2522E+02	2.5500E+00	2.8319E+00	1.6489E-01
-6.8975E+02	-5.5783E+02	2.6500E+00	3.5457E+00	1.4184E-01
-4.1769E+02	-5.9024E+02	2.7500E+00	4.6802E+00	9.8066E-02
-5.7220E+01	-6.2227E+02	2.8500E+00	5.8410E+00	1.4601E-02
4.0495E+02	-6.5376E+02	2.9500E+00	5.0582E+00	-8.8307E-02
9.8315E+02	-6.8454E+02	3.0500E+00	3.4670E+00	-1.5328E-01
1.6929E+03	-7.1440E+02	3.1500E+00	2.4088E+00	-1.8654E-01
2.5507E+03	-7.4318E+02	3.2500E+00	1.7720E+00	-2.0498E-01
3.5744E+03	-7.7067E+02	3.3500E+00	1.3670E+00	-2.1631E-01
4.7831E+03	-7.9667E+02	3.4500E+00	1.0926E+00	-2.2385E-01
6.1971E+03	-8.2096E+02	3.5500E+00	8.9692E-01	-2.2915E-01
7.8379E+03	-8.4335E+02	3.6500E+00	7.5151E-01	-2.3306E-01
9.7286E+03	-8.6360E+02	3.7500E+00	6.3998E-01	-2.3603E-01
1.1893E+04	-8.8150E+02	3.8500E+00	5.5221E-01	-2.3835E-01
1.4358E+04	-8.9680E+02	3.9500E+00	4.8169E-01	-2.4019E-01
1.7150E+04	-9.0927E+02	4.0500E+00	4.2404E-01	-2.4169E-01
2.0297E+04	-9.1866E+02	4.1500E+00	3.7621E-01	-2.4292E-01
2.3831E+04	-9.2473E+02	4.2500E+00	3.3605E-01	-2.4395E-01
2.7782E+04	-9.2721E+02	4.3500E+00	3.0195E-01	-2.4481E-01
3.2185E+04	-9.2585E+02	4.4500E+00	2.7274E-01	-2.4555E-01
3.7073E+04	-9.2037E+02	4.5500E+00	2.4750E-01	-2.4617E-01
4.2485E+04	-9.1050E+02	4.6500E+00	2.2554E-01	-2.4671E-01
4.8458E+04	-8.9596E+02	4.7500E+00	2.0630E-01	-2.4718E-01
5.5033E+04	-8.7645E+02	4.8500E+00	1.8936E-01	-2.4759E-01
6.2251E+04	-8.5169E+02	4.9500E+00	1.7435E-01	-2.4795E-01
7.0157E+04	-8.2318E+02	5.0500E+00	1.6099E-01	-2.4826E-01

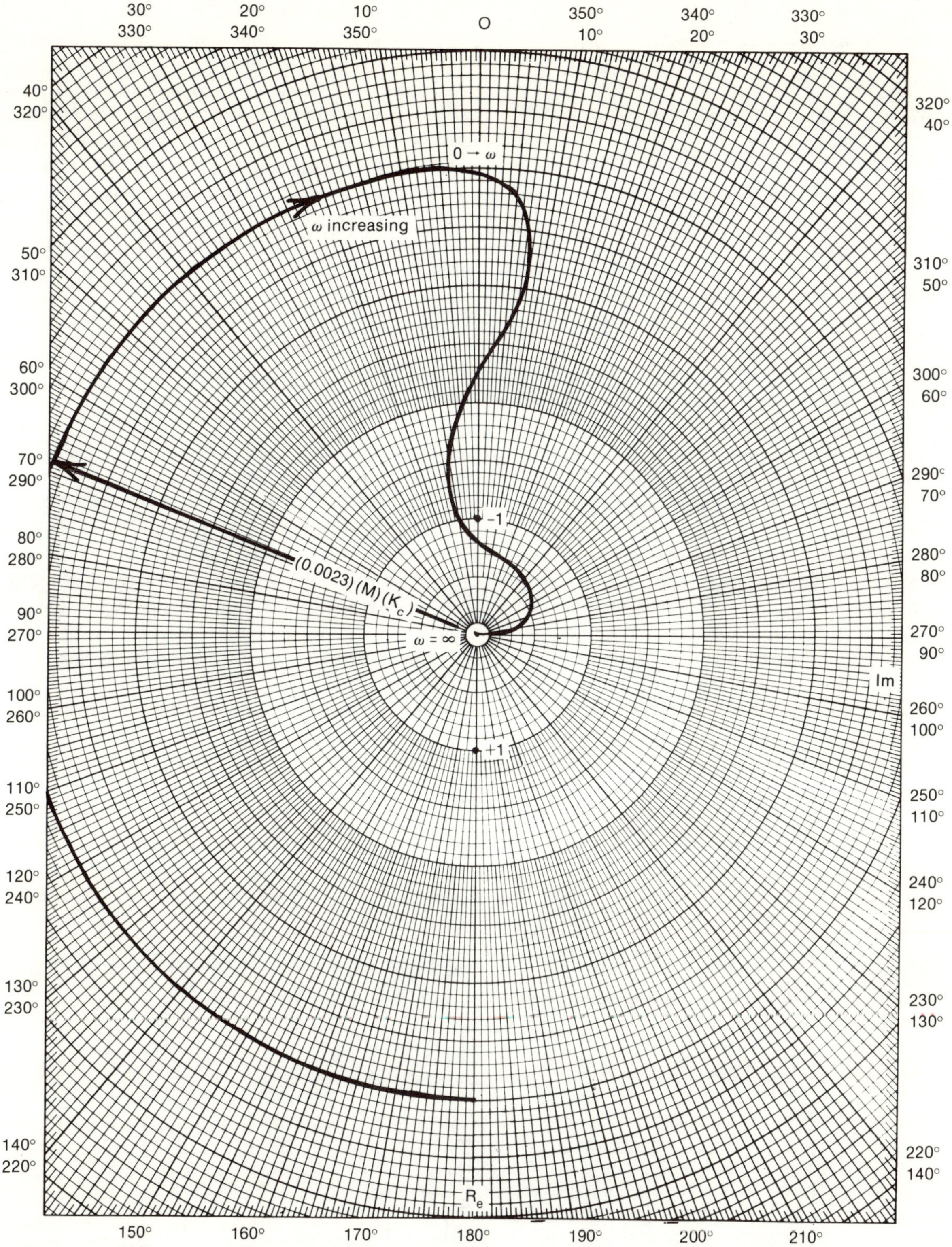

Figure D-10. Nyquist Diagram from Block Diagram of Figure D-7 and Table D-2, Showing Conditional Stability gain (K_c) is increased and also if it is decreased. Reference to the listing, Table D-2, shows that the gain must be between the following values for stability:

Maximum value of gain

$$(0.0023\ K_c)\ (5.81) = 1$$

$$K_c = 75.0\ (\text{P.B.} = 1.4\%)$$

Minimum value of gain

$$(0.0023\ K_c)\ (31.9) = 1$$

$$K_c = 13.6\ (\text{P.B.} = 7.4\%)$$

The condition illustrated by Figure D-10 is called conditional stability.[(106)] It is quite obviously a troublesome situation, as shown by the foregoing narrow range of stable proportional bands. However, with regard to the level control system, notice that conditional stability was instigated by the addition of derivative action to the controller and by the abnormally fast reset action in the controller. Good level regulation is attained by removing the derivative mode from the controller and following the inequality guideline of Section 8.

Wilson and Zoss[(107)] have discussed the information available from the Nyquist diagram. The conditional stability phenomenon is an example of a control problem that is immediately apparent from the Nyquist diagram but might otherwise go unnoticed. However, a glance at the foregoing formidable equations and a little mental gymnastics is convincing evidence of the travail necessary to plot a Nyquist diagram. Thus Tables D-1 and D-2 are computer-generated, and a computer is necessary to plot the number of diagrams necessary to tune a controller. The alternate graph developed by Dr. Hendrick Bode simplifies the procedure to the simple addition of curves traced from templates, as discussed in Section 7.

Figure D-11 is a Bode plot of the level control system as shown in Figure D-6, without the controller. The left-hand side of Equation D-5 can be quickly sketched on the Bode plot paper using the template that conforms to the paper. Permitting the gain of each term to be "one" for the moment, the transfer function of the level transmitter and float cage has been sketched on Bode plot Figure D-12, using the relationships:

$$\tau = \sqrt{0.12} = 0.36 \text{ sec}$$

$$\zeta = \frac{0.035}{2T} = 0.05$$

Likewise the integrating action of the tank has been plotted. Finally, the transfer function of the control valve has been plotted using a template, and the plots combined into the open-loop magnitude ratio and phase angle curves. The resulting magnitude ratio and phase angle plots are shown in Figure D-12, which compares to Nyquist plot Figure D-9 but is much more useful because the component curves can be easily moved to observe their effect on controllability. The strength of the Nyquist diagram can be observed by comparing Figures D-10 and D-12, where the conditional stability is immediately apparent from the Nyquist diagram but could be easily missed on the Bode diagram.

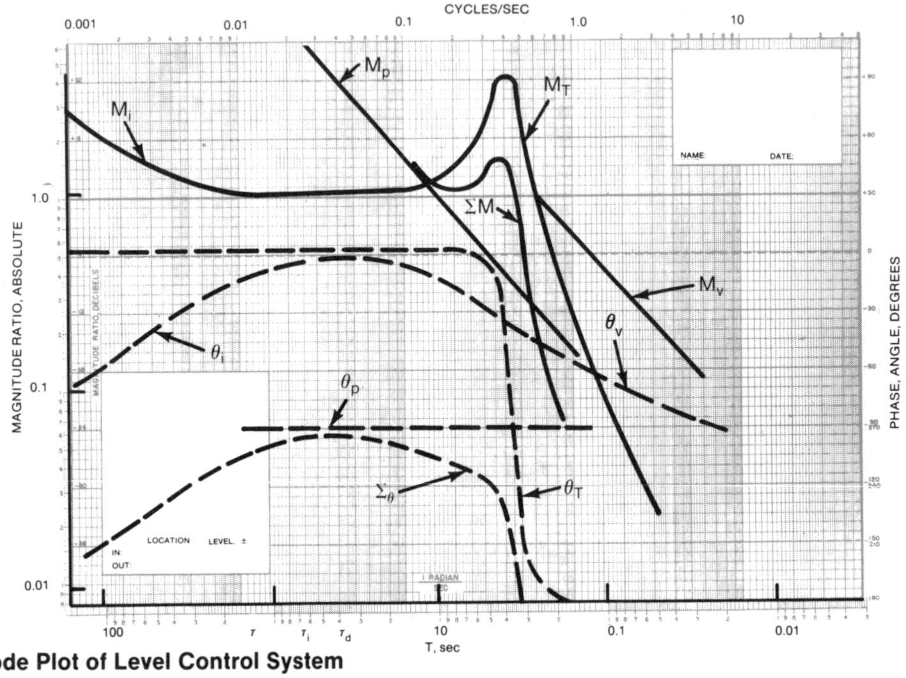

Figure D-11. Bode Plot of Level Control System

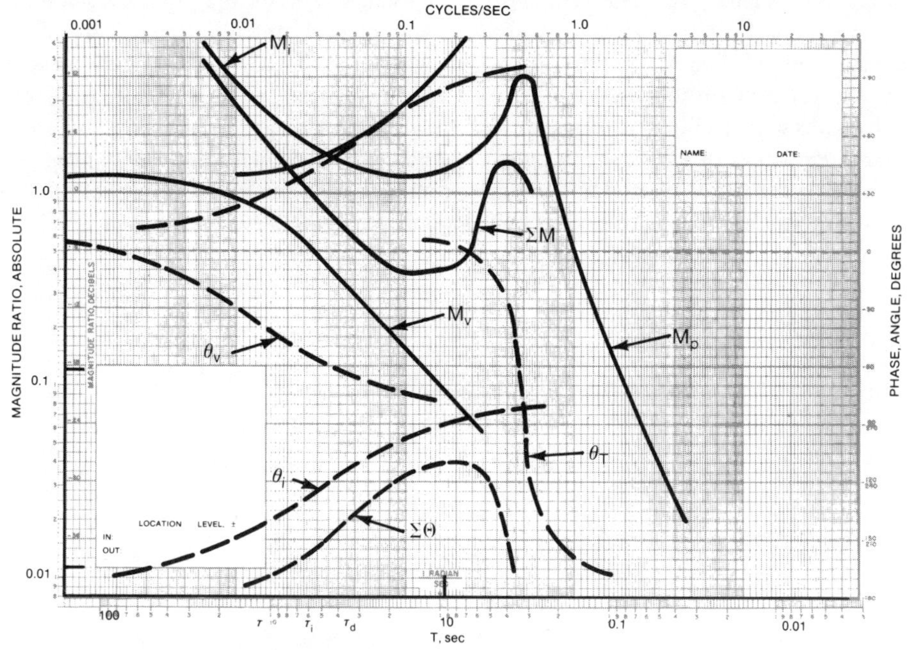

Figure D-12. Level Control System with Conditional Stability

The open-loop Bode plot has been established by combining the individual transfer functions describing the hardware in the control loop. The controller gain (K_c) has been left as a degree of freedom by permitting the open-loop gain, (0.014 K_c in Equation D-5), to arbitrarily be equal to 1. But it has been established that the system is unstable if the magnitude (M) of the open-loop gain is 1 at −180° phase angle. Thus the trick is to move the open-loop magnitude curve vertically on the Bode plot to assure a magnitude of less than 1 at 180° phase lag. There are three guides for this maneuver:

Phase Margin. The phase margin has been defined from the Nyquist diagram as an angle that is positive when measured counterclockwise from 180°. Thus it results in a phase lag of less than 180°. The generally used phase margin of 45° would then say that the gain should be 1 at a phase angle of 135°. Reference to a book with Nyquist diagrams for many systems[106] indicates that the magnitude is generally decreasing as it passes the 180° phase angle point, and thus setting it equal to 1 at 135° should mean that it will be less than 1 at 180°. This criterion has been applied to the Bode plot for the level system shown in Figure D-11. The combined open-loop Bode plot for the process indicates that the 135° phase lag occurs at a frequency of 0.42 cps, where the magnitude is 1.45. Thus K_c becomes

$$0.014\ K_c = \frac{1}{1.45}\ K_c \qquad K_c = 47.5 \text{ or P.B.} = 2.1\%$$

Gain Margin. The gain margin merely takes the open-loop gain at the 180° phase lag and divides it by 2. Thus the magnitude at 180° in Figure D-11 occurs at a frequency of 0.48 cps and is 1.35. The controller gain then becomes

$$0.014\ K_c = \frac{1.35}{2} \qquad K_c = 48.2 \quad \text{P.B.} = 2.1\%$$

Gain margin and its relationship to phase margin is discussed in the publication by Lloyd.[108]

Alternate Criterion. Another criterion sometimes used is that the rate of change of open-loop gain shall not be greater than 2:1 at the point where the gain is equal to one. This criterion is based on the knowledge that a system with a rate of change of 1:1 must be stable, a change of 2:1 can be underdamped but must be stable, and 3:1 can be unstable. On Figure D-11, the 2:1 slope point is 0.48 cps, and the controller gain is

$$0.014\ K_c = 1.35 \qquad K_c = 96.4 \qquad \text{P.B.} = 1\%$$

Thus the stability criteria applied to the level system give remarkably consistent results. Unfortunately, the results are consistently unsatisfactory. The level controller with a P.B. of 2 percent would be quite unstable. Physically, a proportional band of 2 percent would never be used, but typically a very wide P.B. would be selected in order that the tank act as a surge vessel and reduce the amplitude of flow changes through the vessel. This is called "averaging" level control, and has lead to the thought of "nonset point" control,[63] where the level transmitter is connected directly to the valve with no controller whatsoever. This results in an equivalent P.B. of 100 percent (K_c = 1.0) and no set point. But who needs a set point? The water closet on a toilet operates year in and year out without one.

The foregoing analysis has shown a liquid level process, and, in fact, all nonself-regulating processes, to require different precepts in selecting controller modes and in adjusting those controller modes. While the analysis explains many of the observed anomalies in the performance of level control systems, it also shows that the classical guidelines for calculating controller settings lead to unusable results. A bit of consideration of the experimental tuning procedures in Section 8 leads quickly to the conviction that they will also be of little use. The unequality shown as Equation 11-3 in Section 11 provides the only reliable guideline.

APPENDIX **E**

DYNAMIC ANALYSIS OF A FLOW CONTROL SYSTEM

Signal flow diagrams are developed for the analysis of liquid level, thermal, gas pressure, and liquid flow processes. In each case, the controlled variable is fed back through an appropriate transmitter to a controller, and the controller repositions a control valve to compensate for any deviation in the controlled variable. Closed-loop control can thus be applied to each of these typical processes, utilizing a block diagram such as Figure 6-16 for the dynamic analysis of each process.

The control of a liquid flow could seemingly be easily accomplished by measuring the flow and automatically positioning a valve to pass the desired flow rate. Actually, the analysis of the flow control system is one of the most difficult.[50,51] The difficulty lies in the number of disturbances that tend to upset the control system, the nonlinearities used to describe both the measuring element (differential pressure type) and the valve, and the essential lack of capacitance in the system. Since the process remains full of liquid at all times, it cannot accumulate liquid, and thus it has negligible capacitance. The lack of capacitance means that the process has no dynamic lag ($T = RC = 0$), and thus that it responds instantaneously.

Consider the flow process shown in Figure E-1.

The "hydraulic" equation constitutes the mathematical model for the liquid flow through the control valve:

$$q = C_1 x \sqrt{p_2 - p_3} \qquad \text{(E-1)}$$

where C_1 = constant, for a linear installed valve characteristic.

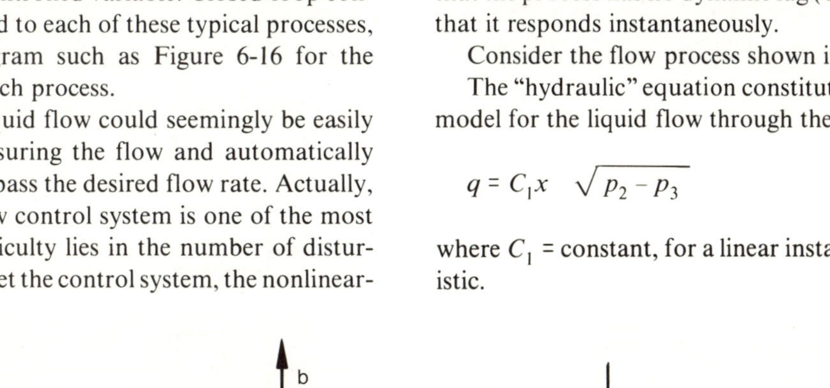

where q — Liquid flow, lb/sec
p_1 — Pressure at upstream measuring tap, in. water
p_2 — Pressure at downstream tap, in. water
p_3 — Valve downstream pressure, in. water
p — Signal from controller, psi
b — Signal to controller, psi
x — Valve opening, fraction

Figure E-1. Flow Process

The orifice-type measuring element can be described by the same type of Equation:[20]

$$q^2 = \frac{2gA_2^2\rho^2(p_1 - p_2)}{1 - \left[\frac{A_2}{A_1}\right]^2} \qquad \text{(E-2)}$$

where A_1 = Area of pipe, sq in.

A_2 = Area of orifice bore, sq in.

ρ = Fluid density, lb/in.^{3}

g = Gravitational constant, in./sec^2

$$\text{Let } C_2 = \sqrt{\frac{2gA_2^2\rho^2(p_1 - p_2)}{1 - \left[\frac{A_2}{A_1}\right]^2}} = \text{constant}$$

Then,

$$q = C_2\sqrt{p_1 - p_2} \qquad \text{(E-3)}$$

The assumption has been made that the pressure drop across the orifice is a permanent head loss. Actually, a part of this loss is recovered, and the pressure upstream from the valve is somewhat higher than p_2. The expected pressure recovery is available as plotted curves[39] and can be included if significant to the analysis.

Equations E-1 and E-3 can be linearized:

$$q = f(x, p_2, p_3) \qquad \text{(E-4)}$$

$$q = \frac{\partial f}{\partial x}x + \frac{\partial f}{\partial p_2}p_2 + \frac{\partial f}{\partial p_3}p_3 \qquad \text{(E-5)}$$

$$q = \left.\frac{q}{x}\right|_{ss} x + \left.\frac{q}{2(p_2-p_3)}\right|_{ss} p_2 - \left.\frac{q}{2(p_2-p_3)}\right|_{ss} p_3 \qquad \text{(E-6)}$$

and

$$q = g(p_1, p_2) \qquad \text{(E-7)}$$

$$q = \frac{\partial g}{\partial p_1}p_1 + \frac{\partial g}{\partial p_2}p_2 \qquad \text{(E-8)}$$

$$q = \left.\frac{q}{2(p_1-p_2)}\right|_{ss} p_1 - \left.\frac{q}{2(p_1-p_2)}\right|_{ss} p_2 \qquad \text{(E-9)}$$

The pressure p_2 can be eliminated by combining Equations E-6 and E-9:

$$q = \left.\frac{q}{x}\right|_{ss} x + \left.\frac{q}{2(p_2-p_3)}\right|_{ss} p_1 - \left.\frac{p_1-p_2}{p_2-p_3}\right|_{ss} q - \left.\frac{q}{2(p_2-p_3)}\right|_{ss} p_3 \qquad \text{(E-10)}$$

In terms of the controlled variable q:

$$q = \left.\frac{q(p_2-p_3)}{x(p_1-p_3)}\right|_{ss} x + \left.\frac{q}{2(p_1-p_3)}\right|_{ss} p_1 - \left.\frac{q}{2(p_1-p_3)}\right|_{ss} p_3 \qquad \text{(E-11)}$$

It is now evident that the tacit assumption has been made that the upstream pressure (p_1) and the downstream pressure (p_3) are independent of the flow (q). This assumption would not be true if a centrifugal pump were upstream from the flow process. Neither would it be true if a vessel that acted as a restricting device were downstream from the process. In either case, the pressure-flow relationship would have to be determined and the intermediate pressures eliminated by the combination of equations. This would lead to substantially more complex relationships.

Define the gain terms:

$$K_1 = \frac{g(p_2 - p_3)}{x(p_1 - p_3)}$$

$$K_2 = \frac{q}{2(p_1 - p_3)}$$

and Equation E-11 becomes:

$$q = K_1 x + K_2 p_1 - K_2 p_3 \qquad \text{(E-12)}$$

Equation E-12 is directly amenable to block diagram presentation as shown in Figure E-2.

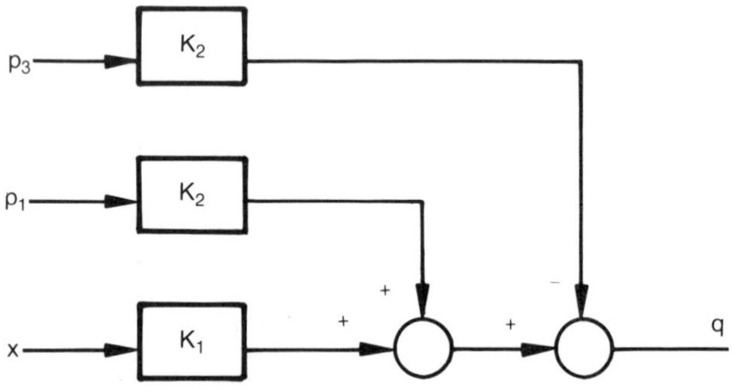

Figure E-2. Flow Process Signal Flow Diagram

The difference between the flow process and the liquid level, thermal, and gas pressure processes formerly described is now apparent. The relationships between the variables in the flow process are steady-state gains, which means that the response is instantaneous. The responses of the other processes are exponential relationships denoted by the τs+1 factor in the denominators. The difficulty in the control of the flow system lies essentially in the condition that the dynamic responses of the control system components dominate in the absence of any dynamic lag in the process.[51]

The foregoing process portion of the block diagram can be incorporated into the closed-loop diagram in a manner similar to the level system described in Section 6.

The dynamic response of the closed-loop control system shown in Figure E-3, when written in accordance with Section 6-6, is dependent entirely on the dynamic responses of the control valve and transmitter. The responses of the valve and the transmitter are dependent on the values of their respective time constants (τ_v and τ_T). Since these two instruments and most pneumatic instruments in general are composed of a series of tubes and chambers filled with a fluid (usually air), the time constants can be determined analytically by the techniques of Section 4-1 for gas pressure systems. Similarly, electrical instruments are analyzed by the techniques of Section 4-4.

Practically, however, more reliable results are obtained from the physical testing of the instruments. The development of Section 8-2 indicated that if a step upset were applied to the input of an instrument, the time constant could be determined directly from the record of the output. This is the transient testing technique. Frequency response testing is a steady-state testing procedure where a sine wave is applied to the input of the instrument. Appropriate plots of the input and the output allow the time constant to be determined.[33] Data on the dynamic response of instruments is generally published in the technical literature,[50] by the manufacturers,[56] or by the users of the instruments.

The analytical description of the process itself is therefore developed by techniques similar to the foregoing and those in Sections 3, 4, and 6. The analytical description of the instruments is generally determined by physical testing. The combination of the resultant formulation into the closed-loop block diagram allows analysis of the controllability of the system.

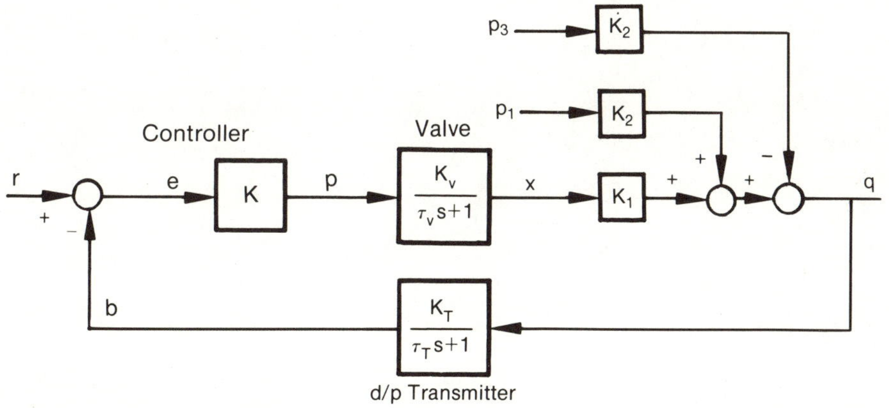

Figure E-3. Flow Control System

APPENDIX **F**

NOMENCLATURE AND DEFINITIONS

NOMENCLATURE

a	Dead time, time
A	Area, ft^2
b	Length of conductive path, ft
c	Specific heat, Btu/lb°F
C	Capacitance
D	Pipe diameter, ft
e	error signal
E	Voltage
f	Frequency, Hz (cps)
g	Gravitational constant, 32.2 ft/sec^2
h	Head, feet
I	Electrical current flow, coulombs/sec
k	Thermal conductivity, Btu-ft/hour-ft^2-°F
K	Gain
K_c	Controller gain, dimensionless
l	Pipe length, ft
L	Liquid level, feet
m	molecular weight
M	Weight, lb
n	Polytropic constant, ratio of specific heats
p	Pressure, psi
P	Period of sine wave, time/cycle
q	Volumetric flow, gpm
Q	Heat flow rate, Btu/hr
r	Control efficiency factor

R	Resistance
$\bar{R}$	Gas constant
s	Laplace operator
t	Time, seconds or minutes
T	Temperature, °C or °F
v	Velocity, ft/sec
V	Volume, ft^3
w	Mass flow, lb/min
W	Disturbance
x	Valve stem movement, normalized 0 to 1
X	Liquid mol fraction
Y	Gas mol fraction
Z	Sampling time, time
ϵ	Emissivity
θ	Angle, radians or degrees
λ	Latent heat, Btu/lb
μ	Viscosity, lb/ft sec
ρ	Density, lb/ft^3
σ_w^2	Variance of disturbance
σ_m^2	Variance of measurement error
τ	Time constant, time
τ_i	Reset time, minutes
τ_m	Measurement time constant, time
τ_w	Disturbance time constant, time
ω	Rate of change of angle, radians/second

Definitions

Btu	British thermal unit, the quantity of heat necessary to raise the temperature of 1 pound of water 1°F.
PCU	Pound centigrade unit, the quantity of heat required to raise the temperature of 1 pound of water 1°C.

Note:

$$\frac{\text{Btu}}{°\text{F}} = \frac{\text{PCU}}{°\text{C}}$$

APPENDIX **G**

REFERENCES

1. Nixon, F.E., *Principles of Automatic Controls*, Prentice Hall, Inc., New York, 1953, p. 1.
2. Wing, P., "Plain Talk on Valve Rangeability," *Instrumentation Technology*, April 1978, p. 57.
3. McKinney, A.H., *An Automatic Control Philosophy*, DuPont Co. Publication, 1951.
4. Janssen, J.M.L., "A Practical Guide to Plant Controllability," *Control Engineering*, November 1955, p. 58.
5. Nourse, T.M., "Decision Making in Systems Engineering," *Instruments and Control Systems*, April 1966, p. 89.
6. Sherwood, G.F.E., and Taylor, E., *Elementary Differential Equations*, Prentice Hall, N.Y., 1943, p. 15.
7. Middlemiss, R.R., *Differential and Integral Calculus*, McGraw-Hill, New York, N.Y. 1946.
8. Williams, T.J., "University Research Results for Industrial Process Control," *Instrumentation Technology*, November 1967, p. 49.
9. Thomas, R.P., and Tong, P.P., "Discrete State Feedback — A Route to Minimum Control," *Instrumentation Technology*, April 1967, p. 59.
10. VanDerGrinten, P.M.E.M., "Control Effects of Instrument Accuracy and Measuring Speed," *ISA Journal*, December 1965, p. 48 and January 1966, p. 58.
11. Hoel, P.G., *Introduction to Mathematical Statistics*, John Wiley & Sons, New York, 1954.
12. Zalkind, C.S., and Shinskey, F.G., "The Application of Statistical Methods in Evaluating the Accuracy of Analog Instruments and Systems," ISA Paper No. 41.2.63, ISA Conference & Exhibit, Chicago, Ill., 1963.
13. VerDerGrinten, P.M.E.M., "Determining Plant Controllability," *Control Engineering*, October 1963, p. 87.
14. Goff, K.W., "Process Studies for Control System Design and Evaluation," *Ceramic Age*, April 1963.
15. Myron, T.J., "Guidelines for Control System Design," *Instrumentation Technology*, January 1972, p. 47.
16. Moore, C.F., Smith, C.L., and Murrill, P.W., "Gaging Performance Costs of Control Loop Hardware," *Instrumentation Technology*, November 1968, p. 69.
17. Buckley, P.S., "Cost vs Performance in Process Control Systems," Paper No. 54-20-3, First ISA International Instrument Congress and Exposition, Philadelphia, Pa., 1954.
18. Badavas, P.C., "Control Only If Necessary — A New Concept," *Control Engineering*, January 1973, p. 54.
19. Fannin, D.B., "Stability of COIN Systems," *Control Engineering*, July 1973, p. 52.
20. Silve, R., "Plant Savings and the Control Heirarchy," *Instruments and Control Systems*, June 1968, p. 85.
21. Stout, T.M., and Cline, R.P., "Control System Justification," *Instrumentation Technology*, September 1976, p. 51.
22. Eckman, D.P., *Automatic Process Control*, John Wiley and Sons, New York, N.Y., 1958, p. 12.
23. Levy, Ezra C., *New Table of Laplace Transformation Pairs*, 3rd ed., Book Page, P.O. Box 1188, Los Angeles, Calif. 90028, 1964.
24. McAdams, W.H., *Heat Transmission*, McGraw-Hill Book Company, New York, N.Y., 1942, p. 63.
25. Lefkowitz, I., "Methods of Dynamic Analysis," *ISA Journal*, June 1955, p. 203.

26. Hutchison, J.W., *ISA Handbook of Control Valves*, 2nd ed., Instrument Society of America, Research Triangle Park, N.C., 1976.

27. Driskell, Les, *Introduction to Control Valves and Other Final Control Devices*, Instructional Resource Package, Instrument Society of America, Research Triangle Park, N.C., 1982.

28. Moore, R.L., "Flow Characteristics of Valves," *ISA Handbook of Control Valves*, Chapter 5, Instrument Society of America, Research Triangle Park, N.C., 1976.

29. Moore, R.L., "Flow Characteristics," ISA Audio Cassette Study Package Topic 4, Instrument Society of America, Research Triangle Park, N.C., 1976.

30. Brainard, H.A., "Some Unique Essentials of Good Liquid Pipeline Control," *ISA Journal*, January 1964, p. 43.

31. Buckley, P.S., "Selection of Optimum Final Element Characteristics," Proceedings of the Fifth National Chemical and Petroleum Instrumentation Symposium, Wilmington, DE, 1964, p. 105.

32. Anderson, N.A., "Frequency Response Analysis," *Instruments and Control Systems*, June 1964, p. 113.

33. Lefkowitz, I., "Frequency Response Analysis," *ISA Journal*, May 1956, p. 170.

34. Lloyd, S.G., "Frequency Response for Process Control Elements," Fisher Controls, Marshalltown, Iowa, p. 3.

35. Vercammen, H., and VanCauwenberghe, A.R., "Parameters Quickly Determined: Second-Order Model with Dead Time," *Control Engineering*, April 1970, p. 79.

36. Hougen, J.D., "Process Dynamics by Pulse Testing," *Chemical Engineering*, June 12, 1961, p. 209.

37. Murrill, P.W., Pike, R.W., and Smith, C.L., "Development of Dynamic Mathematical Models," Part I, *Chemical Engineering*, September 9, 1968, p. 117.

38. VanDerGrinten, P.M.E.M., "Determine Process Characteristics Simply by Filtering Plant Data," *Control Engineering*, July 1964, p. 61.

39. Barber, J.E., "Non-Interacting Time Lags — Frequency Response Analysis," Manual for Taylor Frequency Response Course, Sybron-Taylor, Rochester, N.Y., 1954

40. Murrill, P.W., Pike, R.W., and Smith, C.L., "Frequency Response Data Yield Analytic Equations," Part II, *Chemical Engineering*, November 18, 1968, p. 165.

41. Stoll, H.W., "Non-Interacting Time Lags — Step Response Analysis," Manual for Taylor Frequency Response Course, Sybron/Taylor, Rochester, N.Y. 1954.

42. Anderson, N.A., "Step Analysis Method of Finding Time Constant," *Instruments and Control Systems*, November 1963, p. 160.

43. Oldenbourg, R.C., and Sartorius, H., *The Dynamics of Automatic Controls*, American Society of Mechanical Engineers, New York, N.Y., 1948, p. 276.

44. Murrill, P.W., Pike, R.W., and Smith, C.L., "Transient Response in Dynamic System Analysis," Part IV *Chemical Engineering*, December 16, 1968, p. 103.

45. Eckman, D.P., *Automatic Process Control*, John Wiley & Sons, Inc., New York, N.Y., 1958, p. 41.

46. ISA Standard S26, "Dynamic Response Testing of Process Control Instrumentation," Instrument Society of America, Research Triangle Park, N.C., 1968.

47. Pemberton, T.J., "An Improved Pulse Testing Method," *Instrumentation Technology*, December 1970, p. 61.

48. Banham, J.W., "Obtain Process Dynamics by Pulse Testing," *Control Engineering*, April 1965, p. 83.

49. Murrill, P.W., Pike, R.W., and Smith, C.L., "Pulse Testing Methods," Part VI, *Chemical Engineering*, Feb. 24, 1969, p. 105.

50. Eckman, D.P., "Frequency Response Analysis of a Flow Control System," *ISA Journal*, December 1954, p. 11.

51. Ziegler, J.G. and Nichols, N.B., "Electronic Flow Control," *ISA Journal*, January 1959, p. 58.

52. Schuder, C.B., and Binder, R.C., "The Response of Pneumatic Transmission Lines to Step Inputs," Transactions of the ASME, December 1959, p. 587.

53. Schwent, G.V., McGregor, W.K., and Russell, D.W., "Mathematical Description of Gas Flow Processes for Control Purposes," *ISA Journal*, August 1956, p. 274.

54. Chapin, D., "Transmission Lines," Manual for Taylor Frequency Response Course, Sybron/Taylor, Rochester, N.Y., 1954.

55. Warnock, J.D., "How Pneumatic Tubing Size Influences Controllability," *Instrumentation Technology*, February 1967, p. 37.

56. Buckley, P.S., "Design of Pneumatic Flow Controls," 28th Annual Proceedings, Instrumentation Symposium for the Processing Industries, Texas A & M University, College Station, TX, 1973.

57. Buckley, P.S., "A Modern Perspective on Controller Tuning," Texas A & M Instrumentation Symposium Proceedings, January, 1973.

58. Moore Products Co., "Model 75E/P Positioner Frequency Response," DWG 10400=8, 1956.

59. Ceaglske, N.H., *Automatic Process Control for Chemical Engineers*, John Wiley & Sons, New York, 1956.

60. Hougen, J.D., *Measurements and Control Applications for Practicing Engineers*, Instrument Society of America, Research Triangle Park, NC, 1979.

61. Eckert, E.R.G., *Introduction to the Transfer of Heat and Mass*, McGraw-Hill, New York, 1950.

62. Anderson, N.A., *Instrumentation for Process Measurement and Control*, Chilton Book Co., Radnor, PA, 1975.

63. Buckley, P.S., *Techniques of Process Control*, John Wiley & Sons, New York, 1964.

64. Zoss, L.M., *Applied Instrumentation in the Process Industries, Volume VI, Control Systems: Theory, Trouble-shooting, and Design*, Gulf Publishing Co., Houston, TX, 1979.

65. Moore, R.L., "The Use and Misuse of Pressure Regulators," *Instrumentation Technology*, March 1969, p. 52.

66. Anon., "Getting the Measure of Temperatures in Process Control," *Control and Instrumentation* (British), March 1981, p. 35.

67. Shinskey, F.G., "Controlling Unstable Processes — Part II: A Heat Exchanger," *Instruments and Control Systems*, January 1975, p. 49.

68. Sandford, J., "Understanding Temperature Control," *Instruments and Control Systems*, June 1977, p. 37.

69. Application Notes, "The Relative Merits of Proportional and Switching Type Temperature Control Systems," Magnetic Controls Co., Minneapolis, Minn.

70. Wills, D.M., "Cascade Control Applications and Hardware," Honeywell Technical Bulletin TX 119-1, Honeywell, Inc., Philadelphia, Pa., 1960.

71. Leonhard, W., "Cascade Control, A Versatile Structural Principle for Technical and Cybernetic Control Systems," *Control and Instrumentation* (British), July 1970, p. 28.

72. Chiu, K.C., Corripio, A.B., and Smith, C.C., "Tuning Cascade Control Systems," *Instruments and Control Systems*, January 1974, p. 65.

73. Franks, R.G.E., and Worley, C.W., "Quantitative Analysis of Cascade Control," *Industrial and Engineering Chemistry*, June 1956, p. 1074.

74. Shinskey, F.G., "Avoiding Reset Windup in Cascade Systems," *Instruments and Control Systems*, August 1971, p. 9.

75. Moore, R.L., *Neutralization of Waste Water by pH Control*, Instrument Society of America, Research Triangle Park, N.C., 1978.

76. Naimpally, A., and Santina, M., "Feedforward Control," *Measurements and Control*, June 1981, p. 135.

77. Application Engineering Release AD688, "Model 688 Pneumatic Impulse Relay," Moore Products Co., Springhouse, Pa., July 1979.

78. Lind, L.F., and Nelson, J.C.C., "Feedforward — An Alternative Concept in Control System Design," *Control and Instrumentation* (British), April 1970, p. 39.

79. Master, R.G., "Feedforward Control," *Instruments and Control Systems*, September 1966, p. 107.

80. Shinskey, F.G., "Feedforward Control Applied," *ISA Journal*, November 1963, p. 61.

81. Cadman, T.W., et al., "Linear Feedforward Control of a Continuous Flow, Stirred-Tank Reactor," *ISA Transactions*, July 1967, p. 234.

82. MacMillan, E.C., and Shinskey, F.G., "Feedforward Analog Computer Control of a Superfractionator," *Control Eng.*, March 1964, p. 69.

83. Woolverton, P.F., and Murrill, P.W., "An Evaluation of Four Ideas in Feedforward Control," *Instrumentation Technology*, January 1967, p. 35.

84. Luyben, W.L., "Feedforward Control of Distillation Columns," *Chem. Eng. Prog.*, August 1965, p. 75.

85. Ho, G.E., and Walsh, B.W., "Feedforward Control of Simple Process Systems," *Control* (British).

86. Bollinger, R.E., and Lamb, D.E., "Multivariable Systems — Analysis and Feedforward Control Synthesis," *Chemical Engineering Progress*, Symp. Ser. No. 55, 1965.

87. Luyben, W.L., "Nonlinear Feedforward of Chemical Reactors," *AIChE Journal*, January 1968, p. 37.

88. Luyben, W.L., and Gerster, J.A., "Feedforward Control of Distillation Columns," *I.&E.C.*, October 1964, p. 375.

89. Cadman, T.W., and Carr, N.L., "Feedforward Control Synthesis for Multicomponent Distillation: I. Theoretical Development of Tower Transfer Functions," *ISA Transactions*, October 1966, p. 386.

90. Harman, G.L., "Anticipator-Computer Control System," *Automatic Control*, April 1963, p. 14.

91. Mika, H.S., "Modern Control Theory Applied to Power Plant Analysis," *Instrumentation Technology*, July 1973, p. 53.

92. Rekoff, M.A., "State Variable Techniques for Control Systems," *Electro-Technology*, May 1964, p. 36.

93. Gura, I.A., "State Variable Approach to Linear Systems," *Instruments and Control Systems*, October 1967, p. 106.

94. Wang, J.C., "Relative Gain Matrix for Distillation Control — A Rigorous Computational Procedure," *ISA Transactions*, Vol. 20, No. 1, Instrument Society of America, Research Triangle Park, N.C., 1981, p. 3.

95. Moore, R.L., and McKinney, A.H., "Determining Control Variables," *ISA Journal*, September 1962, p. 81.

96. Bristol, E.H., "Adaptive Control Odyssey," ISA Paper No. 561-70, Proceedings of ISA Annual Conference and Exhibit, Philadelphia, PA, 1970.

97. Bristol, E.H., "On a New Measure of Interaction for Multivariable Process Control," IEEE Transactions on Automatic Control, AC-11, 1966, p. 133.

98. Woolverton, P.F., "How To Use Relative Gain Analysis in Systems with Integrating Variables," *Instrumentation Technology*, September 1980, p. 63.

99. Witcher, M.F., and McAvoy T.J., "Interacting Control Systems: Steady-State and Dynamic Measurement of Interaction," *ISA Transactions*, Vol. 16, No. 3, p. 35.

100. Shinskey, F.G., *Process Control Systems*, McGraw-Hill Book Co., New York, N.Y., 1967.

101. Schuder, C.B., "Control Valve Characteristics," *Instruments and Control Systems*, March 1967, p. 79.

102. Schuder, C.B., "The Dynamics of Level and Pressure Control," Fisher Controls Co., Marshallton, Iowa.

103. Shunta, J.P. and Fehervari, W., "Nonlinear Control of Liquid Level," *Instrumentation Technology*, January 1976, p. 43.

104. Zalkind, C.S., "Practical Approach to Non-Interacting Control," *Instruments and Control Systems*, March 1967, p. 89.

105. Valstar, J.E., "Step Procedure for Analyzing Control Systems," *Instruments and Control Systems*, May 1963, p. 132.

106. Chestnut, H., and Mayer, R.W., *Servomechanisms and Regulating System Design*, John Wiley & Sons, Inc., New York, N.Y., 1951, p. 152.

107. Wilson, H.S., and Zoss, L.D., *Control Theory Notebook*, Instrument Society of America, Research Triangle Park, N.C., 1962.

108. Lloyd, S.G., "Generalized Control Theory," Publication No. TM8, Fisher Controls, Marshalltown, IA, p. 10.

109. Truxal, J.G., *Control System Synthesis*, McGraw-Hill Book Co., Inc., New York, N.Y., 1955, p. 44.

110. Ziegler, J.G., and Nichols, N.B., "Optimum Settings for Automatic Controllers," Transactions of the ASME, 1941, p. 759.

111. Sood, M., and Huddleston, H.T., "Tuning PID Controllers for Random Disturbances," *Instrumentation Technology*, February 1977, p. 61.

112. Ziegler, J.G., and Nichols, N.B., "Process Gage in Automatic Control Circuits," Transactions of the ASME, 1942, p. 433.

113. Anon., "Control Systems, Elements," *Measurements and Control*, January 1982, p. 186.

114. Tivy, V.V., "Notes on the Practical Application of Controllers with Rate Action," Talk to New Jersey Section, ISA, Newark, N.H., April 5, 1949.

115. Bradner, M., "Pneumatic Transmission Lag," Proceedings of the Annual Conference of the Instrument Society of America, Philadelphia, PA, 1948.

116. Buckley, P.S., and Luyben, W.L., "Designing Long Line Pneumatic Control Systems," *Instrumentation Technology*, April 1969, p. 61.

117. Toyama, K., Runstader, P.W., and Dean, R.C., "An Experimental Study of Surge in Centrifugal Compressors," *Centrifugal Compressor and Pump Stability, Stall and Surge*, ASME, 1976.

APPENDIX **H**

ANSWERS TO PROBLEMS

Section 2

1. (a) 2; (b) 3 or 1; (c) 1
2. (a) Inlet water valve; (b) Fuel to furnace; (c) Valve in outlet flow pipeline.
3. (a) Regulation — maintaining controlled variable at set point

   (b) Servomechanism — controlled variable follows changing set point

   (c) Manual — a human being senses a deviation in a variable and takes action to return it to the desired value

   (d) Automatic — a sensor senses the deviation of a variable and transmits the deviation to a controller that acts to return it to the desired value
4. A cause
5. The diaphragm
6. (a) 0.9 (b) 0.000045

Section 3-5

1. (a) $q_1 = q_2 = CsL$

   (b) $E = I/Cs$

   (c) $As^2x + Bsx = 0$

   (d) $(dx/dt)^2$ has no transform — a squared term is nonlinear
2. $$\frac{p}{w_1} = \frac{R}{RCs + 1}$$
3. $$\frac{M}{e} = K\left(\frac{\tau s + 1}{\tau s}\right)$$
4. $$\frac{x}{p} = \frac{A/K}{\frac{M}{K}s^2 + \frac{B}{K}s + 1}$$

Section 3-6-2

1. (a) 1000 volts/sec

   (b) 10 volts/sec

   (c) 0.001 volts sec
2. dT/dt: (a) 15°C/hr

   (b) 30°C/hr

   (c) 30°C/hr
3. Q = 10,000 PCU/hr
4. $$\frac{T}{Q_1} = \frac{1/K}{\frac{M}{K}s + 1}$$

Section 3-6-4

1. dL/dt: (a) 1.0 ft/min

   (b) 1.43 ft/min

   (c) 0.34 ft/min
2. C: (a) 0.00345 in.^{2}

   (b) 0.0500 in.^{2}

   (c) 0.384 in.^{2}
3. $C_2 = C_1/2$
4. V = 53.2 ft^3

Section 3-6-5

1. C = 0.234 in.^{2}
2. V = 98 ft^3
3. Independent of number of positions on stream
4. dp/dt = −37 psi/sec

Section 3-6-8

1. 3.12%/hr

   (a) no

2. w = 33.3 lb/min
3. dX_2/dt = 10%/hr
4. dX_o/dt = −0.167/hr

Section 3-7-2

1. ΔT: (a) 12.0°C
   (b) 40.9°C
   (c) 1.64°C
2. R: (a) 0.00333 hr°C/PCU
   (b) 0.01135 hr°C/PCU
   (c) 0.000455 hr°C/PCU
3. $R_2 = R_1 (v_1/v_2)^{0.8}$
4. R = 0.0241 hr°C/PCU

Section 3-7-3

1. (a) R = 295,000 sec/in.^{2}
   (b) Δp = 82.0 lb/in.^{2}
2. R = 68.5 sec/in.^{2}
3. R = 500 sec/in.^{2}
4. R = 0.517 sec/in.^{2}

Section 3-7-5

1. (a) Δp = 0.735 lb/in.^{2}
   (b) R = 2.95 sec/in.^{2}
2. (a) R = (1.2) (10^6) sec/in.^{2}
   (b) w = 0.135 lb/hr
3. (a) w = 3240 lb/hr
   (b) R = 80 sec/in.^{2}
4. R = 86 hr/mol

Section 4-3

1. System is controlled but not self-regulating.
2. System is self-regulating.
3. Slope = $dD/dt = (A - KD/C) = 0.368A/C$
4. (a) 1.82% incomplete
   (b) 0.66% incomplete

Section 4-4-2

1. (a) $E/I = R/(RCs + 1)$
   (b) τ = 18 ms
2. (a) τ = 1.53 hrs
   (b) τ = 0.23 hr
3. (a) τ = 6.5 sec
   (b) τ = 97.5 sec
4. (a) τ = 1.46 hrs
   (b) τ = 0.685 hr

Section 4-4-4

1. (a) τ = 0.2 hr
   (b) τ = 0.4 hr

2. (a) ΔL = 1.155 ft
   (b) ΔL = 2.31 ft
3. $$p_1 = \frac{20w_1 + p_2}{0.556s + 1}$$

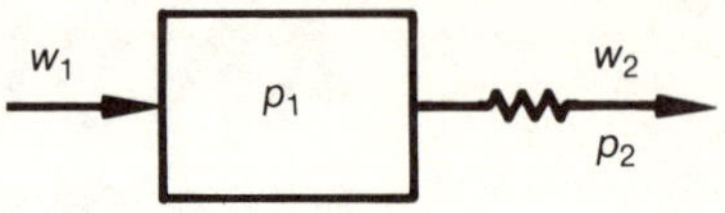

4. V = 2.1 ft^3

Section 4-5

1. (a) $$X_{A3} = \frac{(w_1/w_3)\,X_{A1} + (w_2/w_3)\,X_{A2}}{(M/w_3)\,s + 1}$$
   (b) $\Delta X_{A3} = (w_1/w_3)\,\Delta X_{A1}$
2. $$X_{A2} = \frac{X_{A1}}{\frac{M}{w}\,s + 1}$$
   τ = 2 hrs
   X_{A2} will reach 95% inert gas in 3τ = 6 hrs
3. τ = 1 hour
4. (a) τ = 3 min
   (b) K = 0.0005 hr/mol
5. (a) a = 21.7 sec
   (b) a = 4.34 sec
6. (a) a = 15 min
   (b) No. System is liquid full, hence flow changes simultaneously at all points in the caustic system.

Section 5

1. (a) w = 12.5 lb/sec $K_v = K(p_1 - p_2)^{1/2} = 50$
   (b) w = 12.5 lb/sec $K_v = 50$
   (c) w = 12.5 lb/sec $K_v = (30)(w) = (3.4)(12.5) = 42.5$
   (d) w = 37.5 lb/sec $K_v = (3.4)(37.5) = 127.5$
2. (a) $$K_o = (K_c)\;K_v\,\frac{\text{lb/sec}}{\text{fraction}}\;K_p\,\frac{°\text{C}}{\text{lb/sec}}\;K_T\,\frac{\text{fraction}}{°\text{C}}$$
   (b) Open-loop gain K_o is dimensionless
   1(a) $K_c = 1/K_vK_pK_T = 1/(50)(5)(0.02) = 0.2$
   1(b) $K_c = 0.2$
   1(c) $K_c = 1/(42.5)(5)(0.02) = 0.23$
   1(d) $K_c = 0.078$
3. (a) The equal percentage characteristic should be used.
   (b) $$K_c = \frac{1}{(30)(10)(0.02)} = 1.47$$

Section 6-6

1. (a) Force gain: $X = (A_1/A_2)y$
   (b) Displacement gain: $X = (A_2/A_1)y$
2. $$K_T = \frac{15\text{-}3 \text{ psi}}{10\text{-}7 \text{ pH}} = 4 \text{ psi/pH}$$

3. (a)

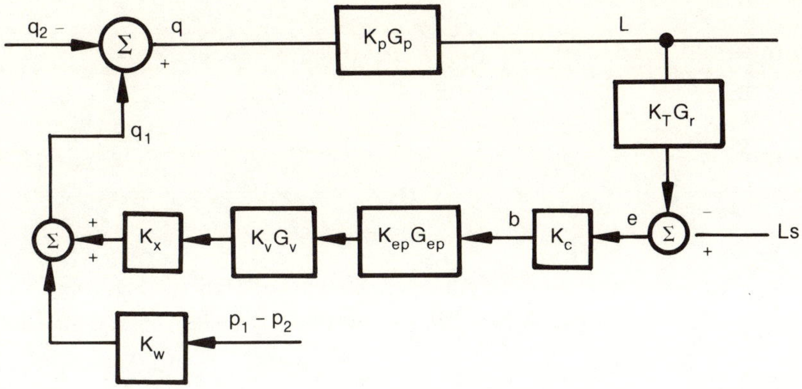

(b) $\left(K_T \dfrac{\text{Vac}}{\text{ft}}\right)\left(K_c \dfrac{\text{mAdc}}{\text{Vac}}\right)\left(K_{ep} \dfrac{\text{psi}}{\text{mAdc}}\right)\left(K_v \dfrac{\text{frac}}{\text{psi}}\right)\left(K_x \dfrac{\text{cfm}}{\text{fract}}\right)\left(K_p \dfrac{\text{ft}}{\text{ft}_3}\right)\left(G_p\, min\right)$

4. (a)

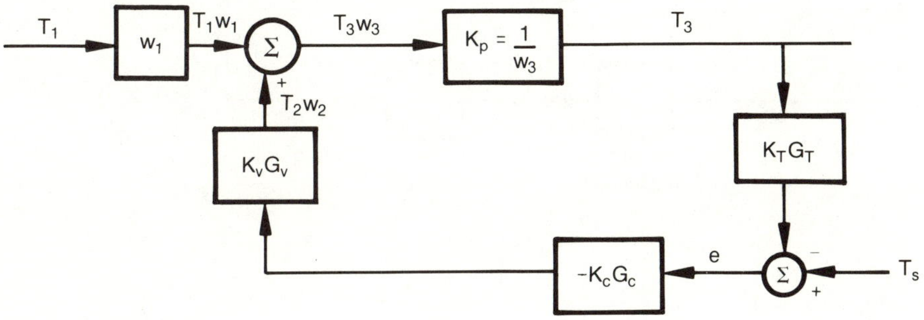

(b) $K_T G_T$ becomes $K_T G_T e^{-as}$

Section 6-8

1. (a) K_c = 0.1
   (b) P.B. = 1000%
2. (a) K_c = 0.05 P.B. = 2000%
   (b) K_c = 0.2 P.B. = 500%

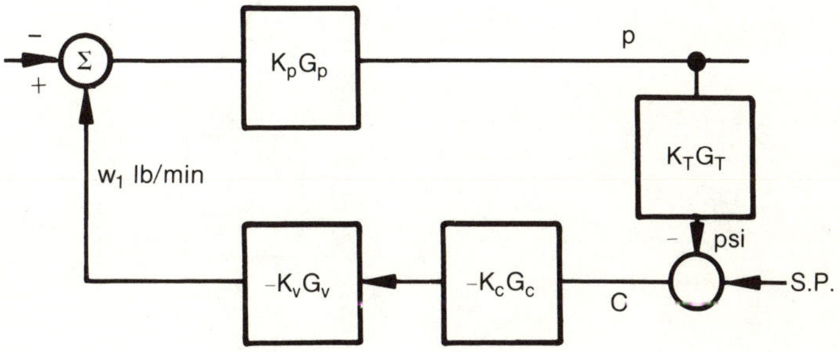

3.

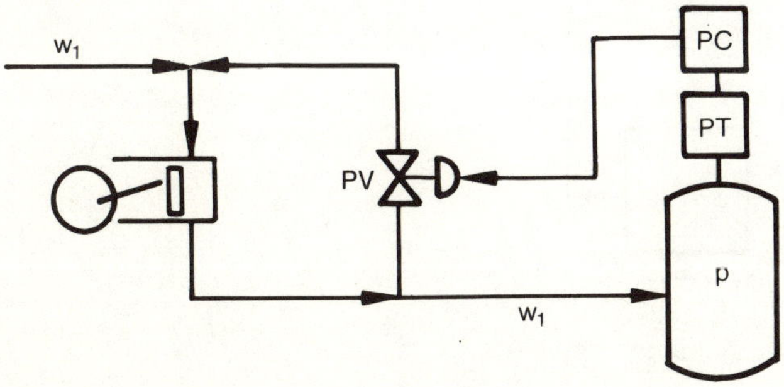

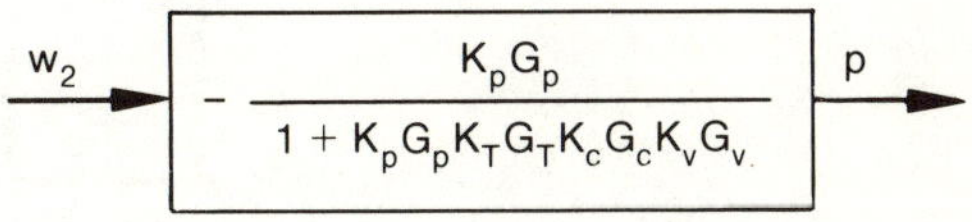

$$w_2 \rightarrow -\frac{K_pG_p}{1 + K_pG_pK_TG_TK_cG_cK_vG_v} \rightarrow p$$

4. (a) $K_pG_p = \frac{\text{atm}}{\text{lb/min}} = \frac{\text{atm - min}}{\text{lb}}$

$K_p = \frac{\text{atm}}{\text{lb}} \quad G_p = \text{min} = \frac{1}{s}$

5.

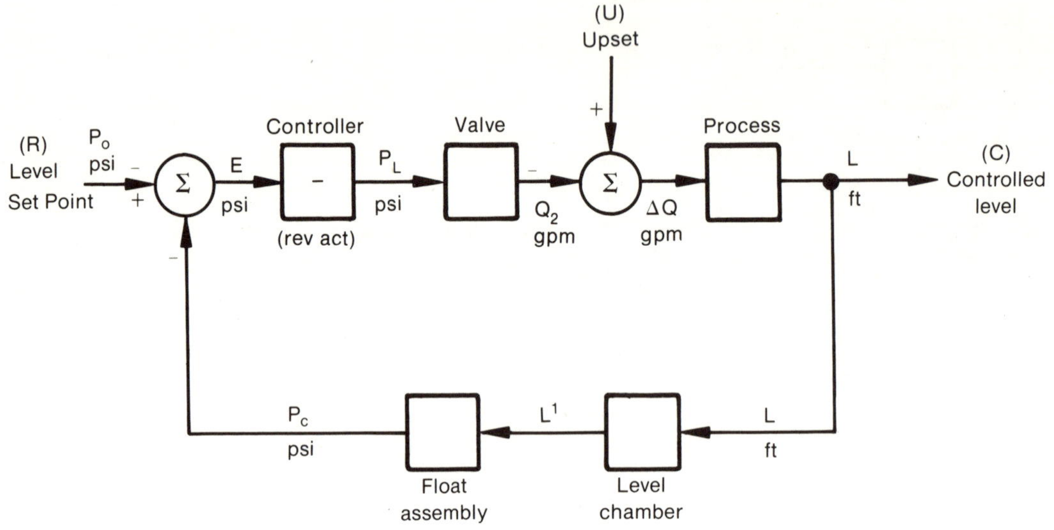

6.

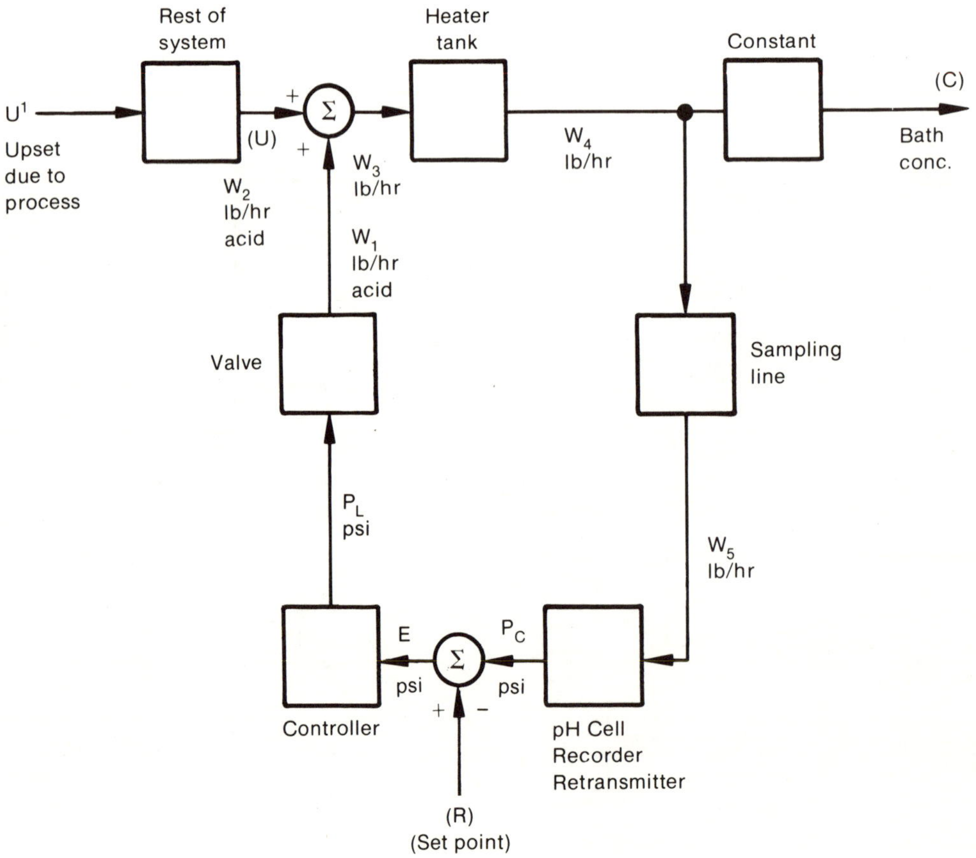

Section 7-4

1. (a) $\omega = 1/\tau = 1/0.159 = 6.28$ radians/min
   (b) $\omega = 2\pi f$ or $f = \omega/2\pi = 6.28/2\pi = 1.0$ cycles/min

2. (a) $M = \left[1^2 + \left(\frac{1}{-\omega\tau}\right)^2\right]^{1/2} = \left(1 + \frac{1}{\omega^2\tau^2}\right)^{1/2}$

   $\theta = \tan^{-1}\left(\frac{1}{-\omega\tau}\right)/1 = \tan^{-1}\left(\frac{1}{-\omega\tau}\right)$

   (b) $\omega = 0$ $\quad M = (1 + 1/0)^{1/2} = (1 + \infty)^{1/2} = \infty$
   $\theta = \tan^{-1}(1/-0) = \tan^{-1}(-\infty)$
   $= -90°$
   $\omega = 1/\tau$ $\quad M = (1 + 1/1)^{1/2} = (2)^{1/2} = 1.414$
   $\theta = \tan^{-1}(1/-1) = \tan^{-1}(-1) = -45°$
   $\omega = \infty$ $\quad M = (1 + 1/\infty)^{1/2} = (1 + 0)^{1/2} = 1$
   $\theta = \tan^{-1}(1/-\infty) = \tan^{-1}(-0) = 0°$

3. (a) $M = [(1)^2 + (\omega\tau)^2]^{1/2} = (1 + \omega^2\tau^2)^{1/2}$
   $\theta = \tan^{-1}(\omega\tau)/1 = \tan^{-1}(\omega\tau)$
   (b) $\omega = 0$ $\quad M = (1 + 0)^{1/2} = 1.0$
   $\theta = \tan^{-1}(0) = 0°$
   $\omega = \frac{1}{\tau}$ $\quad M = (1 + 1^2)^{1/2} = (2)^{1/2} = 1.414$
   $\theta = \tan^{-1}(1)$
   $\omega = \infty$ $\quad M = (1 + \infty)^{1/2} = \infty$
   $\theta = \tan^{-1}(\infty) = +90°$

4. (a) Reset $M = \infty$ at $\omega = 0 \rightarrow$ high gain at low frequencies
   Rate $M = 1$ at $\omega = 0 \rightarrow$ unity gain at low frequencies
   (b) Reset $M = 1$ at $\omega = \infty \rightarrow$ unity gain at high frequencies
   Rate $M = \infty$ at $\omega = \infty \rightarrow$ high gain at high frequencies

Therefore, reset helps at low frequencies, rate at high frequencies. Also: Reset θ starts at −90, ends at zero, i.e., introduces phase lag. Rate θ starts at 0, goes to +90°, i.e., introduces phase lead.

Section 7-5-3

1. (a) $\omega_1 = 1/\tau_d = 1/0.5 = 2.0$ rad/min
   $\omega_2 = \alpha/\tau_d = 6/0.5 = 12.0$ rad/min
   (b) $\omega_1 = 1/5 = 0.2$ rad/min
   $\omega_2 = 6/5 = 1.2$ rad/min

2. (a) Low frequency gain = 1.0
   High frequency gain = α = 6.0

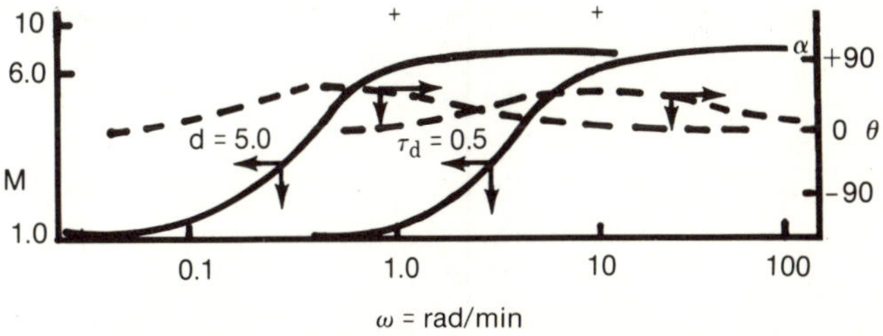

   (b) Moves breakpoints to lower ω; saturates at lower ω

3. (a) $\omega_1 = K_c/200\,\tau_i = 2/200 \times 0.1 = 2/20 = 0.1$
   $\omega_2 = 1/\tau_i = 1/0.1 = 10$
   (b) $\omega_1 = 2/200 \times 0.5 = 2/100 = 0.02$
   $\omega_2 = 1/0.5 = 2.0$
   (c) $\omega_1 = 10/200 \times 0.5 = 0.1$
   $\omega_2 = 1/0.5 = 2.0$

Section 7-5-3, cont'd

4. Low ω, gain = K_i = 200
   High ω, gain = K_c = 2 or 10

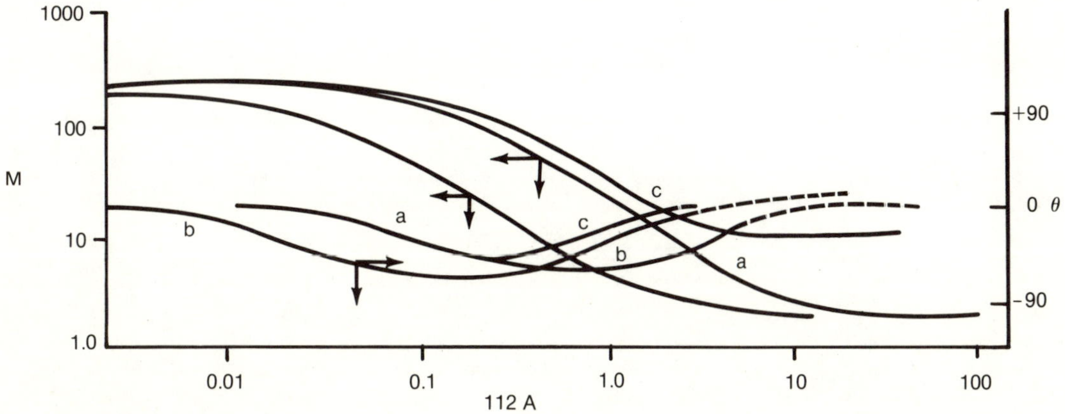

112 A

Section 7-6-2

1, 3.

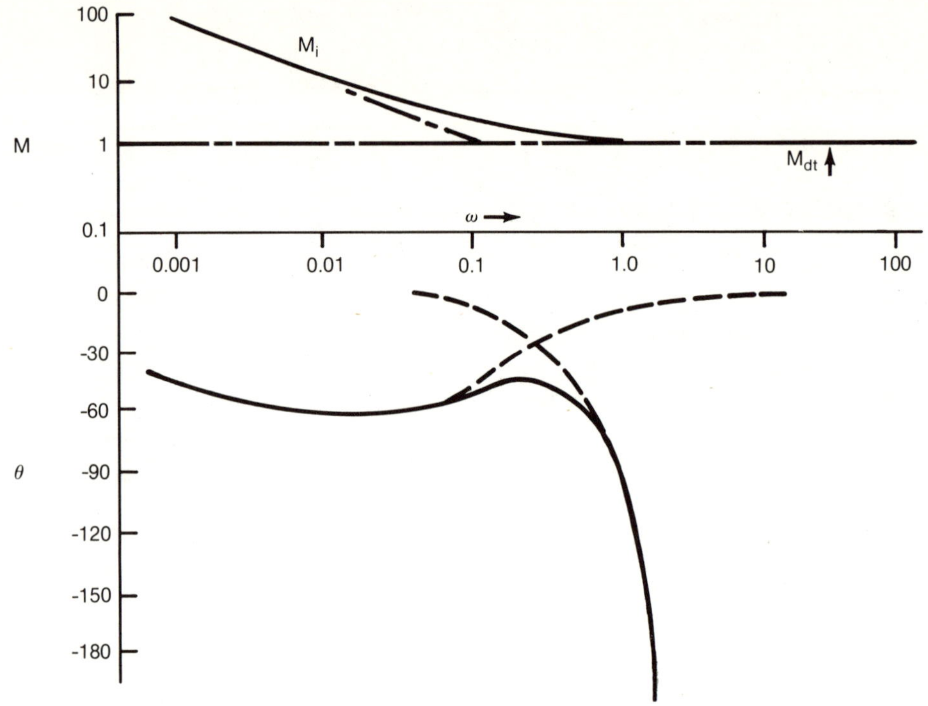

2. $\tau_i = \dfrac{1}{\omega_i} = \dfrac{1}{0.1}$ = 10 min; maybe 7-10 min.

4. Cannot use rate action, since gain is greater than unity after $\theta > -180°$ — unless $K_c < \alpha$. Actually, derivative will eventually die out to unity gain or less at higher frequencies but phase lag will occur also.

Section 7-7

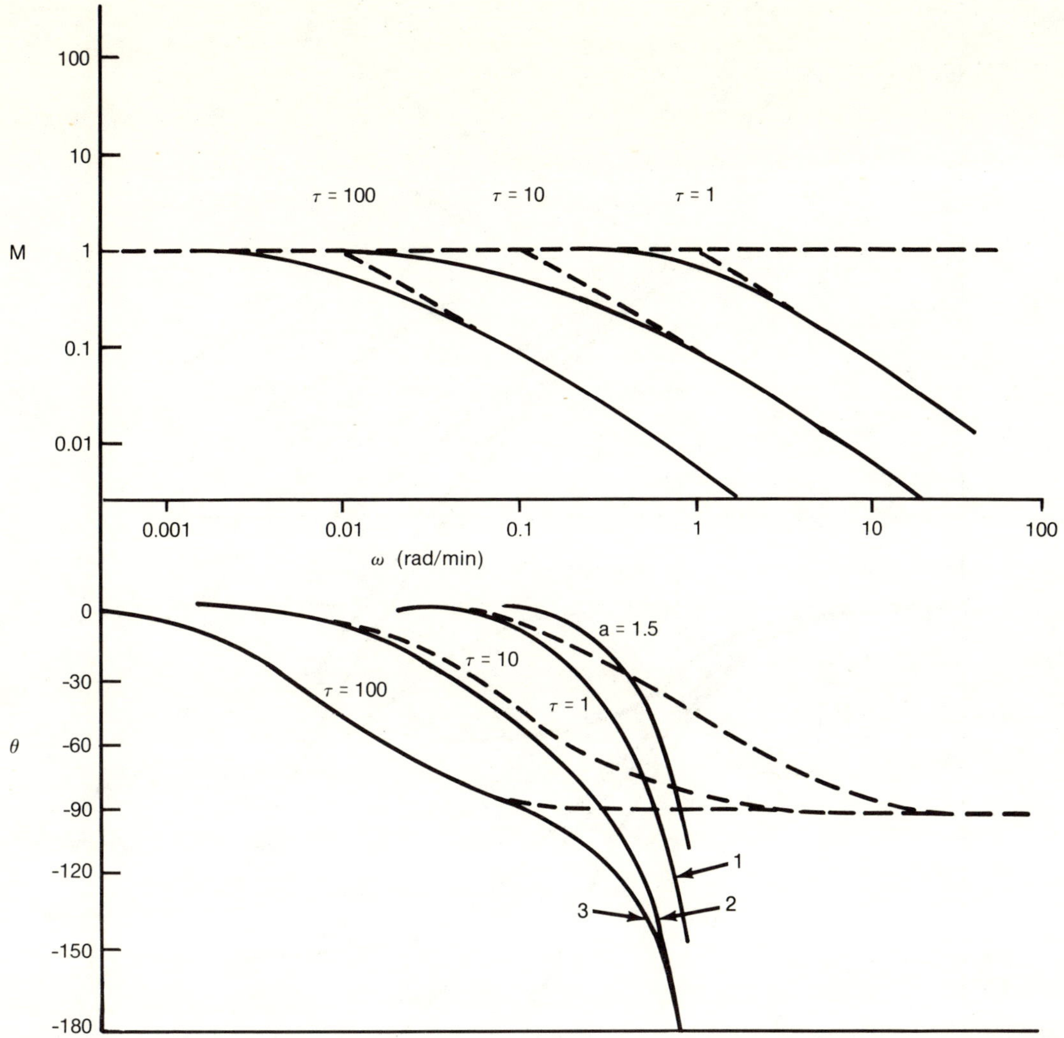

4. (a) For a $= 15$, $\tau_1 = 10$, $\tau_2 = 100$, $\tau_3 = 1000$
   (b) $\theta = -180$ at $\omega \cong 0.1$ rad/min

Section 7-7-4

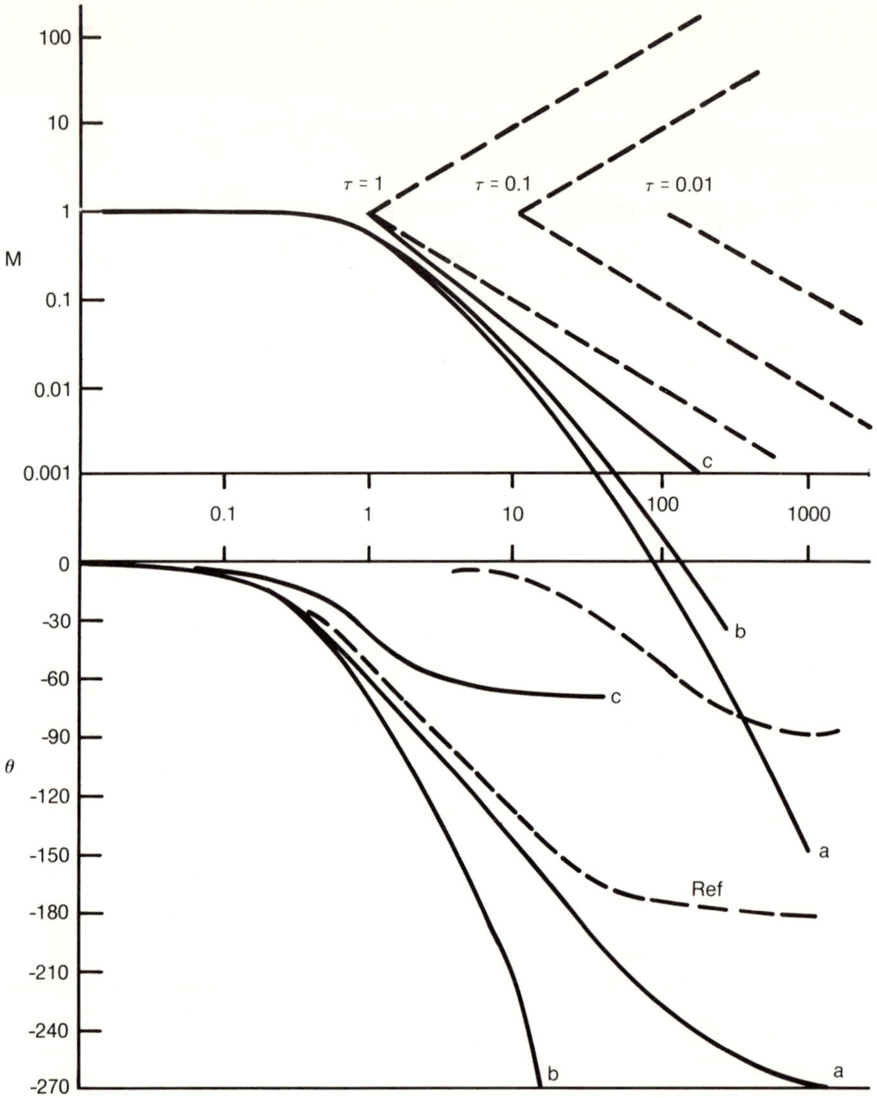

Figure 7-19. Two First-Order Lags

2. (a) M = 0.009
   (b) M = 0.009
   (c) Never gets to −180°

Section 7-8-3

The temperature control valve passes 0.5 lb/min of steam for each psi of input air signal. If the tank temperature is at set point of 160° F, and a 2-psi change is made in valve stem position, increasing heat flow by 1000 Btu/min, the tank temperature would start to rise at 1000/200 or 5° F/min. Assuming that the temperature increase in the tank contents is 25° F after steady state has again been attained, then the controller gain for a PI controller would be

$$K_c = \frac{0.9\ \Delta F}{a\ r} = \frac{(0.9)\ (25°\,F)}{(a)\ (5°\,F/min)}$$

The dead time (a) of the process will be the time necessary for the water flow of 20 lb/min to displace the 12 lb of water in the pipe between tank and sensor.

$$a = \frac{12 \text{ lb}}{20 \text{ lb/min}} = 0.6 \text{ min}$$

And the sensitivity becomes

$$K_c = \frac{(0.9)\,(25)}{(0.6)\,(5)} = 7.5$$

Section 8-4

1. $\tau = 9.5$ sec; $\zeta = 0.15$
   $a = 12$ sec; $\tau = 20$ sec

Section 9

1. $-KT_2$
2. Stabilizing
3. By the sign at the summing junction

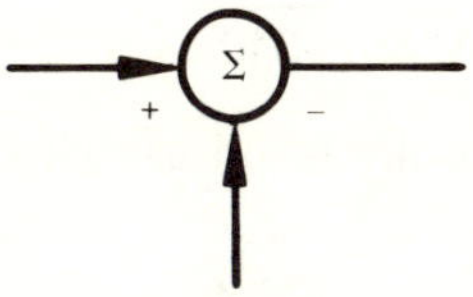

4. An error in measured temperature
5. $v \rightarrow$ small, $h \rightarrow$ small, $\tau \rightarrow$ large (slow response)
6. Yes
7. Offset
8. Decrease → Less stable
9. PI
10. Response time X decreases
    Overshoot Y increases

Section 10

1. Characteristic equation:

$$1 + \frac{K_c K_v K_p K_T}{(\tau_T s + 1)\,(\tau_v s + 1)}$$

Multiplying out:

$$\tau_T \tau_v s^2 + (\tau_T + \tau_v)\, s + 1 + K_c K_v K_T$$

Let $K_o = K_c K_v K_p K_T$

$$\tau_T \tau_v s^2 + (\tau_T + \tau_v)\, s + 1 + K_o$$

$$\frac{\tau_T \tau_v}{1 + K_o} s^2 + \frac{\tau_T + \tau_v}{1 + K_o} s + 1$$

Comparing to the generalized equation:

$$T_c^2 s^2 + 2T_c \zeta s + 1$$

Substituting:

$$2 \sqrt{\frac{\tau_V \tau_T}{1 + K_o}} \qquad \zeta = \frac{\tau_T + \tau_V}{1 + K_o}$$

Let $\zeta = 0.9$

$$1.8 \sqrt{\tau_V \tau_T (1 + K_o)} = \tau_T + \tau_V$$

Divide through by τ_V:

$$(3.24) \left(\frac{\tau_T}{\tau_V}\right) (1 + K_o) = \left(\frac{\tau_T}{\tau_V} + 1\right)^2$$

Then,

τ_T / τ_V	0.1	0.3	0.5	0.7	1.0
K_o	2.7	0.7	0.4	0.3	0.25

2. There are no dynamics between the disturbance and the flow, permitting instantaneous deviations of a magnitude set by K_1 and K_2.

$$C = V \frac{d\rho}{dp} = (0.6 \text{ ft}^3)\,(3)\,(10^{-8})\,\frac{\text{in.}^2}{\text{ft}^3}$$

$$= (1.8)\,(10^{-7}) \text{ in.}^2$$

$$\frac{dp}{dt} = \left(\frac{180 \times 10\%}{3600} \frac{\text{lb}}{\text{sec}}\right)(1.8)\,(10^{-7}) \text{ in.}$$

$$= (9)\,(10^{-10}) \text{ psi/sec}$$

4. PB = 100% is $K_c = 1$

The new K_c:

$$K_c = 1\left(\frac{0.3}{2.7}\right) = \frac{1}{9}$$

P.B. = 900%

Section 11

1. Yes.
2. Install equal percentage value.
3. T_i = 50 minutes (or maximum for controller

$$K_c = \frac{10000 \text{ gallons}}{10 \text{ gpm } T_i \text{ (min)}} = 20 \;(\text{P.B.} = \frac{100}{K_c} = 5\%)$$

illustrates reset problem with large tanks.)

4. Install a baffle.
5. Add a restricting orifice liquid impulse line L_1.
6. Run the level above the weir.
7. The condensing temperature does not change during condensation.

8. Decreasing process temperature causes more heat to be transferred from the steam, causing more condensation, causing a major contraction in steam (water) volume, resulting in a pressure decrease.

9.

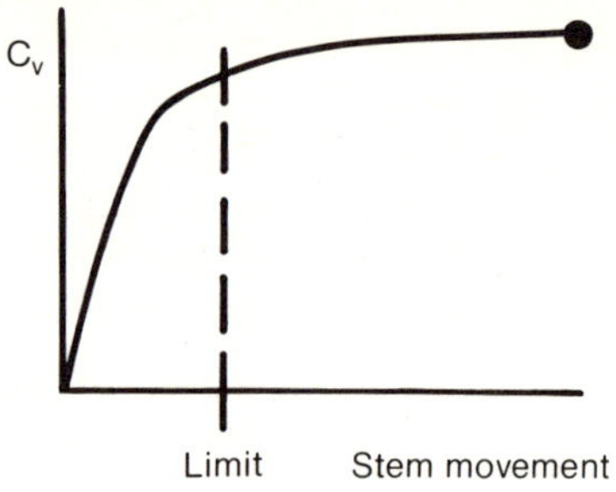

10. The pressure control, since antisurge is protecting a very expensive machine.

Section 12

(See Figure 12-8 and the equation that describes heat balance.)

$$(M_M c_M + Mc)\frac{dT_2}{dt} = w_1 cT_1 - w_2 cT_2 + Q$$

And for a cooling process,

$$Q = w_c c_c (T_{c2} - T_{c1})$$

Then,

Reset Time =

$$\tau_i = \frac{(M_M c_M + Mc)}{wc} = \frac{110 + 160}{160} = 1.7 \text{ min}$$

Derivative Time =

$$\tau_d = \frac{(2)(.11)}{(200)(.5)} = 0.002 \text{ hr} = 0.13 \text{ min}$$

Open-Loop Gain:

$$K_c K_v K_T K_p \leq 3$$

Assume EQ % Valve: $K_v = (\text{Ln}30)\, w_c = 3.4\, w_c$

$$K_p = \frac{C_c (T_{c2} - T_{c1})}{wc} = \frac{(1)(40)}{160} = 0.25$$

$$K_T = \frac{1}{100} = 0.01\ {}^\circ\text{C}^{-1}$$

Then,

$$K_c = \frac{(3.0)(2)}{(3.4)(80)(0.25)(.01)} = 1.76$$

(Assuming coolant flow at half capacity)

$$\text{P.B.} = \frac{100}{1.76} = 57\%$$

Section 13

1. Enclose the major time constant (10 minutes) in the secondary loop. The time constant of 0.5 minute is so small that another cascade loop around it is hardly justified. The small time constant is insignificant when compared to the 10-minute time constant and can be included in the same secondary loop with little effect.

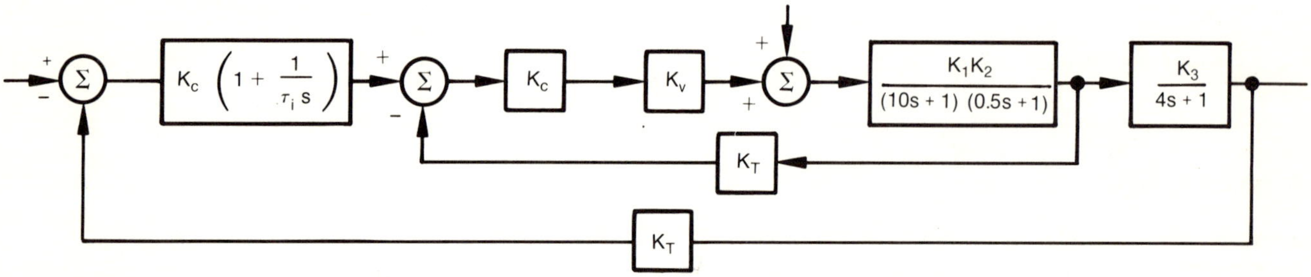

2. Offset doesn't matter because the value of the cascaded variable doesn't matter.

Proportional-only control will not saturate when driven past the end of its span by the primary controller.

3. Three time constants for complete reaction = 4.5 minutes.

Length = (4.5 minutes) (60 second/minute) (1 ft/second)
= 270 feet

4. $KG(s) = -\dfrac{K_w(\tau s + 1)}{K_v K_1}$

INDEX

Accuracy, 17, 81
Adaptive control, 203
Adiabatic gas process, 35
Adsorption process, 41
Amplifier, operational, 55
Amplitude, 97
Analogy, 2
Anticipatory, *see* Derivative
Attenuation, 80, 103

Black-Nichols Chart, 126
Block diagram, 14, 16, 80, 147
  heat transfer, 32
  thermal system, 58, 151, 149, 191
  gas pressure, 64, 168
  controller error, 81, 114
  closed-loop, 82, 89
  liquid level control, 84, 170, 220
  cascade control, 187
  composition control, 112
  flow control, 164, 233
  feedforward, 194
Bode plot, 100, 107, 147, 191, 228

Capacitance, 181, 231
  electrical, 31
  thermal, 32
  liquid, 34
  gas, 35
  vapor, 37
  composition, 40
  mass transfer, 41
  adsorption, 41
  flow system, 159
Cascade control, 155, 187
Compressor, centrifugal, 177
  anti-surge control, 179
  pressure control, 177
Composition process, 40, 66, 118
Concentration, *see* Composition
Conditional stability, 228
Control valves, 50
  flow characteristic, 50,73
    quick-opening, 74
    linear, 74, 76
    equal %, 74, 76, 176, 185
  installed characteristic, 74, 176
Controlled variable, 13, 15, 18
Controller tuning, 88, 126, 128
Correlation, 134
Critical flow, 50

Damping, *see* Zeta
Dead time, 12, 16, 69, 108, 112,.117, 153
Decoupled control, 207
Dependent variable, 13
Derivative control, 9, 105, 114, 122, 127, 153, 184
Diffusion, *see* Mass transfer systems
Distillation column, 41, 201
  bottoms level control, 171
  overhead pressure control, 172
  condenser control, 185
  feedforward, 196
  control strategy, 201
Disturbance variable, 13, 18, 219
Driving force equation, 27, 43
Dynamic analysis, definition, 25

Electrical system, 31
Error, 15
Error, controller, 80, 194, 219

Feedback, 1, 10
Feedforward, 10
Final value theorem, 28, 63
Fired vessels, 176
  pressure control, 176, 202
  temperature control, 189
  level control, 195
Floating control, 154, 165, 175, 185
Flow control, 231
Fluid flow systems, 47
Force balance, 11
Forcing function, 13, 133
Fourier analysis, 100, 139, 144, 223
Frequency response, 97, 126, 134
  nonlinear, 162

Gain, 28, 74, 80, 232
  partial derivative, 74
  open-loop gain, 75, 84, 150, 161, 184
Gain margin, 126, 229
Gas constant (R), 35

Heat balance, 32, 58, 148, 184, 211
Heat exchanger, 175
  temperature control, 175, 184
Heat transfer process, 28, 32, 211
  conduction, 44
  convection, 44
  radiation, 45

Ideal gas law, *see* Perfect Gas Law
Identification,
  adaptive control, 205
Impulse response, *see* Pulse testing
Independent variable, 13
Initial value theorem, 28
Integrating process, 32, 33, 36, 169
Interaction, 206, 215
Isothermal gas process, 35

Lag network, 107
Laminal flow, 47
Laplace notation, 27, 69, 147, 222
Lead network, 106
Level control, 171, 219
  float displacer dynamics, 172
  boiler drum, 195
Linearization, 48, 81, 232
Liquid level, 27, 62
  liquid level control, 170
  multi-capacity, 215

Mach number, 161
Magnitude ratio, 100
Manipulated variable, 9, 13, 18
Manual control, 10, 12
Mass balance, 27, 34, 37, 40, 167, 216, 219
Material balance, *see* Mass balance
Mass transfer systems, 41, 51
Mathematical models, 17
Matrix, control, 202
Measured variable, 13
Mols, 41
Motion blance, 11

Nyquist diagram, 223

Offset, 10, 150, 182
Ohm's Law, 27
Oldenbourg-Sartorius method, 140
On-off control, 7
Open loop, 81, 88
Optimization, 14
  optimal control, 203

Percent incomplete method, 140
Perfect Gas Law, 35, 50, 65
Period, 97
Perturbation, *see* Linearization
Phase, 98

Phase margin, 126, 229
Pipeline, cross country, 165
Pneumatic transmission lines, 161
  Time constant, 163
Polymer system, 38
Polytropic gas process, 35
Potential, 31
Pre-act, *see* Derivative
Pressure control, 168
Proportional control, 10, 19, 126, 150, 155, 182
  proportional-reset, 150, 183, 222
  proportional-reset-derivative, 153, 184, 191, 223
Process variable, 26
Pulse testing, 144

Quantity stored, 31

Radian, 98
Raoult's Law, 68
Reactor control, 148
Recycle system, 67
Regulator, 8
Regulator, self-contained, 15, 150, 176
Relative gain matrix, 206
Reset control, 10, 19, 106, 114, 127, 150, 164
Resistance, 27
  electrical, 27
  thermal, 44,
  laminar flow, 47
  turbulent flow, 48
  compressible fluid flow, 50
  gas diffusion, 51
Response, first-order, 55, 69, 101, 117
  electrical, 57
  thermal, 59
  liquid level, 62
  gas pressure, 64
  composition, 66
  diffusion, 67
  distillation column tray, 68
Routh Criterion, 126

Sample data, 12
Saturation, reset, 183, 191
Self-regulation, 53
  thermal, 54
Separator, liquid, 171
Servomechanism, 8
Set point, 7, 15
Speed of response, 9
Step testing, 28, 139
Streamline flow, *see* Laminar flow
Superposition, 98
System, 1

Temperature sensor, 44, 59, 150, 213
Thermal system, 32
  conductive, 58
  convective, 59
  radiation, 59
Thermostat, 7
Time constant, 16, 53, 54, 134
  electrical, 57, 209
  thermal, 59, 211
  liquid level, 62
  gas pressure, 64
  composition, 66
  diffusion, 67
  distillation column tray, 68
Total derivative, 37, 48, 74, 82, 232
Transfer function, 28
Transient response, 139
Tuning, controller, *see* Controller tuning

Ultimate value, *see* Final value theorem

Vapor pressure, 68
Vapor pressure process, 37

Wind-up, *see* Saturation

Zeta (damping coefficient), 138, 152, 160, 169
Ziegler-Nichols Method, 128, 160